Handbook of Polymer Blends and Composites

Volume 3A

Editors:

C. Vasile
and
A.K. Kulshreshtha

Rapra Technology Limited

Shawbury, Shrewsbury, Shropshire, SY4 4NR, United Kingdom
Telephone: +44 (0)1939 250383 Fax: +44 (0)1939 251118
http://www.rapra.net

First Published in 2003 by

Rapra Technology Limited

Shawbury, Shrewsbury, Shropshire, SY4 4NR, UK

©2003, Rapra Technology Limited

A catalogue record for this book is available from the British Library.

Every effort has been made to contact copyright holders of any material reproduced within
the text and the authors and publishers apologise if any have been overlooked.

ISBN for Volume 3: 1-85957-303-7
ISBN for Complete Set: 1-85957-309-6

Typeset, printed and bound by Rapra Technology Limited
Cover printed by The Printing House Ltd

Contents

Contributors

Georgios Bokias

Department of Chemical Engineering, University of Patras, GR - 265 04 Patras, Greece

Mihai A Brebu

Romanian Academy, 'Petru Poni' Institute of Macromolecular Chemistry, 41A Gr. Ghica Voda Alley, RO 6600 Iasi, Romania

Gina G Bumbu

Romanian Academy, 'Petru Poni' Institute of Macromolecular Chemistry, 41 A Gr. Ghica Voda Alley, RO 6600 Iasi, Romania

Sossio Cimmino

Istituto di Chimica e Tecnologia dei Polimeri - CNR, Via Campi Flegrei 34, 80078 Pozzuoli (Na), Italy

Aurelia Grigoriu

'Gh. Asachi' Technical University of Iasi, Faculty of Textiles and Leather Technology, 53 Mangeron Boulevard, RO 6600 Iasi, Romania

George E Grigoriu

'P. Poni' Institute of Macromolecular Chemistry, 41A Gr. Ghica Voda Alley, RO 6600 Iasi, Romania

Leonard Ignat

'P. Poni' Institute of Macromolecular Chemistry, 41A Gr. Ghica Voda Alley, RO 6600 Iasi, Romania

Elena V Koverzanova

Institute of Biochemical Physics of Russian Academy of Sciences, Russian Academy of Sciences, 4 Kosygin Street, Moscow 117334, Russia

Anand K Kulshreshtha,
Polymeric Materials Group & Internet & Information Centre, Research Centre, Indian Petrochemical Corporation Limited, Vadodara – 391 346, India

Sergei M Lomakin
Institute of Biochemical Physics of Russian Academy of Sciences, Russian Academy of Sciences, 4 Kosygin Street, Moscow 117334, Russia

Mihaela C Pascu
Biophysics Department, 'Gr. T. Popa' Medicine and Pharmacy University, 16 University Street, RO 6600 Iasi, Romania

Yusaku Sakata
Department of Applied Chemistry, Okayama University, Tsushima Naka, Okayama 700-8530, Japan

Hans A Schneider
Institut für Makromolekulare Chemie, 'Hermann-Staudinger-Haus', & Freiburger Materialforschungszentrum, FMF der Universität, Stefan-Meier-Strasse 21, D-79104 Freiburg, Germany

Clara Silvestre
Istituto di Chimica e Tecnologia dei Polimeri - CNR, Via Campi Flegrei 34, 80078 Pozzuoli (Na), Italy

Georgios Staikos
Department of Chemical Engineering, University of Patras, GR - 265 04 Patras, Greece

Aurelian Stanciu
'Petru Poni' Institute of Macromolecular Chemistry, 41A Gr. Ghica Voda Alley, RO 6600 Iasi, Romania

Azhar Uddin
Process Safety and Environment Protection Group, Chemical Engineering, School of Engineering, The University of Newcastle, Callaghan, NSW 2308, Australia

Cornelia Vasile

Romanian Academy, 'Petru Poni' Institute of Macromolecular Chemistry, 41A Gr. Ghica Voda Alley, RO 6600 Iasi, Romania

Guennardi E Zaikov

Institute of Biochemical Physics, Russian Academy of Sciences, 4 Kosygin Street, Moscow 117334, Russia

Preface

There are not many things to add to the preface of the entire handbook presented in the first volume for the preface of the third and fourth volumes. It is intended that these volumes will provide an overview of the theory and practice of polymer blends, from fundamental thermodynamics of mixing to polymer blends processing, from experimental research to modelling developments, although many other very specialised and superb reviews and monographs are frequently published in a relatively short period of time. The choice of subjects has changed over the years as science and technology have progressed. New aspects are constantly being explored. The polymer blends and alloys had a rapid growth in importance during 1980s, although the initial interest in them was much earlier. The prospect of producing new materials with technologically useful properties by mixing different polymers, to form the equivalent of the metal alloys has long been an attractive goal. It is a field characterised by practical needs, basic research, advanced technological development, industrial involvement and legislation. The reasons for blending are: economy, extending engineering resin performance by diluting it with a low cost polymer, adjusting the composition of the blend to customer specifications and polymer waste recycling. About 65% of polymer alloys and blends are produced by resin manufacturers, about 25% by compounding companies and the remaining by transformers who have their proprietary blend formulations for certain products.

There are over 40 chapters (each volume containing 15 chapters) coming from more than 30 distinguished specialists. The 15 chapters of the third volume deal with general aspects of polymer blend morphology, properties and behaviour in various conditions, while the fourth volume is mainly concerned with the various chemical classes of polymer blends.

Unfortunately many polymer blends, have poor mechanical properties because most pairs of chemically different polymers are mutually immiscible and mixing usually results in materials which are phase separated and have weak polymer-polymer interfaces. The thermodynamics of polymer mixing have throughout the years seen many developments and it is still a very fascinating domain because of its attractive and exciting theoretical aspects in solving problems related to structure, conformation and configuration of macromolecules and also due to the direct practical implications. Bulk demixing of binary polymer mixtures has been studied extensively, both theoretically and experimentally, during the last few decades. The presence of an interface, however, may lead to domain morphologies and domain growth kinetics, which differ significantly from those in the bulk. Interface phenomena are still being studied, and a chapter is devoted to this. The third volume includes the topics related to the criteria of selection of the components of the blends, thermodynamic fundamentals of mixing, methods of compatibilisation, general characteristics of the blends such as glass transition, crystallisation behaviour, ageing and thermal degradation, etc. The surface treatment methods to improve some properties such as polarity and biocompatibility, preparation and properties of the interpenetrating polymer networks, etc., are also discussed. The selection of the blend components is the most difficult task in the development of a material with a full set of the desired properties. Material selection works on the principal that the main advantages of one of the component will compensate for the deficiencies of the second and *vice versa*. The trial-and-error approach used to obtain a good blend consists of: define the desired physical and chemical properties of the blend, list the properties of several resins which may provide some required characteristics, tabulate the advantages and disadvantages of various candidates, determine the miscibility and method of compatibilisation, examine the economics including cost of components and operations, define the morphology which assures the optimal performance, select the characteristics of resins (mainly rheological properties, molecular weight, concentration of ingredients, amount of compatibilising agent, etc.), necessary to generate the desired morphology, determine the method of stabilisation of morphology (controlled cooling rate, crystallisation, chemical reaction, irradiation, etc., and select the optimum fabrication method which assures the final morphology. All these aspects are considered in the third volume as well as the up-to-date theoretical aspects regarding the selection of the components of a blend with tailored properties. An enhanced optimisation of material performance through a better understanding of theoretical basis and development in computer simulation of all dependences between composition-miscibility-processing parameters and properties is necessary. Studies show many more examples of polymer mixtures existing which exhibit regions of miscibility than were previously anticipated. The science of complex materials is still at an early stage but there is good reason to expect it to grow dramatically in the near future thanks to the creative efforts of dedicated researchers.

In recent years, much effort has been devoted to the study of polymer blends and the factors, which have an effect on their miscibility. The phenomenology of chemical reactions between polymers at interfaces is only just now being examined in depth from both the experimental and theoretical viewpoints, but much remains to be done and understood.

In Volume 3, there are chapters which highlight the problems related to the morphology of the blends and its direct implications in glass transition and crystallisation phenomena. Chapters 5 and 6 focus on water-soluble polymers, which were a very attractive subject of research in the last decades because of their utilisation in medicine and pharmacy and for environmental protection. Chapters 11-14 also feature recent advances on the effect of radiation and ageing on polymer blends and also behaviour during degradation. Topics of IPN, heterofibres and reactive blending are also included.

The fourth volume contains mainly the descriptive aspects of the various polymer blends classified on their main component, although there is some overlap because sometimes even the minor component plays an important role in blend properties. The following classes are presented: polyolefin, poly (vinyl chloride), styrene polymers, ionomers, polyamides, polyesters, poly (vinyl alcohol), polyacrylates, silicones, polyurethanes, lignocellulosics, liquid crystalline polymers, thermostable polymers, environmentally friendly polymer blends, rubber toughened epoxies/thermosets, etc.

Each chapter is based on a rich and up-to-date literature in the field and the personal research of the authors.

We hope that the selected topics will be readily appreciated by the readers and that they will acknowledge the quality of the contributions.

This handbook will offer to scientists and also to R&D engineers useful updated information on the present developments in the polymer blends area of science and technology.

The editor thanks to all contributors to the handbook because without their efforts this publication should not be possible. Many thanks are due to the Rapra editorial staff, Claire Griffiths (Editorial Assitant), Sandra Hall (Graphic Designer and Typesetter) and in particular Frances Powers for her professionalism and wholehearted support during entire working period.

Cornelia Vasile
'Petru Poni' Institute of Macromolecular Chemistry
Romania

Anand Kulshreshtha
Indian Petrochemical Company
India

2002

1 Terminology

Cornelia Vasile, Anand K. Kulshereshtha and Gina G. Bumbu

1.1 Miscibility and Compatibility

There are some controversial definitions of the terms used in the field of polymer mixtures, therefore their meaning has to be clarified. Most terms are based on the thermodynamics and include the analysis of the phase diagrams of polymer systems as the basis of differentiation, see **Figure 1.1**.

Utracki defines the **polymer blend** as a mixture of at least two polymers or copolymers comprising more than 2 wt% of each macromolecular component. Depending on the sign of the free energy of mixing, the components of the blends are miscible and immiscible [1, 2, 3, 4].

A **miscible polymer blend** is a homogeneous polymer blend at molecular level, thermodynamically associated with the negative value of the free energy of mixing, see **Figure 1.1a**, bottom curve, while the **immiscible polymer blend** is associated with positive value of the free energy of mixing – **Figure 1.1**, upper curve. Blends of two high molecular weight polymers usually produce lower critical solution temperatures (LCST), which means that combinations of those two polymers may be miscible at lower temperature, but phase separate at a higher temperature. The opposite case of phase separation by lowering of the temperature means blends characterised by an upper critical solution temperature (UCST). This is a direct result of the very small entropy of mixing of two polymers – see Chapter 2.

The most important feature of a true miscibility is its **thermodynamic stability** or **equilibrium state.** A system is thermodynamically stable if its formation is accompanied by a decrease in the Gibbs (or Helmholtz) free energy. The Gibbs free energy (G) decreases to a definite equilibrium value which does not change subsequently with time.

When significant interactions exist between the two initial components, ΔG_m becomes negative and the blend is miscible. However, the miscibility is not as complete as in the case of small molecules. Nevertheless, this 'nanometric' phase separation cannot be easily

1

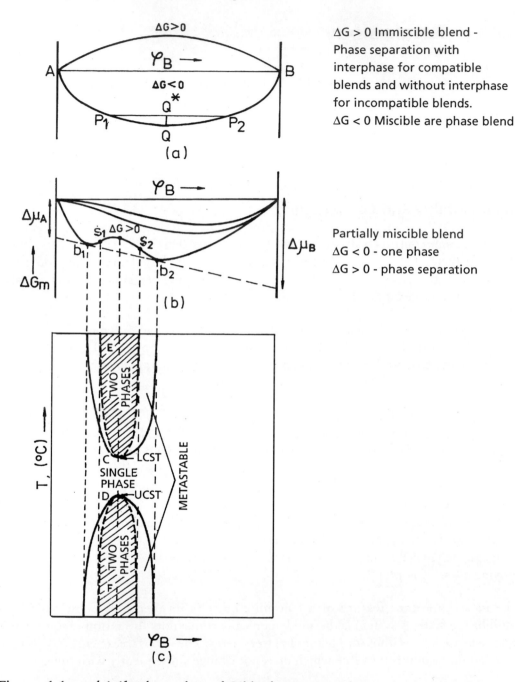

Figures **1.1a** and **1.1b,** show plots of Gibbs free energy of mixing (ΔG_m) for blends with differing miscibility, which lead to correspondingly various morphologies. **Figure 1.1c** is a schematic representation of a phase diagram for a system exhibiting both LCST and UCST

detected by most methods of characterisation. For example such a blend is characterised by one glass transition temperature (T_g).

The complete miscibility phenomenon is rare for polymers. For example it is present in the case of polyphenylene oxide and polystyrene (Noryl from GEP or Vestoblend from Hüls) which is a miscible blend of amorphous polymers or polyvinylchloride (PVC) and polycaprolactone (PCL) which are miscible in the amorphous state. For concentrations of PCL higher than 50%, PCL crystallises. A biphasic blend is observed mixing a crystalline phase of PCL and an amorphous phase of miscible PVC/PCL.

An **immiscible polymer blend** is a subclass of polymer blends referring to those blends that exhibit two or more phases on entire composition and temperatures range.

Partially miscible polymer blends are a subclass of polymer blends including those blends that exhibit a 'window' of miscibility, i.e., they are miscible only at certain concentrations and temperatures, see **Figure 1.1b**.

Miscibility has been also defined in morphological terms of homogeneous domain size. The strict definition of miscibility requires a single phase over the entire composition range at a particular temperature [5].

The miscibility defined in the terms of the equilibrium thermodynamic ($\Delta G_m = 0$) must be considered only within the range of independent variables (temperature T, pressure P, molecular weight, chain structure, etc). The rate at which the thermodynamic equilibrium can be achieved depends on the driving force, i.e., the polymer-polymer interaction parameter, and the resisting rheological forces, i.e., the diffusivity. Only detailed studies of the tendency of these blends to mix or to phase separate over a long period of time can answer the question of their true thermodynamic miscibility. In thermodynamics 'miscibility' means single phase down to molecular level. In a pragmatic sense miscibility means that the system appears to be homogeneous in the type of test applied in the study, i.e., it is defined in terms of degree of dispersion. In the literature one finds conflicting reports on miscibility of a given polymer/polymer blend. In a global sense, polymer/polymer miscibility does not exist, observed miscibility is always limited to a 'miscibility window' also for a range of independent variables namely, composition, molecular parameters (as molecular conformation and configuration, molecular weight, molecular weight distribution), temperature, pressure and others. More than 1600 of these 'miscibility windows' have been identified for two, three or four component blends. However, immiscibility dominates the polymer blends technology [1-4, 6].

For the blends with a 'miscibility window' the miscibility is obtained only for a limited domain of concentrations. Outside the 'miscibility window' interactions are not strong enough to prevent the demixing of the blend. Nevertheless, no phase consists of a pure polymer, each one must be considered as a miscible blend of major component with a small amount of a minor component. In this case two glass transition temperatures T_g are observed, but each one is different from the values reported for the virgin polymers. As an example, terpolymer acrylonitrile (AN)/butadiene/styrene (ABS) and polycarbonate (PC) with strong interactions between acrylonitrile and PC give a **miscible alloy** for a certain range of concentrations.

Miscible or not, most polyolefin (PO) blends are compatible, with the enhancement of their physical performance responsible for the predominance of blending in PO industry.

Compatibility can be defined on the basis of the property-composition curve, where properties are taken as macroscopic characteristics such as mechanical or rheological properties, etc. **Figure 1.2** plots the qualitative trends of property-composition curves for blends which are compatible, partially compatible or incompatible. The synergistic effect is found only in blends with strong interactions between the two phases. Instead, the antagonistic effect is typical of those of the two components, partially compatible or semi-compatible although, most of them show values lower than expected on the basis of an additive law. For blends where the components form separate phases, properties depend on the arrangement of these phases in space and the nature of the interface between the phases. Immiscible blends behave like composite materials in many respects.

For improved performance, the immiscible blends usually need compatibilisation.

Compatibilisation is any physical or chemical process of modification of interfacial properties of an immiscible polymer blend, resulting in formation of the interphase and stabilisation of the desired morphology thus leading to the creation of a polymer alloy.

The compatibilisation must not only ensure a certain morphology and improvement in performance. The compatibilised blends must be stable and reproducible, insensitive to forming stresses and repeated processing. Compatibilisation prevents the phase separation. The 'compatibilisation' does not mean that immiscible mixtures become miscible at thermodynamic equilibrium of the system. The immiscible blends become more transparent by adding a block copolymer (compatibilising agents) due to the reduce of the interfacial tension that facilitates dispersion therefore the decrease of domain size, that stabilises morphology against high stress and strain processing and enhance adhesion between the phases in the solid state. By compatibilisation systems with acceptable properties can be obtained from mixtures of immiscible polymeric components. As results improved mechanical properties results.

A **compatibilising agent or compatibiliser,** can either be added to the polymer mixture as a third component or generated '*in situ*' during reactive compatibilisation process. It can be a grafted copolymer or, most often, a block copolymer, see **Figure 1.2**. The added

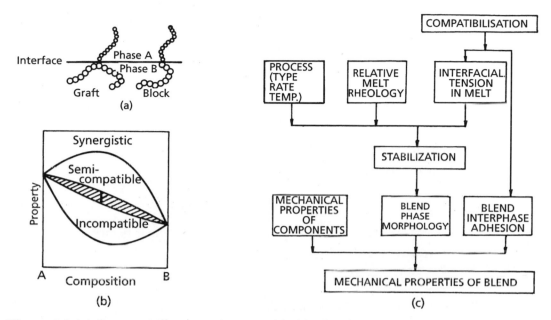

Figure 1.2 (a) Compatibilisation of immiscible blends of polymers A and B by block or graft copolymers (b) the subsequent modification of property responses and (c) Role of compatibilisers in determining end-use properties

compatibiliser should migrate to the interface, reducing the interfacial tension coefficient and the size of the phase domains (a finer dispersion of the minor component), as well as creating improved adhesion in the solid state. Intensified phase interactions and a controlled phase morphology lead to improvements in mechanical properties of the blends. An additional function of compatibilisers is stabilisation of the blend morphology against coalescence and agglomeration of the dispersed particles that can take place during the following processing and forming steps. Different groups of compatibilisers are known and used for such applications. As an example the thermoplastic elastomer Kraton (Shell) is a typical compatibilising agent. This compatibilising agent is a copolymer consisting of a central elastomeric block (polybutadiene, polyisoprene, hydrogenated styrene-butadiene rubbers) with two lateral glassy polystyrene blocks. It is used to compatibilise and to improve the impact properties of PO/polystyrene (PS) and PO/polyethyleneterephthalate (PET) blends. The elastomeric block is miscible with the olefinic component and the glassy styrenic block is miscible with PS or with PET.

It is important to know that miscibility *per se* may not be required or even desired in optimising the properties of a given blend. However what is important is that the miscibility-immiscibility boundary also delineates the regions of compatibility **Figure 1.3(a)**.

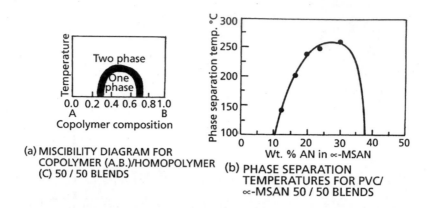

Figure 1.3 Phase diagrams for the homopolymer/copolymer binary blends

Blends of PVC and α-methylstyreneacrylonitrile copolymer (α-MSAN) exhibit a miscibility window that stems from an LCST-type phase diagram. **Figure 1.3(b)** shows how the phase-separation temperature of 50% PVC blends varies with the AN content of the copolymer.

A **compatible polymer blend** indicates a commercially attractive polymer mixture, normally homogeneous to the eye, frequently with some enhanced properties compared to those of the individual components. This type of blend is characterised by phase separation and two glass transition temperatures (T_g). However a thick interphase exists which is composed of a mixture of entangled blocks of the two kinds of macromolecules. These blocks are linked by a certain level of molecular interactions. Consequently, the morphology of this blend consists of a minor component well dispersed in the other. Most of the commercial alloys belong to this category (for example: in the blend of PC and styrene/acrylonitrile (SAN) copolymer, depending of SAN composition, interactions occur between PC and acrylonitrile) and the properties of this products are reasonably good.

Incompatible blends. This is the most frequent case because polymers have not often good affinity for each other. Differences in crystallisation increase repulsive interactions. This results in a very coarse morphology. The interphase is very thin corresponding to a very low adhesion between the components. The blend is characterised by two T_g with values corresponding to those of the virgin polymers and by one or two melting temperatures (T_m) if crystalline polymers are present.

Usually **miscibility (in fact the compatibility)** is assessed against the method or technique of study [7]. Often compatibility is judged by analysing the T_g temperature of the blends by differential scanning calorimetry (DSC) or dynamic mechanical analysis (DMA). These techniques are sensitive to approximately 10 nm in domain size. The existence of one T_g of the blend between those of the constituent materials indicates miscibility, while presence of two T_g in the original position is an evidence of a completely immiscible blend. If two T_g are found, that are shifted towards each other from the value of the components, the blend is termed partially miscible.

The term **pseudo-compatibility** is also used to differentiate between real miscibility and compatibility [8]. Compatibility being found only for a narrow composition range.

Generally, a **polymer alloy** is an immiscible polymer blend having a modified interface and/or morphology. On the contrary, using metals as an analogy, Lipatov and Nesterov [9], proposed the term 'alloy' for binary or multicomponent systems that when they are mixed in the molten state they are situated in the region of the phase diagram corresponding to the mutual miscibility of components, forming of one-phase system. By cooling the melt, a different morphology could be developed depending on the thermodynamic state at a given temperature and on the type of system. The binary system with UCST enter in the region above spinodal forming a two-phase system while the systems with LCST, below spinodal so one-phase structure is preserved by cooling, a miscible system is maintained.

Rätzsch and Haudel [10] define the alloys using interface properties. If no new phases occur, i.e., no miscibility in the interfaces is demonstrable then the term of '**blend**' is used, if an interface (a new phase) is present it is denoted as an '**alloy**'. A polymer alloy is also a specific sub-class of polymer blend. They could be subdivided into two categories: 1) those in which the compatibilisation leads to very fine dispersion so the molded part will show neither streaking not excessive weld-line weakening, and 2) those where some compatibiliser is added in order to facilitate the formation of the desired morphology is a subsequent processing step.

1.2 Related Terms used for Polymer Blends

The term '**compound**' will be used as generic term for blends, alloys, filled and reinforced polymers.

A special case of polymer blends are **semi- and full interpenetrating polymer (SIPN and IPN) networks** and other systems that are formed from two or many polymers during the chemical reactions. In IPN both components form a continuous phase and at least one is synthesised or crosslinked in the presence of the other [11, 12]. Such networks interdisperse two immiscible polymers down to a fine scale of phase separation, and usually use crosslinking to stabilise this morphology. This can produce a remarkable synergism of properties.

An **engineering polymer** is a processable polymeric material capable of being formed to precise and stable dimensions exhibiting high performance at the continuous use at temperature above 100 °C, and having tensile strength in excess of 40 MPa.

An **engineering polymer blend** is a polymer blend or alloy either containing or having the properties of the engineering polymer.

A **homologous polymer blend** is a subclass of polymer blends limited to mixture of two of chemically identical polymers usually of two narrow molecular weight distribution fractions.

Thermoplastic Elastomers (TPE) are materials that possess, at normal temperatures, the characteristic resilience and recovery from the extension of crosslinked elastomers and exhibit plastic flow at elevated temperatures. They can be fabricated by the usual techniques such as blow moulding, extrusion, injection moulding, etc. This effect is associated with certain interchain secondary valence forces of attraction, which have the effect of typical conventional covalent crosslinks, but at elevated temperatures, the secondary bonds dissociates and the polymer exhibits thermoplastic behaviour.

Elastomeric Alloys are a special subclass of thermoplastic elastomers, generated from the synergistic interaction of two or more polymers possessing better properties than those of a simple blend. In elastomeric alloys the elastomer is crosslinked and dispersed in a continuous matrix of thermoplastic under dynamic conditions resulting in a fine dispersion of fully crosslinked rubber particles in the alloy.

Block Copolymers (BC) and Graft Copolymers In a block or graft copolymer, several incompatible components are linked each other by chemical bonds, and the bulk of the copolymers form a stable structure even if the two phases are separated. Usually they show the character or function of each component independently.

Techniques used to retard or eliminate demixing of the polymer blends are called **compatibilisation procedures** [13]. They are:

1. Use of non-reactive compatibilising agents such as statistical, graft or block copolymers;

2. Techniques rely on slow diffusion rates by mixing high molecular weight polymers and co-crystallisation;

3. Techniques to prevent segregation by: crosslinking, forming IPN or mechanical interlocking of components;

4. The use of a reactive polymer or functionalised polymer: One of the components of the blend has been modified, by functionalisation or by copolymerisation, in order to offer reactive groups which are able to react chemically with the second component of the blend.

5. *In situ* Reactive Compatibilisation: Another possibility is offered by the addition of a small molecular compound with a 'double' functionality giving it the ability to react with the two components of the blend. A small amount of solvent is used to carry the small reactant at the strategic location, the interface between the two polymers. It is expected to chemically bind the two polymers.

6. Compatibilisation by Transesterification: Exchange reactions are possible, especially in the melt, in a blend of two polycondensates. As an example melt blending of PC and polybutylene terephthalate (PBT) enhances some exchanges between carbonate and ester functions leading to block copolycondensates. This kind of block copolymers act as a compatibiliser for the corresponding blend;

7. Mechanical Procedures [14]: Methods are studied to promote the chemical compatibilisation during processing without addition of any reactant. During extrusion and injection of polymers, processing temperatures, high shears and strains of the materials generate '*in situ*' macromolecular radicals by mechanochemical action. In the case of a blend of incompatible polymers the recombination of these radicals may produce some macromolecules binding blocks of the original components. These new macromolecules have the structure of compatibilisers and are expected to act as such.

8. Dynamic Vulcanisation: An impact modifier is added to the polymer matrix. Compatibilisation between the matrix and the modifier occurs in the extruder allowing a finer dispersion of the latter. This results in a greater adhesion between the two phases. Simultaneously, another reaction stabilises the morphology of the impact modifier by crosslinking.

9. 'Tapered' Block Copolymer as a Compatibiliser [15]: During blending experiments on a two roll mill, low density polyethylene and polystyrene in the presence of 9 percent two-block styrene-(ethylene-butene) (S-EB) polymers showed improved dispersion of the two polymers and its retention on subsequent compression moulding. Macroscopic examination showed that polystyrene particles as small as 1 mm existed in the polyethylene rich region and a fine semicontinuous to continuous two phase structure in polystyrene rich blends. Two-block polymers were found to be more effective than three-block polymers. A 'tapered' two-block polymer (in which some mixed sequences occur at the block centre) was more effective than a pure two-block polymer of the same molecular weight. It was suggested that the tapered block polymer,

because of its structure, stabilises a diffuse interface between pure polyethylene and polystyrene phases, acting more as solubilising agent for the homopolymers than an anchoring agent. In all cases, however, enhanced strength was observed in the blends, particularly in the polystyrene rich composition region where the ductility was also improved. Factors that favour compatibilisation are: an increased polarity, decrease of molecular weight, specific group attractions, presence of block and graft copolymers

Elastomer toughening: In this 'nonchemical' technique, addition of a low-modulus elastomer to a two-phase polyblend may create a 'soft' deformable interphase, which reduces the tendency of the 'hard incompatible' interface to fracture under stress.

Co-crystallisation: Certain crystallisable polymers are thermodynamically miscible and truly miscible in the melt; but upon cooling, each polymer separates and forms its own unique crystal structure. Occasionally two polymers that have such similar isomorphous crystalline structures that both can enter the same crystal lattice and co-crystallise, thus forming a single homogeneous solid product. This behaviour has important consequences in several polyolefin polyblends.

Plasticisation consists of addition of various liquids or solids (**plasticisers**) to rigid polymers to improve their elasticity and to make them much resistant to frost and easier to process, increasing mobility of supermolecular structures. The efficiency of a plasticiser depends on its compatibility with the polymer and formation of a thermodynamically stable polymer-plasticiser system. Phase separation may lead to the loss of plasticiser by its migration to the surface of polymer articles.

An **interface** is an imaginary plane without thickness between the two phases, a mathematical convenience.

An **interphase** is a third phase in a binary system (blend, composites or nanocomposites) that controls their performance. It is increased by adding a compatibilising agents, see **Figure 1.4**.

The **interface** is characterised by a two-dimensional array of atoms and molecules which is impossible to measure, while the **interfacial layer** has a large enough assembly of atoms or molecules to have its own properties such as modulus, strength, heat capacity, density, etc. According to Sharpe [16], the **interphase** is 'a region intermediate for two phases in contact, the composition and/or structure and/or properties of which may be variable across the region and which may differ from the composition and/or structure and/or properties of either of the two contacting phases. Lipatov [17-19] considers that terms of **interfacial layer** and **interphase** are equivalent.

Rheopectic is a rheological term used to denote an increase of apparent viscosity with time at a constant rate of agitation or shear.

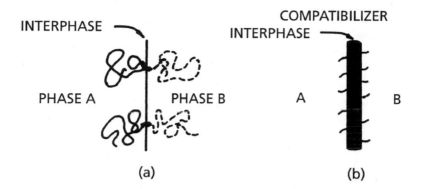

Figure 1.4 (a) Ideal configuration of a block copolymer at the interface between polymer phases A and B; (b) Formation of an interphase between them

Pseudoplastic is a rheological term used to describe material which has a yield stress above which flow occurs and where the flow rate increases sharply with further increases in stress. Printers ink and paper (cellulose) pulp are pseudoplastic in behaviour.

Biodegradation is the gradual degradation of the molecular structure of materials induced by environmental conditions. Polysaccharide derivatives and polyalkanoates have been exploited in the development of **environmentally friendly materials.**

Bioplastics are biodegradable materials with high water vapour permeability, a good oxygen barrier, not electrostatically chargeable and usually environmentally friendly materials. As an example they are obtained by blending thermoplastic starch (TPS), with biodegradable polymers like polylactic acid (PLA) [20] or poly(vinyl alcohol) PVA and plastifying agents (glycerol and its derivatives, sorbitol and its derivatives, etc). Bioplastics can be processed with the existing plastic processing industrial production lines [21, 22, 23, 24]. Bioplastics reduced the waste amounts in an ecological manner by closing the natural material cycles [25]. Commercial biodegradable plastics, used for packaging and agricultural mulch films, are Mater-bi AF10H, prepared from a blend of starch (60%) and natural additives with modified PVA (40%) and plasticisers [26, 27].

Biocompatibility is one of the most important criteria involved in the selection of a biomaterial for use in long-term biomedical applications. Biocompatibility by definition is [28, 29] the ability of a material to perform with an appropriate host response in a specific application. The definition of biocompatibility will vary with the application of the microsystem but in general requires the use of surfaces with low protein adsorption, non-leaching materials, and other non-interactive features [30].

Blood compatibility means that a foreign surface placed in contact with blood should not induce adverse responses such as thrombosis, destruction of the blood cellular components, alteration of the plasma proteins, damage to adjacent tissue and toxic or allergic reactions [31].

A **biopolymer** is a biologically active protein, peptide, carbohydrate or nucleic acid produced by and purified from naturally occurring or recombinate plant or animal organism, tissues or cell lines or synthetic analogs of such molecules.

Special terms are used for highly interacting polymer systems, such as water-soluble ones: An **interpolymer complex** is a chemical compound obtained by mixing solutions as a result of the interaction of two or more compatible polymers. It can occur both between two oppositely charged polyelectrolytes due to an ionic bond and through hydrogen bonding between a polyacid and a proton-acceptor groups of a noncharged polymer [32] or two components having electron donor-electron acceptor groups [33-39]. An interpolymeric complex is characterised by a **decomplexation temperature** at which phase separation occurs. Interpolymeric complex formation depends on the association-dissociation equilibrium.

The **complex coacervation** means a spontaneous liquid/liquid phase separation that occurs when oppositely charged colloids are mixed as a results of electrostatic attractions.

Polymers which do not dissociate into ions, form solutions of **polymeric non-electrolytes** or **neutral polymers**. Polymers which undergo electrolytic dissociation when dissolved form solution of **polymeric electrolytes** or **polyelectrolytes**. A **polyelectrolyte** is a macromolecule in which a substantial portion of the constitutional units have ionisable or ionic groups, or both [40]

Hydrocolloids are colloids which can interact with water to form new textures and perform specific functions [41]. Polysaccharides such as starches, carrageenan, pectin, xanthan, arabic gum, agar gum, locust beam gum, guar gum, carboxymethyl cellulose (CMC), gellan, konjac mannan and alginates are known from the literature as hydrocolloids.

A **gel** is a nonfluid two-component system formed by molecular dispersion of a micromolecular liquid in a polymer with crosslinked chains [42]. It is the best described as a continuous three-dimensional network that is held together by chemical or physical bonds. Sufficient space exists within the network, and solvent molecules can become trapped and immobilised, filling the available free volume. The gels can be divided into two major categories based on their types of bonds. These include chemical gels and physical gels [43]. Gels with chemical bonds between their structural elements (limited swollen network polymers) are single-phase thermodynamically stable systems, in which the content of the low molecular liquid at given temperature and pressure depends on the nature of the liquid and the polymer, as well as on the crosslink density of the latter. A gel with strong intermolecular bonds between its structural elements (gelatine gel) is a thermodynamically unstable system, the state of equilibrium of which corresponds to the formation of two phases. Such gels form only because the energy of interaction of the polymer chains with each other is higher than the energy of their interaction with the molecules of the low molecular liquid. The latter must be a poor solvent for polymer. If the liquid is very polar, it cannot disturb the interaction between polymer chain fragments of low polarity. If the liquid is nonpolar or of low polarity crosslinks appear as a result of interaction between the polar groups of polymer. Since the affinity of polymer for the liquid is low, the system should separate into phases, but this separation takes time. They often have a biphasic structure induced by **segregative interactions** (thermodynamic incompatibility) between the constituent polymers [44]. Systems phase separation occurs in the pre-gel solution state or during cooling [45] or one component gels first from a single phase solution and the other then forms a second (dispersed) phase by gelling within the pores of the original (continuous) network [45-48]. The ultrastructure of the resulting co-gels can vary widely depending on the time-temperature course of gelation in relation to the rate of the segregation and network formation [49].

The term **hydrogel** is used to describe materials that are hydrophilic in nature and exhibit the characteristic structure of a gel [43].

The **gelation process** is the beginning of phase separation. It could induce the differences between phase relationships and pre-gel solution state. In the early state of gelation after the onset of intermolecular association but before development of a continuous network there will be a massive reduction in the number of species free to move independently which will reduce the relative importance of entropy of the mixing and hence promote further segregation therefore the assumption of complete de-mixing could be not entirely unrealistic (especially when the rate of gelation is slow enough) [48].

The overall physical properties of such **co-gels** depends on the relative deformation [12, 49].

Syneresis. The process of gradual phase separation of a gel and segregation of the low molecular weight liquid is called syneresis. The seepage or release of water from gels, particularly on storage is also known as syneresis. Syneresis in food applications may be avoided by choosing a suitable chemically stabiliser.

The transition of a nonfluid gel into a fluid solution is called 'melting' of the gels.

The gelatinisation and **retrogradation** were described as non-equilibrium processes. It is generally found with starch and it is defined as irreversible insolubilisation of starch paste with formation of a precipitate or gel depending on concentration. Also known as 'set back.' Retrogradation is mainly due to the presence of amylose in the starch. The linear amylose molecules are attracted to each other and form bundles of parallel polysaccharide chains by the formation of hydrogen bonds between hydroxyl groups on neighbouring molecules [50].

References

1. L.A. Utracki, *Polymer Alloys and Blends*, Hanser Publishers, Munich, Germany, 1989.

2. *Encyclopedic Dictionary of Commercial Polymer Blends*, Ed., L.A. Utracki, ChemTech, Toronto, Canada, 1994.

3. A. Ajji and L.A. Utracki, *Progress in Rubber and Plastics Technology*, 1997, **13**, 3, 153.

4. A. Ajji and L.A. Utracki, *Polymer Engineering and Science*, 1996, **36**, 12, 1574.

5. M.M. Coleman, J.E. Graf and P.C. Painter, *Specific Interactions and the Miscibility of Polymer Blends: Practical Guides for Predicting & Designing Miscible Polymer Mixtures*, Technomic Publishing Company Inc., Lancaster, PA, USA, 1991.

6. O. Olabisi, L.M. Robeson and M.T. Shaw, *Polymer-Polymer Miscibility*, Academic Press, New York, NY, USA, 1979.

7. C. Wästlund, *Free Volume Determination in Polymers and Polymer Blends*, Department of Technology, Chalmers University of Technology, Goteborg, Sweden, 1997. [PhD Thesis]

8. C. Vasile and I.A. Schneider, *European Polymer Journal*, 1973, **9**, 10, 1063.

9. Y.S. Lipatov and A.E. Nesterov, *Thermodynamics of Polymer Blends*, Technomic Publishing Company Inc., Lancaster, PA, USA, 1997.

10. M. Ratzsch and G. Haudel, *Macromolekulare Chemie - Macromolecular Symposia*, 1990, **38**, 81.

11. M.A. Winnik, *Polymer Engineering and Science*, 1984, **24**, 2, 87.

12. J.A. Manson and L.H. Sperling, *Polymer Blends and Composites*, Plenum Press, New York, 1976.

13. A. Rudin, *Journal of Macromolecular Science*, 1980, **C19**, 2, 267.

14. R.D. Deanin and M.A. Manion in *Handbook of Polyolefins*, Second Edition, Ed., C. Vasile, Marcel Dekker Inc., New York, NY, USA, 2000, Chapter 23, p.633.

15. S. Bywater, *Polymer Engineering and Science*, 1984, **24**, 2, 104.

16. L. Sharpe, *Journal of Adhesion*, 1989, **29**, 1.

17. Y.S. Lipatov, *Polymer Science, USSR*, 1978, **2**, 1-18.

18. Y.S. Lipatov, *Interfacial Phenomena in Polymers*, Naukova Dumka, Kiev, Ukraine, 1980.

19. Y.S. Lipatov, *Polymer Reinforcement*, ChemTec Publishing, Toronto, Canada, 1995.

20. P. J. Wuk, L. D. Jin, Y. E. Sang, I. S. Soon, K. S. Hyun and K. Y. Ha, *Korea Polymer Journal*, 1999, 7, 2, 93.

21. I. Tomka, inventor; Fluntera AG, assignee; EP 0542 155, 1993.

22. I. Tomka, inventor; no assignee; US 5,280,055, 1994.

23. I. Tomka, inventor; Fluntera AG, assignee; EP 596 437, 1994.

24. J. Lörcks, W. Pommeranz and H. Schmidt, inventors; Biotech Biologische Naturverpackungen GmbH, assignee; WO 96/19599, 1996.

25. J. Lörcks, *Polymer Degradation and Stability*, 1998, 59, 1/3, 245.

26. T. Iwanami and T. Demura, *Japanese Journal of Polymer Science and Technology*, 1993, 50, 767.

27. C. Bastioli, V. Bellotti, L.D. Giudice and G. Gilli, *Journal of Environmental Polymer Degradation*, 1993, 1, 3, 181.

28. *Polymers of Biological and Biomedical Significance*, Eds., S.W. Shalaby, Y. Ikada, R. Langer and J. Williams, ACS, Washington, DC, USA, 1994.

29. D. Klee and H. Hocker, *Polymers for Biomedical Applications: Improvement of the Interface Comaptibility*, Advances in Polymer Science No.149, Springer-Verlag, Berlin, Germany, 1999, 1.

30. B.K. Gale, R.S. Besser, J. Brazzle, I. Papautski and A.B. Frazier, Packaging for Biomedical Analysis Systems, Proceedings of the 198th Meeting of the Electrochemical Society: Microfabricated Systems and MEMS V, Phoenix, AZ, USA, 2000.

31. M. Gheorghiu, G. Popa and O.C. Mungiu, *Journal of Bioactive and Compatible Polymers*, 1991, **6**, 2, 164.

32. O. Nikolaeva, T. Budtova, Y. Brestkin, Z. Zoolshoev and S. Frenkel, *Journal of Applied Polymer Science*, 1999, **72**, 12, 1523.

33. M.C. Piton and A. Natansohn, *Macromolecules*, 1995, **28**, 1197, 1598, 1605;

34. G. Cho and A. Natansohn, *Chemistry of Materials*, 1997, **9**, 148.

35. U. Epple and H.A. Schneider, *Thermochimica Acta*, 1987, **112**, 123.

36. U. Epple and H.A. Schneider, *Thermochimica Acta*, 1990, **160**, 103.

37. M. Bolsinger and H.A. Schneider, *Makromolekulare Chemie*, 1994, **195**, 2683

38. M. Bolsinger and H.A. Schneider, *Journal of Thermal Analysis*, 1998, **51**, 643.

39. M. Bolsinger and H.A. Schneider, *Journal of Thermal Analysis*, 1998, **52**, 115.

40. A.D. Jenkins, P. Kratochvil, R.F.T. Stepto and U.W. Suter, *Pure and Applied Chemistry*, 1996, **68**, 12, 2287.

41. *Hydrocolloids*, Ed., K. Nishinari, Elsevier, Amsterdam, The Netherlands, 2000.

42. A. Tager, *Physical Chemistry of Polymers*, Mir Publishers, Moscow, Russian Federation, 1972, 453.

43. R.J. LaPorte, *Hydrophilic Polymer Coatings for Medical Devices: Structural Properties, Development, Manufacture and Applications*, Technomic Publishing Company, Lancaster, PA, USA, 1997.

44. V.V. Suchkov, V. Ya Grinberg, V.B. Tolstoguzov, *Carbohydrate Polymers*, 1981, **1**, 39.

45. I.S. Chronakis, S. Kasapis and R. Abeysekera, *Carbohydrate Polymers*, 1996, **29**, 137.

46. A.H. Clark, R.K. Richardson, G. Robinson, S.B. Ross-Murphy and A.C. Weaver, *Progress in Food and Nutrition Science*, 1982, **6**, 149.

47. A.H. Clark, R.K. Richardson, S.B. Ross-Murphy and J.M. Stubbs, *Macromolecules*, 1983, **16**, 1367.

48. S. Kasapis, E.R. Morris, I.T. Norton and A.H. Clark, *Carbohydrate Polymers*, 1993, **21**, 269.

49. S. Alevisopoulos, S. Kasapis and R. Abeysekera, *Carbohydrate Research*, 1996, **293**, 79.

50. *Dictionary of Food Starch Terms*, National Starch & Chemical website, New York, NY, USA, www.foodstarch.com.

2 Thermodynamics of Multicomponent Polymer Systems

Cornelia Vasile and Gina G. Bumbu

The thermodynamics of polymer solutions and blends has the same basic principles as classical thermodynamics however, it was developed as a new discipline.

The fundamental scientific questions raised by polymer systems and their commercial significance, as well as their potential for achieving novel materials properties, are among the reasons for continued interest in study of their structural and thermodynamic properties. Rational development of polymer blends' technology has raised many interesting scientific challenges. The answer these questions requires progressively higher levels of sophistication and knowledge about the systems using a thermodynamic theory or model appropriate for the level of detail to be predicted. The reason for the effort is that thermodynamic solubility is an exception and if compatible systems can be found, the probability of obtaining a proprietary compatible blend is high. The thermodynamic study of polymer systems also stimulated the study of the individual polymers. Some of parameters, related only to the pure components, have to be specified through appropriate characterisation or physical properties determination, e.g., pressure-volume-temperature behaviour, as in recent years experimental observations have demonstrated the sensitivity of macrophase and microphase boundaries on local polymer architecture. The knowledge of the polymer-polymer interactions are generally required, too. In many cases, the most practical means, is to reverse the prediction procedure, i.e., the parameters are deduced from observation of phase behaviour. A quantitative prediction of phase diagram should be the most useful for practical purposes.

Because of its attractive and exciting theoretical aspects in solving problems related to structure, conformation and configuration of macromolecules and also due to the direct practical implications, thermodynamics of polymer mixing has seen over the years many steps of developments and it is still a very fascinating domain. Many very specialised and superb reviews and monographs have been published in a relatively short period of time, starting with Flory, Huggins and Staverman books [1-3] and their co-workers papers, who developed the first theory that was commented, contested, revised and again reconsidered in the last year as demonstrated by Sanchez and Stone [4]. Due to the reasonable character of the assumptions underlying the Flory-Huggins-Staverman theory and mainly due to the remarkable simplicity of the final formulas, this theory up to now remains widely popular among investigators dealing with experimental data treatment. During the last ten years important reviews appeared with special reference to polymer blends. The approaches developed in thermodynamic characterisation of polymer systems including polymer blends are summarised in many books and reviews [5-30].

2.1 Approaches Developed in Thermodynamics of Polymer Blends

Two quite divergent approaches are currently in use for describing or predicting the phase behaviour of polymer mixtures. One uses association or quasi-chemical models and is applicable for systems where strong specific interactions, like hydrogen bonding are involved. The other uses a mean field approach which is appropriate when the interactions are given by geometric mean of the interactions between the two like pairs. This last approach leads to endothermic heats of mixing and positive Gibbs free energy (ΔG) values and always predicts immiscibility in the limit of very high molecular weight of the components, see **Figure 1.1a**.

Fundamental studies were started independently by Flory, Huggins and Staverman. The advancement of the theoretical concepts in the field of thermodynamics of homopolymer solutions and blends was due to development by relaxing some of the five assumptions of the Flory-Huggins theory (see subchapter 2.1.2). Further progress in the Flory-Huggins theory was associated with an account for polydispersity of macromolecules and the selection of more complicated dependencies of the enthalpy of mixing on concentration and temperature [31]. The most serious shortcoming of the Flory-Huggins theory is apparently, a disregard of the volume change occurring at dissolution. The effect was taken into account in the framework of the molecular theories of polymer solutions proceeding from their equation-of-state that originate from fundamental studies by Prigogine and co-workers [32, 33] who demonstrated the possibility of applying the principles of corresponding states to polymer systems. The refinement of the Prigogine theory [34-36] enabled to predict the existence of upper and lower critical solution temperatures, to explain the phenomenon of volume change during mixing. The difficulty of the theory resides in finding the values of reduction parameters and the accuracy of their determination [37-39], (because the predicted phase diagram very sensitively depends on them), which was obviated by new Flory's theory [40-45] and lattice liquid theory [46-50]. The new Flory theory underwent a number of modifications mainly connecting with regard for inhomogeneity of mixing and some other factors [51-55]. The construction of phase diagrams – **Figures 1.1b and 1.1c** - was started 40 years ago by Scott, [56], Tompa [57, 58] and Stockmayer [59] who applied Flory's formalism. The theories were experimentally verified first by Koningsveld and co-workers, in a series of works what is undoubtedly of first-rate important for physical chemistry of polymers [16, 31, 60-65] who considered the influence of polydispersity of polymers upon the thermodynamic characteristics and cloud point curves. They also developed a gas-lattice model to describe multi-peak spinodals. Other authors took into consideration this factor by so-called 'continuous thermodynamics' method [23, 24, 66-69]. Solc [67-70] studied binary and ternary systems containing two chemically different polymers. While there is appreciable progress in theoretical description of solutions and blends of homopolymers the quantitative theory heteropolymer systems is still to be developed [27]. Scott seems to be the first who examined the problem of thermodynamic compatibility of copolymers [71]. Almost 30 years later approaching the same problem Leibler [72] managed, in the framework of Flory-Huggins formalism, to obtain the expression for the spinodal and critical point for the copolymer blends with different compositions. In other papers a homopolymer with two copolymer components, two copolymers with one identical type of monomer units [73] or two copolymers with different pairs of monomer units, and other blends containing copolymers [74] were examined. A strong directional interactions model was proposed by ten Brinke and Karasz [75].

Practical formulas have been proposed by several authors [76, 77].

In this chapter the milestones and the progress in thermodynamics of polymer mixing is discussed, emphasising the aspects and relationships currently used in miscibility/compatibility studies and the practically applications.

2.1.1 Basic Principles

The thermodynamic behaviour of polymer blends determines the compatibility of the components, their morphological features, rheological behaviour, microphase structure and in such a way the most important physical and mechanical characteristics of blends. The study of miscibility is also important for judicious selection of compatibilisation strategies.

A polymer blend is a special type of polymer solution. All results developed for polymer solutions, including those for multicomponent mixtures can be applied to polymer blends, allowing that a polymer is distinguished from a small molecule by some size parameters, such as the degree of polymerisation.

The main task of the thermodynamics of binary polymer systems is to establish the theoretical background to find the conditions of mixing and the ranges of mutual solubility of different polymers depending on their chemical nature and molecular mass. The models provide for the specification of the size or length of the macromolecule. The mixing of polymers at elevated temperatures and of amorphous polymers can be considered as mixing of two liquids. According to the general principles of thermodynamics, the state of system can be described by the variation of the Gibbs free energy (ΔG):

$$\Delta G = \Delta H - T\Delta S \qquad\qquad (2.1)$$

The necessary but not sufficient condition for the system stability is that $\Delta G < 0$ which is fulfilled if enthalpy $\Delta H < 0$ and entropic term $T\Delta S > 0$ or if $\Delta H > 0$, $|T\Delta S| > |\Delta H|$ see **Figure 1.1a** and **1.1b**.

At first sight, it appears that for two polymers with a high molecular mass, the contribution of the entropic term is very small, so the change in enthalpy is responsible for the mixing behaviour of two polymers. However the Russian school showed that contribution of entropy cannot be neglected at least for several systems [5, 13, 14].

ΔG is known as a function of pressure, temperature, and composition. The enthalpic part is usually expressed in terms of contact interactions.

Pressure related phenomena are usually treated by means of equation-of-state theories, they led to additional terms in enthalpy and entropy.

The change of molecular packing upon mixing and its effect upon the thermodynamic properties of the mixture, have not, as far as we know, received adequate attention, although several studies are known [78]. The packing affects results because the molecules will arrange themselves in such a way as to shorten the intermolecular distances, i.e., decrease the volume, as much as possible in order to maximise intermolecular interactions. The most favourable arrangement depends on the molecular geometry. The best arrangements are achieved in crystals, but in liquids mutual molecular orientations are still preserved in a rudimentary form. Parallel arrangement generates liquid crystals. Their entropy is very low because their freedom of movement is restricted along their molecular axes.

2.1.2 Flory-Huggins - Simple Mean Field, Rigid Lattice Treatment

For over 50 years the lattice models as Flory-Huggins-Staverman theory dominated the polymer solution thermodynamics. The Flory-Huggins theory, being the original from which the thermodynamics of polymer systems emanated, in its first version has the following assumptions [1-7, 27]:

• Fluctuation effects are excluded from consideration. That is why, like any mean field theory, this one does not work when dealing with dilute solutions or with phenomena occurring in the close vicinity of the critical point.

- The system is supposed to be incompressible. This restriction precludes one from explaining the experimentally observed volume alteration under polymer mixing and leaves out effects induced by pressure change.

- The application of the rigid lattice model. Within its framework, no account is taken of distinction in size and shape of monomer units of different homopolymers or in copolymers.

- Random mixing approximation. This assumption, whereby the local composition of quasicomponents coincides with their average composition in a mixture, would be expected to be violated for some systems where specific interaction (electron-donor-electron acceptor or π-complex formation) are of special importance.

- Polydispersity is ignored.

According to regular solution thermodynamics, Flory-Huggins equation for the Gibbs free energy of mixing (ΔG_m) is:

$$\Delta G_m - (RTV/V_r)\left[\frac{\varphi_A}{r_A}\ln\varphi_A + \frac{\varphi_B}{r_A}\ln\varphi_B + \chi_{AB}\varphi_A\varphi_B\right] \qquad (2.2)$$

where, R is the universal gas constant, T is the absolute temperature, V is total volume of the mixture, r_A and r_B are the numbers of segments (or polymerisation degree) of polymer A and B, $r_i = V_i /V_r$ (V_i is molar volume of component i) φ_A and φ_B are volume fractions of components.

$$\chi_{AB} = (\varepsilon_{ii} + \varepsilon_{jj} - 2\varepsilon_{ij})/k_BT \equiv z\Delta\varepsilon_{ij}\ r_i/k_BTV_i \text{ or } \chi_{AB}\ (\varphi_I,\ T,P,M_w)$$

χ_{AB} is a thermodynamic interaction parameter related to the enthalpy of interaction between different segments (ε) of volume, V_r which is the volume of segment taken equal to the volume of repeating unit of the polymer chain (the same for both polymers) or a reference volume related to the size of the unit cell on the lattice, k_B is Boltzman's constant and z is the coordination number of the lattice.

Koningsveld kept these terms in Equation 2.2, but added an Γ interaction function. By means of this function several improvements of the theory have been accommodated in ΔG_m expression [79]. For example:

$$\Gamma = g_{AB}\varphi_A\varphi_B \text{ With } g_{AB} = a + b(T)/(1-c\varphi_B) \text{ where } b(T) = \chi_S + h/T$$

where χ_S and h/T represent the entropic and enthalpic contribution or the parameter, χ_{AB} is a free energy parameter comprising an energetic χ_H and an entropic part, χ_S:

$$\chi_{AB} = \chi_H + \chi_S.$$

Koningsveld [79] established the dependence of the interaction parameter on concentration and temperature:

$$\chi_{AB}(\varphi, T) = \sum^{2} a_i \varphi_2^i; \quad \text{with} \quad a_i = \sum^{i} a_{ij} T^i \tag{2.3}$$

this relationship being introduced to explain many experimental data as multiple peak spinodals. Other expressions have been also proposed by Koningsveld and other authors [80]. They concluded that four factors determine the polymer/polymer miscibility namely: interacting surface areas of various types of segments, coil dimensions as a function of temperature, molar mass and concentration, molar mass distribution and free volume. The simplest form is:

$$\chi_{AB} = (a_o + a_1/T + a_2 T)(b_o + b_1 \varphi_B + b_2 \varphi_B^2) \tag{2.4}$$

Notwithstanding general acceptance of Equation 2.2 of ΔG_m cannot be used to quantitatively anticipate the phase behaviour of most polymer blends. In practice these deficiencies are usually accounted for empirically by incorporation a temperature-dependent term into χ, as well as composition (ϕ) and/or molecular weight dependencies (N) [81, 82]:

$$\chi(T, N, \phi) = \frac{\alpha(\phi_A, N)}{T} + \beta(\phi_A, N) \tag{2.5}$$

which may be attributed to specific interactions, equation-of-state effect, e.g., mismatched compressibilities, and asymmetric monomer shapes. N is number of segments and a_o, a_1, ...a_I and b_o, b_1, ...b_I are constants.

In this approach the contribution to the entropy of mixing represented by the first two terms in the brackets (Equation 2.2), is purely combinatorial and will become negligibly small as the molecular masses of each component increase, i.e., r_i becomes large. However, the entropy of mixing also has other significant non-combinatorial contributions.

This means that the miscibility becomes increasingly dependent on the interactional free energy, expressed as a segmental interaction parameter, which will required to be either very small or negative, if a one-phase system is to be obtained (ΔG_m negative). The χ_{AB} parameter is then assumed to be a composite term that includes contribution from dispersive forces, specific interactions, non-combinatorial entropy effects and, to a lesser extent compressibility effects. In the recent treatment the χ_{AB} is considered a binary thermodynamic interaction function that depends on concentration, temperature, pressure, molecular weight, molecular weight distribution, etc. The interchange energy for formation of A/B contacts of monomer units is:

$$\Delta\varepsilon = \varepsilon_{AB} - (\varepsilon_{AA} + \varepsilon_{BB})/2$$

equal with algebraic sum of energies of intermolecular interaction. Hildebrand [83] realised that the geometrical average seems to be reasonable:

$$\varepsilon_{ij} = \sqrt{\varepsilon_{ii}\varepsilon_{jj}}$$

It should be realised that the equation of ΔG_m (Equation 2.2) is only a crude approximation because the distribution of the molecules on the pseudo-crystalline lattice is non-random and the molecules usually have different size and shapes. A much more correct analysis replaced the volume fraction φ, by the surface fraction 0 and the composition independent parameter χ, by a concentration dependent function g_{ij}, with a character of a free energy function. All these values result mainly from contact interactions. Various kinds of interactions have to be considered but they must be understood in local terms not in the terms of the whole molecule. While the energies of the homo-contacts, ε_{ii}, may be obtained easily from the properties of the pure components, the hetero-contacts energy, ε_{ij}, are generally derived from the thermodynamic studies of the mixtures themselves, frequently a very difficult task. Being able to predict the value of ε_{ij} from the properties of the pure components would therefore be of great practical importance.

Sometimes it is considered more useful to use the polymer/polymer interaction energy density (B_{ij}) instead of χ_{ij} parameter since the use of arbitrary reference volumes involved in the definition of this last parameter can lead to confusion and serves no useful purpose. The χ_{ij} is used extensively in theories of polymer solutions while B_{ij} is the excess cohesive density, used mainly with bulk polymer and polymer blends. Values B_{ij} and χ_{ij} extracted from experimental data may contain effects other than simple enthalpic contribution, so these quantities are more appropriately regarded as excess free energy parameters. In many cases evaluation of ΔG_m indicates that the systems are not in thermodynamic equilibrium.

The blends tend to reach equilibrium in two ways. Firstly, it tends to lower its Gibbs free energy by further phase separation into a state of smaller mutual solubility and the second effect is the increase of mutual solubility in the case of immiscible blends. This phenomenon was named 'thermodynamic ageing' [79]. This phenomenon appears because the chemical potential gradients that will force the system into some degree of mutual solubility, however slow the change may be that will eventually lead to physical properties deviating from those of the initial blend. Special measures must be taken if this inevitable process of thermodynamic ageing is to be prevented. The problems associated with the production of polymer blends under far-from equilibrium conditions can be very difficult. During the product's lifetime, its properties drift toward the equilibrium. For poorly mixed homogeneous polymer blends this may be detrimental although it may even lead to an improvement of properties. However, the heterogeneous polymer blends, the phase ripening and deterioration of mechanical properties may lead to disastrous results. Vitrification of one of the phases or both will usually take place, but reprocessing of the blend will then start renewed thermodynamic ageing. Crosslinking of one or both phases might offer a better perspective because a network tends to expel free chains.

Alongside the theoretically rigorous approaches mentioned above, polymer scientists have traditionally used a much simpler 'rule-of-thumb' technique for estimating miscibility. The method is based upon comparison of Hidebrand solubility parameters for the polymer of interest.

2.1.3 Solubility Parameter Approach

When attempting to modify or tailor the properties of a polymer by blending, one quickly realises it would be highly desirable to have available a technique that provides guidelines for estimating miscibility or regions (windows) of miscibility. A simple approach that could provide guidelines rather than a rigorous theoretical treatment, particularly useful to the industrial polymer scientist/engineer is solubility parameter approach. It is capable to emphasise trends and not rigorous quantitative results.

The dispersive forces are often approximated from interaction parameters. χ_{AB} can be calculated from the relationship:

$$\chi_{AB} = (\delta_A - \delta_B)^2 (V_S / RT) \qquad (2.6)$$

Equation 2.6 is valid for non-polar substances that means in the presence of only van der Waals or London interactions, where δ_A and δ_B are solubility parameters of components. It can be seen that calculated values of χ_{AB} are always positive. Critical values of interaction parameter and volume fraction are:

$$(\chi_{AB})_{crit} = (1/2)[r_A^{-1/2} + r_B^{-1/2}] \cong 0.5$$

$$(\varphi_A)_{crit} = \frac{1}{[1 + (r_B / r_B)^{1/2}]}$$

Therefore $(\chi_{AB})_{crit}$ as it was shown is close to zero and contribution of combinatorial entropy is negligible small, therefore *the theory predicts that it is almost impossible to find a miscible polymer pair in absence of any specific interactions*. Therefore the components of the blend would have to have closely matching values of δ_i if miscibility is to be achieved. Some small assistance might come from the combinatorial entropy but consideration of the critical conditions for miscibility, namely $\chi_{AB} < (\chi_{AB})_{crit}$, shows that the difference between the solubility parameters of the components is very small indeed.

For low molecular weight compounds, the solubility parameters can be directly calculated from heat of vapourisation measurements $\delta = [\Delta E/V]^{1/2}$ in $(cal/cm^3)^{1/2}$, where ΔE is the energy of vapourisation or $\delta^2 = CED$, cohesive energy density. The polymer solubility parameters are determined by indirect methods as swelling, viscosity or surface tension measurements in various solvents or by calculation from group molar attraction constants (F_i):

$$\delta = \frac{\sum F_i}{V}$$

Hansen [84-86] recognised three types of interactions or δ as a sum of various contributions as dispersive (δ_d) polar (δ_p) and hydrogen bonding (δ_h):

$\delta^2 = \delta^2_d + \delta^2_p + \delta^2_h$ so that the Equation for B_{ij} is [78]:

$$B_{ij} = (\delta_{i,d} - \delta_{j,d})^2 + (\delta_{i,p} - \delta_{j,p})^2 + (\delta_{i,h} - \delta_{j,h})^2$$

Considering hydrogen bonding as a special case of electron donor-electron acceptor interactions it was proposed the introduction of four solubility parameters to characterise the systems [78] and using the molar surface interactions, a reasonable improvement is obtained.

Weight fractions of repeated units are multiplied by their respective individual solubility parameter contributions. $(\delta_A - \delta_B)^2$ is considered a miscibility parameter, denoting miscibility for values 0.1 or lower. To use this parameter to estimate miscibility a bank (or database) of solubility parameter is necessary connected to a program (Matprop from SF Technologies) including polymer properties [10, 87].

Coleman, Graf and Painter [10] produced a practical guide to polymer miscibility based on solubility parameter approach, using the Mac II MATLAB programme, that is used to select the components of a blend, reducing the time used in searching the potential miscible polymer systems and indicating the extent of chemical modification of polymers by incorporation of some functional groups to render them compatible. It allows calculation of the free energy, phase diagrams and miscibility windows. However due to the errors involved in such an evaluation it is not recommended for theoretical purposes. Prediction of compatibility on the basis of solubility parameters using a computer programme known as 'spherical volume of solubility' gave good correlation with mechanical properties for the polycarbonate (PC)/acrylonitrile-butadiene-styrene (ABS) blends [88].

The molar attraction constants tabulated by Small, Barton or van Krevelen, [89-91] have been compared and selected for unassociated and weakly associated groups, see **Table 2.1**, using the same set of model compounds.

Several solubility parameters of common monomers are given in **Table 2.2**.

Table 2.1 Molar attraction constants [90, 91]			
Group	V* (cm³/mol)	F+ [10] (cal.cm³)$^{1/2}$ /mol	F [90] (cal.cm³)$^{1/2}$ /mol
Unassociated groups			
-CH$_3$	31.8	218	214
-CH$_2$ -	16.5	132	133
> CH -	1.9	23	28
> C <	-14.8	-97	-93
C$_6$H$_3$	41.4	562	-
C$_6$H$_4$	58.8	652	658
C$_6$H$_5$	75.5	735	735
CH$_2$ =	29.7	203	190
-CH =	13.7	113	111
> C =	-2.4	18	19
- OCO -	19.6	298	310
- CO -	10.7	262	275
- O -	5.1	95	70
> N -	-5.0	-3	-
Weakly associated groups			
- Cl	23.9	264	260
- CN	23.6	426	410
- NH$_2$	18.6	275	-
> NH	8.5	143	-
V* - molar volume from density measurement			

Table 2.2 Solubility parameters of several monomers [87, 88]			
Monomer	δ_d (cal.cm^3)$^{1/2}$	δ_p (cal.cm^3)$^{1/2}$	δ_h (cal.cm^3)$^{1/2}$
α-Methyl styrene	8.56	0.459	0
α-Methyl styrene-ran-4-(2-hydroxy ethyl α-methyl styrene)	8.311	1.637	5.590
Acrylonitrile	8.51	12.00	3.65
Butadiene	8.01	0	0
Butylene terephthalate	8.747	2.756	4.373
Cyclohexane dimethylene succinate	8.02	2.389	4.084
Cyclohexyl methacrylate	8.493	1.582	3.324
Ethylene	8.022	0	0
Maleic anhydride	9.038	13.962	5.479
Methacrylonitrile	8.038	8.42	3.058
Methyl methacrylate	8.075	2.767	4.396
Methylene phenylene oxide	9.395	1.252	2.068
ε-Caprolactame	8.447	4.599	3.42
Propylene	7.665	0	0
2,2-Propane bis(4-phenyl carbonate)	8.769	1.545	3.358
Phenylene oxide	9.099	1.708	2.457
Styrene	8.88	0.55	0
Vinyl alcohol	7.62	6.80	11.54
Vinyl acetate	7.853	3.317	4.813
Vinyl butyral	7.72	2.90	3.26
Vinyl chloride	8.65	5.95	1.45
Vinyl methyl ether	7.382	3.394	3.527
ε-Caprolactone	7.959	2.241	3.956
Butylene adipate	7.948	2.649	4.301
Butylene terephthalate	8.747	2.756	4.373
Dipropylene succinate	7.724	2.897	4.497
Ethylene adipate	7.931	3.239	4.735
Ethylene orthophthalate	8.378	3.190	4.705
Ethylene succinate	7.905	4.165	5.393
Hexamethylene sebacate	7.974	1.714	3.459
Pivolactone	7.647	2.62	4.277

The errors inherent in the indirect experimental methods used to determine polymer solubility parameters are too large to be useful in prediction of polymer miscibility.

It has been established on the basis of numerous studies that, for high molecular weight polymer blends in the absence of favourable intermolecular interactions, miscibility is only achieved when the solubility parameters of the two polymers are within about ≤ 0.1 $(cal/cm^3)^{1/2}$ of one another for $r_A = r_B = 5000$. Coleman, Graf and Painter [10] found that the errors in calculating solubility parameters are approximately ± 4 times higher (of ± 0.4 $(cal/cm^3)^{0.5}$) than mentioned value of 0.1 and standard error in molar volume determination was ~ 2.8 cm^3/mol. Therefore for such a kind of blends, the solubility parameters cannot be used to predict phase behaviour. For one-phase blends to be obtained it is desirable to have systems in which favourable specific interaction are present. The ability to quantify such contributions, however, is less well-established. Coleman, Graf and Painter [10] simply add an extra term to Equation 2.1 and then scale this up to accommodate varying strength of different specific interactions. The critical values of $(\chi_{AB})_{crit}$, or solubility parameter difference $(\Delta\delta)_{crit}$ change in a particular manner with the blend composition and type, number and relative strength of specific favourable intermolecular interactions. Taking into account the errors involved in calculating solubility parameters to avoid over-interpretation of the results and also considering the equilibrium constants describing both the self-association and the inter-association of the two polymers the following values of $(\Delta\delta)_{crit}$ have been proposed to be used to discern significant trends in phase behaviour of some sets of polymer blends, in respect with the forces involved in mixing process and that can be accommodated if a one-phase blend is to be obtained as follows:

- Dispersive forces or absence of favourable intermolecular interactions:

  $(\Delta\delta)_{crit} \leq 0.1 \ (cal.cm^3)^{1/2}$

- Polar Forces (dipol-dipol) $(\Delta\delta)_{crit} = 0.4 - 0.7 \ (cal.cm^3)^{1/2}$

- Weak, moderate and strong hydrogen bonding (or other specific interactions):

$(\Delta\delta)_{crit} = 1.0, 1.5, 2.0$ and $\geq 3.0 \ (cal.cm^3)^{1/2}$, respectively, see **Table 2.3**.

Favourable intermolecular interactions increase the probability of finding miscible blends compared to non-polar polymer mixtures. They serve to counteract the unfavourable contribution to the free energy of mixing expressed in the χ_{AB} parameter that takes into account only physical interactions. The free energy of mixing of hydrogen bonded system may be calculated by introducing into the Equation (2.1) ΔG:

$$\frac{\Delta G_m}{RT} = \frac{\Delta G_m^H}{RT} + \frac{\Delta G_m^{FH}}{RT}$$

a term that takes into account the specific interaction $\Delta G_m^H/RT$, which has a complicated composition dependence, given the necessity of taking into account inter- and self associations. According to this relationship any immiscible binary system may be transformed into a compatible one by chemical incorporation into the chain of a sufficient number of groups of specific interactions through their favorable contribution to dominate over unfavorable ones,

Hydrogen-bonded groups can be determined by high resolution (~2 cm^{-1}) Fourier Transform Infrared Spectroscopy (FT-IR) or from the isotherm of adsorption [92]. The samples have

Table 2.3 Comparison between contribution to interaction energy of various types of interactions		
Strength of Interaction [10]	$\Delta\delta_{crit}$ (J/cm^3)$^{1/2}$	B_{ij}^{sp} (J/cm^3)
Non-polar	<0.2	<0.04
Polar	0.2 – 1.0	(-0.04) – (-1.0)
Weak	1.0 – 2.0	(-1.0) – (-4.0)
Weak-moderate	2.0 – 3.0	(-4.0) – (-9.0)
Moderate	2.0 – 4.0	(-9.0) – (-16.0)
Moderate – strong	4.0 – 5.0	(-16.0) – (-25.0)
Strong	5.0 – 6.0	(-25.0) – (-36.0)
Associative balance [95]		
Very favourable	0.0	(+36.0) – (+16.0)
Unfavourable	0.0	(+16.0) – (+4.0)
Slightly unfavourable	0.0	(+4.0) – (-0.0)
Slightly favourable	0.0 – 2.0	(0.0) – (-4.0)
Favourable	2.0 – 4.0	(-4.0) – (-16.0)
Very favourable	4.0 – 6.0	(-16.0) – (-36.0)

to be sufficiently thin (usually as films deposited on KBr tablets from diluted solution) to be within the absorption range where the Beer-Lambert law is obeyed [93]. Non-hydrogen bonded carbonyl stretching frequency at ~ 1739 cm^{-1} and hydrogen bonded carbonyl band is observed some 21 cm^{-1} lower at 1718 cm^{-1}. Considering the equilibrium constants describing both the self-association and the inter-association of the two polymers, that describe the stoichiometry of hydrogen bonds in the blend and, assuming that the system is single phase, one can readily calculate the fraction of hydrogen bonded groups as a function of composition. Using this procedure, Painter, Coleman and others [93] compared experimental and calculated phase diagrams for a great number of systems and have established some limit of miscibility as: poly(2,6-dimethyl-4-vinyl phenol) (PDMVPh), forms miscible blends with the homologous series of poly(*n*-alkylmethacrylates) up to poly(*n*-hexyl methacrylates) and also with polyethylene-co-vinyl acetate (EVA) copolymers containing greater than 40 wt% vinyl acetate, or less than 35 wt% vinyl acetate, two phase systems are formed. The accentuation of the 'strength' of interassociation over self-association enhances the favourable contribution from the $\Delta G_m^H/RT$ term and is also a favourable trend for miscibility.

The theoretical improvement considers two dimensional or multidimensional expressions for solubility parameter or contribution of specific interactions as: π-π (bonding as in polyphenylene ether (PPE)/polystyrene (PS) blends), dipol-dipol, hydrogen bonding (as in poly vinyl methyl ether (PVME)/PS or PVC/polycaprolactone (PCL), polyethylene glycol (PEG)/polyacrylic acid (PAA)) with various strength, electron donor-acceptor, ionic interaction (between ions and ion-dipol), etc., has to be considered. In this chapter we are not interested in ionic interactions (see Chapter 5 on Water Soluble Polymer Systems).

For unfavourable polymer-polymer interaction, (i.e., $B_{ij}>0$), the instability or phase separation will occurs when molecular masses of two polymers increased to a critical value. In the limit of *infinite* molecular weight, miscibility exists only when the polymer-polymer interactions are favourable, (i.e., $B_{ij}<0$). For a *finite* molecular weight, the combinatorial entropy is finite and always favours miscibility. The contribution of this term to the free energy becomes greater the higher the system temperature. The Flory-Huggins theory, therefore, naturally forecast upper critical solution temperature (UCST) behaviour or phase separation on cooling see phase diagram of **Figure 2.1b**. If B_{ij} is regarded as a constant independent of temperature, this theory does not predict lower critical solubility temperature (LCST) or phase separation by heating behaviour, which is quite common for polymer blends. The theory can be empirically modified to address this shortcoming by allowing B to be temperature dependent, but this increases the number of parameters to be determined. UCST predominates in the systems containing solvent, but is less frequent phenomenon in polymer blends and alloys, see Chapter 3 on Phase Separation.

The specific interactions between electron donor and electron acceptor groups could be taken into account as a negative contribution to the interaction parameter as: $V_m E_{da}/RTV_{ctc} \varphi_a \varphi_d$, where φ_d and φ_a are volume fractions of donor and acceptor components in the mixture. V_m and V_{ctc} are average molar volumes of the component of the blends and respectively of the charge transfer complex and E_{da} is average cohesive energy of the charge transfer complex usually its value is about ~ 6 kJ/mol [94].

It has been concluded that high molecular mass polymers may be miscible only at the negative values of χ_{AB} when the interactions of the A-B type dominate over A-A and B-B interactions, therefore for specific interactions between two polymers. It is only one example of atactic polyvinylethylene and *cis*-1,4-polyisoprene where no specific interactions are present (only van der Waals) and system is miscible, but χ_{AB} has a very low value of 0.0004 [5]. However, some authors found that because of the self-association/inter-association equilibrium the existence of the groups able to interact is not a necessary and enough condition for miscibility, because they can give positive values of interaction parameter when self-association is dominant [95].

However, when δ_A and δ_B are almost equal, the calculated value of χ_{AB} may be below the critical value. Other authors considered both χ_{AB} or δ arising from various contributions [96, 97] but the findings are still unable to deal adequately with the presence of miscibility windows in copolymer blends and they also fail to recognise the part played by structural factors such as tacticity, sequence distribution in copolymers, and non-combinatorial contributions. Consequently, following analyses developed by various authors [96-98] consider the segmental parameters established initially from experimental phase behaviour measurements are used to predict behaviour of a range of similar systems [98]. The effects of the local environment or micro-structural differences (as tacticity in PVC-based blends, on the miscibility behaviour are described by using χ_{AB} as arising from inter- and intra-segmental interactions and 'co-solvent concept' in explaining the miscibility in blends involving copolymers [11, 12], or by Cantow-Schulz approach [76, 77]. The miscibility maps for various homopolymer/copolymer or copolymer/copolymer blends have been calculated [10-12] both in binary and ternary blends on the basis of solubility parameter approach.

To predict the miscibility in solid state the following thermodynamic criteria can be used:

- δ_A and δ_B very close to each other, with approximately the same values

- $(\chi_{AB}) < (\chi_{AB})_{crit}$

- $\Delta H_{mix} = w_A M_{A_1} \rho_{A1} (\delta_A - \delta_B)^2 [\dfrac{w_B M_2 \rho_B}{(1 - w_B)} - (1 - w_A) M_A \rho_A]$

where M_i, ρ_i and w_i are molecular mass of monomeric unit polymer density and weight fraction, respectively. It was established that for miscible polymers, values of ΔH_{mix} are in the range from 10^{-3} to 10^{-2} J. In spite of the satisfactory agreement between experimental data and theoretical calculations some exceptions are found. Heat of mixing approach is useful for explaining the: 'miscibility windows' of a series of polymer blends as polyamides, polyesters and copolymers. It is not useful in generation the phase diagrams [99, 100]

Jacobson and others have described how molecular modelling studies can give a rapid estimation of whether two polymers will form a miscible blend using a method that can account for specific interaction between polymer segments [101, 102]. The theoretical foundation of their method relies upon the premise that miscibility is determined by thermodynamic factors alone, and furthermore that these thermodynamic factors are dominated by the energetics of local interactions between segments of the polymer chain. Mainly rigid high-performance polymers were studied. Tiller and Gorella, have extended and modified the two-segment approach to miscibility prediction of flexible polymers by a 'flexiblend algorithm' [103]. A computer model was elaborated using high-temperature molecular dynamics or rotational-isomeric-state Monte-Carlo sampling and also amorphous cell approach. The FLEXIBLEND method developed by Tiller and Gorella is useful for initial screening of potentially miscible combination of polymers for a given application and for understanding changes in phase behaviour resulting from modifications in chemical structure. It is also very convenient for the polymers with close solubility parameters. The FLEXIBLEND method is quick, typically requiring only a couple of central processing unit (CPU) hours but only energetics of local interactions are considered. The amorphous cell method is more rigorous but requires greater computing resources, calculations on a polyethylene oxide (PEO)/polypropylene (PP) blend took about three CPU days on a Silicon Graphics 4D/35 workstation and it allows investigation of the effects of concentration by varying the volume fractions of the two components in the mixed cell.

Other criteria established using the experimental data on compatibility/miscibility are: film clarity, display of properties, which are at least a weighted average of the components and a single glass transition

2.1.4 Equation-of-State Theories

Equation-of-state theories are a new step of the development of thermodynamics of polymer solutions adaptable to polymer-polymer system, too. It was developed by Flory [40-44], Patterson [35, 36, 104] and Sanchez [46, 105].

Sanchez's theory predicts lower values for ΔG, therefore predicts a better solubility in polymer systems than other theories.

2.1.4.1 Sanchez-Lacombe Lattice Fluid (Hole) Model

Finite compressibility adds a destabilising influence, and often this is origin of phase separation upon heating of LCST diagrams (see **Figure 1.1c**). The average mutual distances among the molecules depend not only on the interplay of attractive and repulsive forces, but also on the temperature and pressure. Generally the entropy of a system increases with its volume. The expansion of a condensed system is paid for by a decrease in cohesive energy. An equilibrium is established, at which the free energy of the system is minimised and the volume adopts an equilibrium value. With increasing temperature, the entropy effect becomes more important and the equilibrium shifts to larger volumes. At large expansions, and especially close to the critical point, the behaviour of most substances is remarkably similar. Under such conditions, the behaviour of many compounds is described well by the theory of corresponding states. In the equation-of-state theories, the key term is the change in volume upon mixing. Excess contact energies increase the volume when positive and decrease it when negative. The equation-of-state is written in terms of reduced volume, pressure and temperatures.

Among the new developments the most important are equation-of-state models proposed for the first time in 1950 by Prigogine and co-workers and extended by many others [32, 33].

In contrast a hole model has been developed by Sanchez and Lacombe [46, 105] where free volume is represented as vacant lattice sites. To characterise the stability of the systems consisting of two polymers of high molecular mass, Sanchez has developed the theory based on the model of compressible lattice. This theory is capable of describing the volume changes by mixing and predicts the existence of LCST. It is assumed that polymer segments are rigid cores and that four or six parameters operating in the theory can be found from pressure-volume-temperature (PVT) properties of individual components. On the basis of this theory Sanchez derived the expression for chemical potentials, phase stability conditions, and interfacial tension. The lattice-fluid equation-of-state has the following simple closed form:

$$\tilde{\rho}^2 + \tilde{P} + \tilde{T}[\ln(1-\tilde{\rho}) + (1-1/r)\tilde{\rho}] = 0 \qquad (2.7)$$

where reduced properties are defined as $\tilde{P} = P/P^*$, $\tilde{T} = T/T^*$ and $\tilde{\rho} = \rho/\rho^* = v^*/v$ and r is a chain length with P^*, T^*, ρ^*, v^* and M characteristic pressure, temperature, density, hard core volume per structural unit and weight average molecular weight, respectively. The basic relationships among the parameters are: $T^* = \varepsilon^*/k$; $P^* = \varepsilon^*/V^*$ and $\rho^* = M/rv^*$, where ε^* and v^* are the energy and volume parameters of the intermolecular potential. $V^* \sim \sigma_{ij}^3$ is a densely packed volume, where σ_{ij} is the nearest distance between i-th and j-th segments.

The characteristic parameters may be obtained by fitting the equation-of-state to experimental PVT data for the components and from 'mixing rules' for mixtures and copolymers. Mixing rules for the characteristic parameters given by Sanchez and Lacombe [46, 50]. The characteristic pressure for a mixture P* is related to those of the pure components P*$_i$ and the bare interaction energy density ΔP* by relationship:

$$P^* = \phi_A P^*_A + \phi_B P^*_B - \phi_A \phi_B \Delta P^*$$

Where the ϕ_i are close-packed volume fractions. The ΔP* replaces the interaction parameter in the Flory-Huggins theory. The ΔP* = χ when $\tilde{\rho} = \tilde{\rho}_i = 1$.

For a binary system [26]:

$$\chi = [\varphi\, T_A^* + (1-\varphi)\, T_B^* - T^*]/((1-\varphi)\, T) \qquad (2.8)$$

The critical point reduced variables are:

$$\tilde{\rho}_c = 1/(1+r^{1/2}); \quad \tilde{T} = 2r\tilde{\rho}_c^2 \quad \text{and} \quad \tilde{\rho} = \tilde{T}_c[\ln(1+r^{-1/2}) + (1/2 - r^{1/2})/r \qquad (2.9)$$

An examination of these equations shows that T$_c$ increases with r while P$_c$ and ρ$_c$ decrease with increasing r.

The expressions for other thermodynamic functions have been developed but they are not reproduced here.

This class of theories retained forms similar to the Flory-Huggins theory for the combinatorial entropy and the interaction energy but adds free volume contribution to both the entropy and enthalpy of mixing. They predict LCST behaviour even when ΔP* is constant and independent of temperature by volume contribution. In the case of the enough strong specific interactions even ΔP* (or χ) values deduced from this framework will be temperature dependent.

A general result of the lattice fluid theory is that differences in equation-of-state properties of the pure components make a thermodynamically unfavorable entropic contribution to the chemical potential. The difference between the pure component parameters, especially T* values tend to destabilise a solution and make it more susceptible to phase separation. A small change in combinatorial entropy term limits miscibility both at low and high temperatures so both LCST and UCST may appear.

This theory was widely applied to polymer blends [106-109] and some improvements have been proposed [110-112].

Another complication of the theory is that the interaction energy is dependent on blend composition, but for current objectives, it is more important to know the range of temperature where phase separation occurs than the precise phase diagram shape rather than to predict or interpret the phase diagram in fine details. However, sometimes this effect and also the effect of polydispersity on the bimodal curve are necessary to be known. Theories and computer software that allow for these effects are available.

The most important equation-of-state parameter controlling polymer-polymer solubility is the thermal expansion coefficient. Small differences in pure component thermal expansion coefficient are sufficient to lead to an LCST for two molecular weight polymers. When the energies of interaction are negligible, two polymers having high molecular weights (~ 200 000) must have thermal expansion coefficients within 4% of one another to exhibit significant mutual solubility. When interaction parameter exhibits small positive values, simultaneous LCST and UCST behaviour is possible, For larger positive values of this parameter, the LCST and UCST merge to yield hourglass shaped bimodal and spinodal curves see **Figures 1.1** and **3.1**. Negative values of interaction energy parameter should anticipate whether strong hydrogen bonding or acid-base interaction exists.

The Flory and Sanchez-Lacombe 'equation-of-state' thermodynamics has many advantages over the classical Flory-Huggins theory in describing properties of polymer mixtures. McMaster examined the contribution of the state parameters to the miscibility of hypothetical mixtures and showed that the theory is capable of predicting both LCST and UCST behaviour [112]. Also because it abandons a fixed lattice, it is capable of predicting volume changes on mixing and, hence, the effects of pressure on the phase diagram. Olabishi has applied McMaster treatment to PCL/polyvinyl chloride (PVC) blend [113] and Walsh and Rostami applied the same treatment on PMMA or polybutylacrylate/chlorinated polyethylene (PE) blends [114].

Two systems are considered very convenient to test the theory EVA-copolymer/chlorinated PE and polyether sulfone/PEO because their properties are very diverse and they show very large effect of pressure on the phase separation temperature and a very large volume change on mixing. LCST of these systems increases with increasing pressure.

2.1.4.2 Generalised Lattice Fluid (Sanchez) Model

Recently Sanchez and Stone [4] developed a generalised lattice fluid model (with z coordination number) able to explain most of thermodynamic properties of polymer blends including systems with strong interactions. So, both the old Sanchez-Lancombe theory and Flory-Huggins theory appear as special cases, all relationships being deduced in certain conditions. They demonstrated that although the Flory-Huggins model does indeed have limitations and drawbacks, it is still much more general than is accepted by the thermodynamics scientists. They have obtained former Flory-Huggins relations without ever using the lattice description of a fluid.

Sanchez and Stone [4] took into account the finite compressibility of polymer solutions that affects the phase stability and chemical potentials in a significant way. The Gibbs free energy at a T and P constant can be separated into its compressible and incompressible contributions. Searching for phase stability they concluded that a compressible solution is always less stable than the corresponding incompressible solution, because the equation of ΔG contains a term for the compressible contribution that is an unfavorable contribution to phase stability. This term increases with temperature and at high enough temperatures, this unfavourable term becomes large enough to overwhelm the positive constant volume term and a phase instability develops. This phenomenon of thermally induced phase separation (LCST) is well known in polymer solutions and blends [115]. With increasing temperature, the interactions are weakened, causing LCST behaviour, while the noncombinatorial entropy contribution becomes more and more dominant, leading eventually again to phase stability of the blend or UCST behaviour. This phase behaviour is experimentally easily accessible only for blends with component having very different polymerisation degree. The χ parameter has to be negative at low and high temperatures and positive in a temperature range between.

The next step was to consider that thermodynamic behaviour depends on local structure. These effects are generally referred as 'free volume' effects which have been extensively theoretically treated but not entirely successful from a quantitative point of view.

Attempts to quantify the effect of free volume have proceeded by two ways. The first uses a cell model, described initially by Prigogine and Simha [32, 33] and later in a modified form by Flory and co-workers [40-42] which allows the lattice site to vary in size so as that each site, or cell, can contain a portion of the systems free volume. Patterson and Robard [116] using the Flory's model approximated χ_{AB} in the original Flory-Huggins theory to the sum of two terms, one describing contact or interactional energies and the second consisting of so-called free volume terms. The former decreases with temperature through the usual 1/T dependence, while the second increases, see **Figures 2.1a** and **2.1b**. The sum of the interactional and free volume terms has a parabolic dependence for χ_{AB} crossing the line of $(\chi_{AB})_{crit}$ twice, giving both UCST and LCST behaviour with a region of miscibility in-between. The LCST behaviour appears usually for hydrogen bonding materials, where entropy changes due to association (changes in hydrogen bonding) and it is seems to be critical factor and due to the free volume differences observed in mixtures of some non-polar polymers or both factors could contribute. Other theories manage the last effects as compressibility effects as the most common so the phase separation is entropically driven.

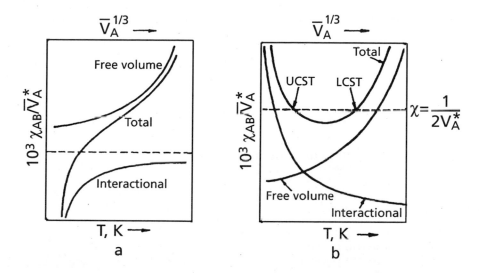

Figure 2.1 Schematic representation of the temperature dependence of binary interaction parameter on temperature and concentration, contribution of the free volume; contribution of the interaction term, sum of the two contributions (a) $\chi_{AB}>0$ (b) $\chi_{AB}<0$. Horizontal dashed line is critical value of χ_{AB}/V_A^* of the blend. [5, 10, 116]

A significant failure of the Flory-Huggins model is its inability to account for the universal property of polymer solutions to phase separate on heating. The simple Flory-Huggins theory does not predict LCST behaviour because it was developed as an incompressible model. For this reason it is useful to use an equation-of-state theory that can predict LCST behaviour stemming from compressible nature of polymers.

The Flory-Huggins theory is useful if one only wishes to know whether miscibility occurs or not, but an equation-of-state theory is generally needed to answer more detailed questions about phase diagram.

The surface in TPx (x denotes concentration) – space that separate the stable (metastable) from thermodynamically unstable region is defined by spinodal condition namely ΔG =0, because $\Delta G < 0$ inside the spinodal and $\Delta G > 0$ outside the spinodal. On the spinodal line we have the following conditions:

$$\frac{\partial Gxx}{\partial T}\bigg|_{P,x}\begin{cases}<0 \Rightarrow \text{LCST}\\>0 \Rightarrow \text{UCST}\end{cases} \text{ or } S_{xx}\begin{cases}>0 \Rightarrow \text{LCST}\\<0 \Rightarrow \text{UCST}\end{cases}$$

Therefore unfavourable entropies drive the phase separation in an LCST system, while unfavorable energetics drive the phase separation in an UCST system.

All known miscible polymer blends appear to phase separately at elevated temperatures, unless thermal decomposition intervenes.

The equation-of-state for the lattice fluid modified by this new approach is [4]:

in absence of attractive interactions is:

$$Pv_o / \beta kT + \ln(1 - \tilde{\rho}) / \tilde{\rho} - (1 - 1/r) \ln(1 - \theta\tilde{\rho}) / \theta\tilde{\rho} = 0$$

or   (2.10)

$$Pv_o / \beta kT + \ln(1 - \tilde{\rho}) / \tilde{\rho} - (1 - 1/r)[1 + \frac{1}{2}\theta\tilde{\rho} + ...] = 0$$

and in presence of attractive interactions:

$$\tilde{\rho}^2 (1 - \theta\tilde{\rho})^{-2} + \tilde{P} + \tilde{T}[\ln(1 - \tilde{\rho}) - (1 - 1/r) \ln(1 - \theta\tilde{\rho}) / \theta] = 0$$ (2.11)

where $\tilde{T} = kT/\varepsilon^*, \tilde{P} = Pv_o /\varepsilon^*$, $\theta = 2/z(1\text{-}1/r)$ v_o intensive volume and z is the number of adjacent sites. In this last case ε^*, r and θ are composition dependent. ρ is number density, β is the compressibility.

For $\tilde{\rho} \rightarrow 0$, the ideal gas law is recovered while setting $\theta = 0$ we obtain the Sanchez-Lacombe equation-of-state (Equation 2.7).

Parameters of equation-of state for polymers can be found in [4, 5] and in the *Polymer Handbook* [117].

2.1.5 Thermodynamics of Ternary Polymer - Polymer – Solvent Systems

Thermodynamics of polymer-polymer-solvent systems were developed by Scott [56], Tompa [118], Patterson [119], Koningsveld and others [64]. The first two authors analysed symmetrical cases when interaction of each polymer with solvent is similar and the polymers have the same monodisperse average molecular weight: $\chi_{SA} = \chi_{SB}$ and $r_A = r_B = r$ and, they established that the phase behaviour of these ternary systems is determined only by the unfavourable interaction between segments of different chains (on χ_{AB}) and shape of bimodal does not depend on χ_{SA} and χ_{SB}. It is considered that the role of solvent is only to dilute the binary mixture of two polymers, decreasing the numbers of contacts between different segments and also the interaction parameter with χ_{AB} (1-j_S) where φ_S is volume fraction of solvent. So when $\varphi_S \to 0$ the equations for a binary system are obtained while for $\varphi_S \to 1$, the $\chi_{AB} \to 0$ and system becomes a single phase one. The critical values of the system do not depend on the solvent nature at the same φ_S. This should means that if two polymers are not miscible in a given solvent, they will not be miscible in any other solvent. This conclusion is not in accordance with many experimental data, being established that the influence of solvent on the phase behaviour is considerable [120-122]. The influence is manifested both in solution and in solid state (films from various solvents exhibits various morphologies). A correlation between critical concentration of mixing and the exponent from well-known Mark-Houwink equation has been also established. Due to this influence the Scott and Tompa theories have been reconsidered by Patterson and Koningsveld [64, 119] whose thermodynamic treatment took into account the role of solvent. According to their theories the spinodal equation is:

$$\sum r_i \varphi_i - 2 \sum r_i r_j (\chi_i + \chi_j) \varphi_i \varphi_j + \varphi r_A r_B \\ (\chi_S \chi_A + \chi_S \chi_B + \chi_A \chi_B) \varphi_S \varphi_A \varphi_B = 0 \tag{2.12}$$

with: $2\chi_S = \chi_{SA} + \chi_{SB} - \chi_{AB}$

The other χ_{ij} are obtained by cyclic interchange of subscripts. where r_i are number of segments in components, φ_i are volume fractions, χ_{ij} correspond to three energies of the formation of contacts between segments. Contrary to their significance in Flory's theory where contacts were related to the molecules of the polymer χ_{ij} (Flory) = $r_i \chi_{ij}$ (Tompa).

The following conclusion could be drawn from theories:

Symmetric bimodals and spinodals are obtained when $\chi_{SA} = \chi_{SB}$ and χ_{AB} are constant. Higher difference in χ_{ij} leads to a great broadening of the immiscibility region. At $\chi_{SB} = \chi_{SB}$ and $\chi_{23} = 0$, the spinodals disappear and therefore the polymers become miscible both in presence and absence of solvent. If $\chi_{SA} \neq \chi_{SB}$ at $\chi_{AB} = 0$ or <0, the region of immiscibility will be present and shifted to lower concentrations. In this case the spinodals have a closed loop shape and do not intersect the axes of triangle diagrams, therefore binary mixtures are fully miscible [119] see **Figure 2.2**. Increasing the quality of the solvent for polymer produces almost the same effect as decreasing it. Increasing the molecular weight of one of the polymers enlarges the region of high concentration of the polymer of lower molecular weight.

When molecular masses of polymers decreased and $\chi_{SA} \neq \chi_{SB}$, closed loops of immiscibility rapidly narrow and at low molecular masses disappear. Koningsveld and others [64] have studied a system, which shows a closed immiscibility loop (high density polyethylene/ isotactic polypropylene (IPP)/diphenyl ether) and pointed out that polymer immiscibility

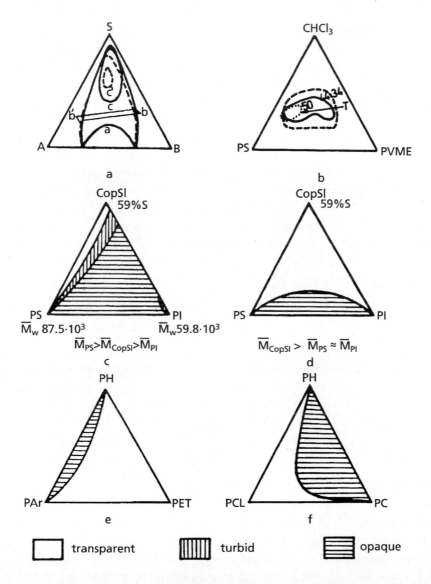

Figure 2.2 Schematic representation of ternary diagrams for systems: polymer A /polymer B/ solvent S a) curve a and b: $\chi_{AB} = 0.0025$; curve 'a' $\chi_{SA} = \chi_{SB}$; curve 'b' $\chi_{SA} \neq \chi_{SB}$; curves c) $\chi_{AB} = 0$, $\chi_{SA} \neq \chi_{SB}$ with increasing differences between χ_{SA} and χ_{SB}. [119] b) PS/PVME/chloroform [116, 123] experimental closed region of incomplete miscibility decreased with increasing temperature; the system is turbid inside the region and clear outside c, d) polymer A/polymer B/copolymer block copolymer with various composition and molecular weight [124-126a] and e, f) polymer A/polymer B/polymer C (phenoxy (PH)/aromatic polyamide/polyethylene terephthalate (PET) and phenoxy resin/PC/polycaprolactone) [95]. The light and shaded zones represent transparent, turbid and opaque films, respectively

in solution does not necessarily imply incompatibility in the absence of solvent. The origin of the closed loop is not completely understood.

Another area where the solvent effect could be important is in the 'segregation' of the blocks within isolated block copolymer macromolecules [127].

It is very clear that the miscibility/immiscibility is mainly related to changes in χ values. The experimental χ values may be obtained by various methods such as: composition of the coexisting phases, thermodynamic cycles, inverse gas chromatography, vapour sorption [1, 5, 7, 13, 14, 16].

For real systems the restriction of miscibility in the region of small concentration is connected with an inequality of polymer-solvent interactions, whereas in the region of high concentration the immiscibility is the result of unfavourable polymer-polymer interactions.

The ternary mixtures of polymer homologues, a phase diagram of closed type may be obtained. Such kind of phase diagrams are also typical for miscible mixture exhibiting LCST, such as the PS/PVME system **Figure 2.2.b**. On the basis of χ_{AB} measurements has been found that $\Delta\chi = \chi_{SA} - \chi_{SB}$ diminishes with decreasing temperature, the immiscibility region narrows. But for this system χ_{AB} should increase with temperature according to the Scott theory [56]. Contrary, Patterson's theory [119] explains the experimental data, emphasising the role of inequality of χ_{ij} on the miscibility of two polymers in common solvent and so the role of solvent in ternary mixtures. Moreover, it was established that polymers immiscible in solution are immiscible in absence of solvent, too. For the systems with UCST as oligostyrene and dimethylsiloxane mixture, χ_{AB} should decrease with increase in temperature. For such systems, the introduction of solvent should rapidly decrease χ_{AB} and change its dependence on temperature from negative to positive.

By increasing the total concentration of polymers in a mixture, the conformational effects begin to play a very important role in changing χ_{AB}. The studies on the systems IPP/PVME and atactic polypropylene (aPP)/PMVE [128] showed that the first system exhibits a narrow miscibility region while the second such region was not seen. This was explained by the difference in gyration radii of both types of PP and due to the fact that in aPP, the probability of the formation of inter-segment contacts is lower than in IPP. For the system PS/polymethyl methacrylate (PMMA) it was found that in dilute regime was no change in coil dimensions at high concentration of PS, whereas at PMMA concentrations higher than cross-over concentration the chains of PS were contracted by a power law [129], and a specific variation of gyration radius was observed. The reason is the exclusion of PMMA from PS coils, surrounded by pure toluene adsorbed on PS coils.

In the other theory based on the Flory's theory of equation-of-state the interaction parameter was obtained function on solvent nature and polymer concentration and the interaction parameters are multiplied with surface fractions [130]. The use of surface fraction instead of volume fraction counts for the molecular contact probability. This theory explained how the affinity of each polymer influences the ternary phase diagram. It was found that if a solvent has a different affinity to the mixture components, the region of immiscibility of the phase diagram of the blends shifts rapidly towards the polymer-solvent axis. For a good solvent for both polymers and small values of $\Delta\chi$, the binodals are symmetric.

The influence of concentration has been studied for PS/PMMA in toluene and bromobenzene. It has been established that in dilute solutions the χ does not change, whereas in a semidilute solution χ increases with concentration. The linear increase of χ_{AB} begins at concentrations exceeding crossover concentration. The extrapolation of the χ_{AB} in solution to $\varphi \to 1$ gives the χ_{AB}, corresponding to polymer-polymer pair.

The morphology of the polymer blends obtained by solvent casting depends of the solvent, temperature, time, concentration, etc. This dependence is important in membrane manufacture.

According to Kleintjens and Koningsveld [131], four different levels of approximation for polymer-solvent interaction may be defined: 1) solubility parameter theory predicts whatever a blend is miscible or not, but does not take into account the dependence on molar mass or concentration dependence, 2) the Flory-Huggins-Staverman model does not consider variation of volume upon mixing, therefore the dependence on temperature and molar mass is considered but the dependence on concentration is not correct; 3) Extended Flory-Huggins model in the equation-of-state terms (mean-field lattice-gas model) and contact statistics predicts phase concentrations and influence of pressure; 4) Extended Flory-Huggins model takes into account non-uniform segment density and chain flexibility.

2.1.6 Thermodynamics of Polymer-Polymer-Polymer Systems

The thermodynamic behaviour of the ternary systems constituted from three different polymers has been described using the same approaches as for ternary polymer-polymer-solvent system. The interaction parameter having the same expression (Equation 2.12). If $\chi_{crit} \to 0$, then the $\chi_{ij} > 0$ therefore the ternary polymer system is immiscible. The general picture of the effect of the third component thermodynamic behaviour of the ternary mixture was developed on the basis of the Flory theory of equation-of-state taking into account the constants between surface by the fraction of surface $\theta_i = (S_i/S)\varphi_i$ where S_i is a segment surface area and $S = \sum_{i=1}^{N} S_i \varphi_i$ is total surface area.

The following situations have been described:

a) The addition of the third component to the two immiscible polymers makes all system miscible if this is miscible with two other examples: PMMA/polyethylene methacrylate (PEMA)/polyvinylidene fluoride (PVDF); PVDF is miscible both with PMMA and PEMA.

b) By adding a component miscible with one of the component of miscible binary mixture, the phase separation proceeds as demixing of single phase regime. With increasing temperature, the miscibility region broadens – example PS/PVME/polypropylene oxide (PPO) system;

c) The following systems are of practical interest: 1) two with LCST and one with UCST (PS/PVME/PCL, PVC/PC/chlorinated polyvinyl chloride (CPVC), PVC/PMMA/ styrene/acrylonitrile copolymers (SAN), *p*-methylstyrene-acrylonitrile copolymers (*p*-MSAN)/PEMA/PMMA, SAN/PMMA/PVA); 2) three with LCST; 3) two with UCST and one with LCST (PS/PC/PCL) [5].

The miscible ternary polymer blends are: PVA/PMMA/PEG, PMMA/PVA/ poly(epichlorohydrine) (PECH), phenoxy/PVME/PCL, PMMA/PEO/PECH. In PEO/ PMMA/PEG system all pair of interaction parameters are negative and the system should have the closed region of immiscibility. Other ternary blends have been studied [95, 132, 133]. The ternary diagrams show various miscibility regions as a function of blend composition, see **Figures 2.2e** and **2.2f**.

2.1.7 Some Thermodynamic Aspects of Homopolymer/Copolymer and Copolymer/Copolymer Mixtures

Block and graft copolymers display unique properties, in solution as well as in solid state, which can essentially be attributed to the incompatibility of the dissimilar blocks in their molecule. Due to their possessing two incompatible blocks within a molecule, block and graft copolymers exhibit surface activity and colloidal properties [124-126a].

The phase behaviour of homopolymer/copolymer and copolymer/copolymer mixtures has been studied both by mean-field theory by Karasz, MacKnight and others [134-138] and other [139-142] and also by modified equation-of-state theory Prigogine-Flory-Patterson [143-146]. However, the first theory does not take into account the changes in the free volume by mixing and dependence of the χ_{ij} on composition. It reveals several characteristics of the thermodynamic behaviour using the binary interaction model that involved both intra- and inter-molecular interactions. The experimental investigation of such blends showed that the miscibility of homopolymers with copolymers is not necessary connected with specific interactions. The transition from miscible to immiscible mixtures occurs vary rapidly even at small changes of the copolymer composition, appearing so-called 'miscibility windows' which have been explained by the effects of repulsion between comonomeric units in copolymer chain. The miscibility is strongly influenced by distribution of monomeric units which is described by introducing a order parameter (Θ) and a composition parameter (Ω) by whose variations it is possible to differentiate between random ($\Theta = 1/2$), alternating and block copolymers ($0 \le \Theta \le 1$). A model taking into account the sequence distribution was proposed by Balazs [139-142]. She analysed the effect of sequence distribution and established a relation for critical compositions at which the mixtures become miscible. A copolymer acting as a compatibiliser has an effect of dilution of unfavourable interactions between two homopolymers and formation of a homogeneous phase. The expression of Gibbs free energy contains terms depending either on composition and sequence distribution.

The transition from immiscibility to miscibility and again to immiscibility with appearance of miscibility windows has been theoretically assessed by equation-of-state theory, the theoretical variation of the interaction parameter with temperature and composition proving this behaviour because in function of copolymer composition (x), the χ takes both positive and negative values, see **Figure 2.3**.

Kammer [147] introduced a new parameter to the generalised χ that reflects the difference in the segment dimension, the final equation predicting both LCST and UCST.

Because the aim of this handbook is mainly practical, the expressions for thermodynamic functions are not given. Several compatibility windows for hopolymer/copolymer mixtures of potential practical interest are given in **Table 2.4**.

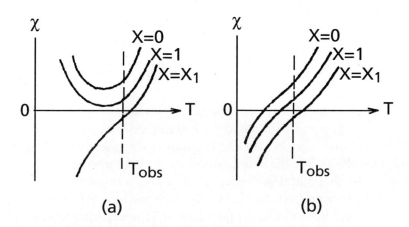

Figure 2.3 Schematic representation of the variation of the χ with temperature and copolymer blends whose miscibility changes with copolymer composition x. T_{obs} is temperature where the system changes from immiscible to miscible [5]

Table 2.4 Compatibility windows in several homopolymer/copolymer mixtures selected from reference [5]		
Component A	Component B	Compatibility range (wt%)
PMMA	SAN	10 – 38 AN
PMMA	PMSAN	11 - 32 AN
PMMA	SMAN	8 - 63 MAN
PCL	SAN	8 – 28 AN
PVME	SAN	0 - 11 AN
PVME	SMA	0 - 15 MA
PVC	SAN	11.5 – 26 AN
CPVC	EVA	35 - 37 E
TMPC	SAN	0 – 13 AN 0 – 20 AN
TMPC	SMA	0 – 8 MA
	SMMA	0 - 40 MMA
SAN	MMA-co-EMA	22 AN in SAN; 30 MMA in MMA-co-EMA
SAN	SMMA (18% MMA)	8.5 – 30 AN
PphMA	SAN	11.5 – 32 AN
PS	PMSAN	0 - 5 AN
PS	SDMS	60 - 90 S
PPMA	SAN	2 – 25 AN
PPMA	PMSAN	8 - 29 AN
PEMA	PMSAN	12 - 32 AN
PBMA	SVPh	0 - 97 Ph
PMSAN	MMA-co-EMA	22 AN in PMSAN; 60-30 MMA in MMA-co-EMA
PMSAN	MMA-co-BMA	22 AN in PMSAN; 70 BMA in MMA-co-BMA

AN: *acrylonitrile*
E: *ethylene*
MMA: *methyl methacrylate*
SDMS: *poly(styrene-co-dimethyl siloxane)*
SVPh: *styrene-co-vinyl phenylene*
PphMA: *poly(phenyl methacrylate)*

BMA: *butyl methacrylate*
MAN: *poly(methacrylate-co-acrylonitrile)*
S: *styrene*
SMAN: *styrene methacrylonitrile*
TMPC: *tetramethyl polycarbonate*
PPMA: *poly(propyl methacrylate)*
PMSAN: *poly(f-methylstyrene-co-acrylonitrile)*

From the miscibility maps the following values of the segment-segment interaction parameters have been determined [11, 12], see **Table 2.5**.

Random copolymers offer a power tool for achieving the desired results since they permit adjustment of both the intermolecular interaction between the mixture components and

Table 2.5 The segment-segment interaction parameters for a $V_r = 100$ cm³/mol (selected from references [11, 12])					
Segments i/j	B_{ij} (J/cm³)	χ_{ij}	Segments i/j	B_{ij} (J/cm³)	χ_{ij}
S/MMA	0.71	0.03	MMA/itaconic anhydride	8.5	0.34
S/AN	22.8	0.92	MMA/AN inter	13.3	0.54
S/B	2.7	0.11	MMA/AN intra	9.3	0.38
S/VME	0.14	0.006	MMA/methacrylonitrile	2.9	0.12
S/α-MS	0.09	0.004	EMA/EA	0.8	0.03
S/itaconic anhydride	12.0	0.48	B/chloroethylene	1.4	0.06
S/maleic anhydride	40.0	1.16	B/AN	38.0	1.53
S/phenylene oxide	-1.6	-0.06	B/vinyl chloride	1.0	0.04
S/phenyl itaconimide	4.0	0.16	B/E	4.8	0.19
S/*m*-nitrostyrene	3.6	0.15	AN/E	20.9	0.84
S/*p*-nitrostyrene	4.2	0.17	AN/chloroethylene	17.0	0.69
α-MS/MMA	0.44	0.02	AN/MA	1.0	0.04
α-MS/AN	22.3	0.9	AN/itaconic anhydride	6.5	0.36
α-MS/E	5.0	0.20	AN/vinyl chloride	12.2	0.49
α-MS/chloroethylene	1.1	0.04	Vinyl chloride/E	3.3	0.13
MMA/MA	1.1	0.04	Vinyl chloride/chloroethylene	1.0	0.04
MMA/EA	0.85	0.03	E/chloroethylene	9.6	0.39

α-MS: *alpha-methylstyrene*
EA: *ethylacrylate*
B: *butadiene*
V_r: *reference volume related to the size of the unit cell on the lattice*

MA: *methylacrylate*
EMA: *ethylemthacrylate*
VME: *vinyl methyl ether*

within components. In fact random copolymers are more often found to be miscible with other polymers than are homopolymers because of this. A thermodynamic treatment of random copolymers is based on a mean field binary interaction model in which interaction energies B_{ij} (or ΔP^*) between monomer unit pair is considered as a sum of terms the positive terms reflecting intermolecular interactions and negative terms intramolecular interactions [148]. A copolymer composition mapping is useful in establishing the compatibility windows. Formal theories have been proposed for dealing with inductive effects but no theory of inductive effect itself has appeared and unit sequence distribution is not yet considered.

A small amount of block copolymer is quite often used as compatibiliser in many polymer blends for miscibility enhancement between immiscible polymer pairs, leading to increased film transparency and improved mechanical properties. The transparency is related to the presence of domains with dimensions less than 1000 Å. The considerable effect of the copolymer on the reduction of domain sizes is clearly observable. The effect of such copolymers depends on their composition and molecular weight as is easily remarked from diagrams c, d of **Figure 2.2** [124-126a] The block copolymers polystyrene – polyisoprene are much effective than PS PMMA block copolymers in improving compatibility of corresponding homopolymers.

According to the binary interaction model for a blends of two copolymers composed of units 1 and 2 and 3 and 4, respectively, the net interaction energy parameter is written as a sum of at least 6 terms [149]. The interaction parameter is considered a segmental interaction one. Some information about them can be deduced from the experimental miscibility maps. The theory was applied for copolymer blends based on glycidylmethacrylate (GMA)/methylmethacrylate or GMA/S copolymers with S/AN copolymers or with PPO and tetramethylcarbonate, [120] establishing the maps of copolymer compositions where miscibility exists. These help to model the GMA use as a functional monomer in reactive processing.

The microphase boundaries and ordered phase symmetries for diblock copolymer met in the weak segregation regime were examined by mean field theory [150, 151], incompressible field theoretical approaches [152] and by fluctuation theory [153-155] within highly coarse grain. Theoretically has been predicted in a quantitative manner the interfacial and emulsifying activity of block copolymers A-b-B in immiscible homopolymers A and B. The authors [150-155] claimed that the compatibilising effect depends not only on the ratio of molecular weight of block copolymer to that of homopolymers but also on the composition $f = N_c/N$ (where N is the chain length of block copolymer and N_c is the chain length of block C) of the block copolymer. A maximum compatibilising effect can be observed when the block copolymer has the same composition as that of homopolymer blend.

Computer simulations of the dynamics (Monte Carlo simulation) of phase behaviour of A/B immiscible polymer blends in presence of various C-b-D block copolymers have been performed [156, 157]. It was found that the growth rate of phase separated domains was significantly retarded especially in the later stages of phase separation and the reduction of the interfacial energy [158, 159] due to localisation of block copolymer at the interface and that this retardation effect is dependent upon both the interaction energies between the copolymer blocks and homopolymers and the chain lengths of block copolymers. The reason for this observation is the curvature properties of the interface, i.e., the bend elasticity and the spontaneous radius of curvature of the interface formed during the phase separation. According to Balazs and co-workers [139], the interaction parameter is a sum of the contribution of the composition χ_{comp} and the sequence distribution χ_{dist} as follows:

$$\chi_{tot} = \chi_{com} + \chi_{dist} \qquad\qquad (2.13)$$

$$\chi_{com} = f_A \chi_{AC} + f_B \chi_{BC} - f_A f_B \chi_{AB} \qquad\qquad (2.14)$$

$$\chi_{dist} = (f^2{}_{AA}/f_A)\, \Delta\chi_{AC} + (f^2{}_{BB}/f_B)\, \Delta\chi_{AC} \qquad\qquad (2.15)$$

$f_{AA(B)}$ are the pair probabilities of AA, AB and BB in a single chain so that $f_A = f_{AA} + f_{AB}$, etc.

The dependence of interaction parameter on composition was found for several systems as isotopic blends (protonated and deuterated components), but for some systems such as polyethylbutylene and polymethylbutylene, the dependence was not found [160] either theoretically nor experimentally (by small angle neutron scattering; SANS).

2.1.8 Some Experimental Data on Thermodynamic Properties of the Polymer Blends

The experimental investigation of thermodynamic properties of binary blends consists of establishing their miscibility and their dependence on polymer component chemical nature, molecular mass, temperature, composition of the blends, etc.

The best method of estimating the phase state of polymer mixtures consists of determining their phase diagrams. Unfortunately most studies deal with establishing only the region of composition having a better compatibility or determination of properties of the blends.

The earlier investigations used very simple method as visually observation of the turbidity of polymer mixture solutions or films [5, 161]. Other criteria for compatibility have been developed based on dynamic mechanical measurements (DMA) and differential scanning calorimetry (DSC). If a binary blend has only a single glass transition temperature the system is considered compatible, depression of the melting point and the viscosity of solutions and melts was also used to estimate compatibility. Heat effects of mixing have been directly related to compatibility and later the role of mixing entropy has been also established in determining compatibility as being due to the negative values of excess volume of mixing, but positive values of excess volume of mixing have been also found (PMMA/PVDF).

Many efforts have been dedicated to the estimation of the thermodynamic interaction parameter χ_{AB} using different methods as SANS, melting point depression, heat of mixing, cloud points, light scattering, osmotic pressure, inverse gas chromatography, FT-IR, etc.

It has to be mentioned that the extrapolation of the results obtained in ternary polymer/polymer/common solvent system for binary polymer/polymer system must be made very carefully. To sustain this, it was supposed that in ternary systems two polymers do not fully mix, i.e., the number of contacts between them is much smaller than it should be for random mixing, in order to avoid the coil overlapping should lead to their repulsion and to segregation of the two different macromolecular species.

The majority of data on interaction parameters are obtained by measurements of melting point depression according to the equations:

$$\frac{\Delta H_A V_A}{R V_B}\left(\frac{1}{T_{m,A}} - \frac{1}{T_{mA}^0}\right) + \ln \varphi_B / N_B (1/N_B - 1/N_A) = \chi_{AB}\varphi_A^2$$

where V_i and ΔH_i are molar volume of monomeric unit of i-th component and enthalpy of mixing per segment correspondingly, T_m and $T_m^{\,\circ}$ are melting temperature of sample and of totally crystalline polymer. This equation does not include the dependence of χ_{AB} on concentration and temperature. This deficiency can be solved by using inverse gas-chromatography, but the dependences obtained by this methods depend on the molecular probe used. The Prigogine-Flory-Patterson theory corrects in part this shortcoming.

To minimise the free volume effects, polymers with similar T_g values should be selected.

2.1.9 New Models for Thermodynamic Description of Polymer Mixtures

During the past two decades, several groups have attempted to incorporate the dependence of interaction parameter on concentration, pressure, molecular weight, etc., into predictive microscopic theories as follows:

Statistical Associating Fluid Theory (SAFT) is an equation-of-state developed from Wertheim's perturbation theory and computer simulation results. It provides very accurate equations of state for individual components but is necessary to demonstrate its application to the mixtures [162-167].

Polymer Reference Interaction Site Model (PRISM) allows certain structural features of polymers to be incorporated. The well-known Flory-Huggins theory has been remarkably successful in describing many aspects of polymer mixing. This theory is known, however, to suffer from several deficiencies that can be traceable: it requires that both components to have the same volume, a mean field or random mixing approximation is made which effectively ignores chain connectivity. It does not include packing effects and can not be used to make quantitative molecular engineering calculations. Recently Curro and Schweizer developed a new approach for treating these blends that is PRISM theory. This is based on a continuous space description of a polymer liquid, which includes chain connectivity and non-random mixing effects in a computational tractable manner [168-174]. It is considered that each polymer has a certain structure factor that includes details on polymer structure. By using of a series of coupled integral equations that take into account the repulsive interactions while the attractive interactions are handled by a perturbation theory. The equations-of-state obtained are inferior to those obtained by other theories, but PRISM model has ability to consider polymer blends. Solubility parameters can be calculated directly from the radial distribution functions, by integrating the pair-pair energy function. It can be also applied to polymers with different architectures as block copolymers, star polymers and ring polymers because subtle structural effects can be systematically investigated both in one polymer component system or mixtures. Applying the density functional theory thermally induced phase separation both UCST (even in isotopic blends) and LCST behaviour has been explained [172].

Lattice Cluster Theory (LCT) is an extension of the Flory-Huggins lattice model, but it is able to distinguish polymer structural details [175-179] because a superior solution is developed. This includes contributions to thermodynamic properties from packing and induced local correlations. In LCT, χ has an energetic and an entropic component and can be developed in a compressible and an incompressible form. Chain stiffness is introduced in model by using a bend or flex energy. The advantage of LCT is that it can be used to investigate in a systematic way the detail effect of structure on polymer blend miscibility. The differences in the miscibility of different polyolefin (PO) systems have been explained using this model.

A Perturbed Hard-Sphere Chain (PHSC) theory appears rather prospective when performing the thermodynamic calculations of actual polymer systems because it allows for, in a natural way, distinction in size of different type of monomer units [180], including copolymer-containing blends [181, 182]. The first two theories conclude that quantitative measurements of phase behaviour in model A-B block copolymer and A/B homopolymer blend would yield different functions for $\chi(T, \phi, M)$. This conclusion is supported by

Monte Carlo simulation of diblock copolymers and corresponding homopolymers that indicate non-random mixing even in the athermal case.

Other models and theoretical treatments proposed are able to describe multicomponent polymer systems with arbitrary distribution of both polymerisation degree and composition, based on variational principles [183], molecular dynamic simulation [184], computer simulation using bond fluctuation model [185-187], surface enrichment [188]. The effects of non-local mixing entropy have been also taken into account [189]. It was found that these effects are significant near the glass transition of the blend.

References

1. P.J. Flory, *Principles of Polymer Chemistry*, Cornell University Press, Ithaca, NY, USA, 1953 and 1981.

2. M.L. Huggins, *Physical Chemistry of High Polymers*, Wiley Interscience, New York, NY, USA, 1958

3. A.J. Staverman, *Recueil des Travaux Chimiques des Pays-Bas*, 1941, 60, 76, 640.

4. I.C. Sanchez and M.T. Stone in *Polymer Blends*, Volume 1, Formulation, Eds., D.R. Paul and C.B. Bucknall, John Wiley & Sons, New York, NY, USA, 2000, 16.

5. Y.S. Lipatov and A.E. Nesterov, *Thermodynamics of Polymer Blends*, Technomic Publishers Co. Inc., Lancaster, PA, USA, 1997.

6. *Polymer Blends*, Eds., D.R. Paul and J.W. Marlow, Academic Press, New York, USA, 1978.

7. O. Olabishi, L. Robeson and M. Shaw, *Polymer-Polymer Miscibility*, Academic Press, New York, USA, 1979.

8. A. Manson and L.H. Sperling, *Polymer Blends and Composites*, Plenum Press, New York, NY, USA, 1976.

9. L.A. Utracki, *Polymer Alloys and Blends*, Hanser Publishers, Munich, Germany, 1989.

10. M.M. Coleman, J.E. Graf and P.C. Painter, *Specific Interactions and the Miscibility of Polymer Blends: Practical Guides for Predicting & Designing Miscible Polymer Mixtures*, Technomic Publishers Co. Inc., Lancaster, PA, USA, 1991.

11. J.M.G. Cowie, *Die Makromolekulare Chemie - Macromolecular Symposia*, 1994, 78, 15

12. J.M.G. Cowie, and J.H. Harris, *Polymer*, 1992, 33, 4592.

13. A. Tager, *Physical Chemistry of Polymers*, Mir Publishers, Russian Federation, Moscow, 1972.

14a. N.V. Kuleznev, *Smesi Polymerov. Structura I Svoistva, Himia*, Moscow, Russian Federation, 1980.

14b. *Fisico-Himia Mnogokmoponentnîh Polimernîh Sistem*, 2 Volumes, Ed., Y.S. Lipatov, Naukova Dumka, Kiev, Ukraine, 1986.

15. C. Vasile, *Studii si Cercetari de Chimie*, 1972, 20, 393.

16. R. Koningsveld, *Pure and Applied Chemistry*, 1989, 61, 1051.

17. R. Koningsveld, W.H. Stockmayer and E. Nies, *Die Makromolekulare Chemie - Macromolecular Symposia*, 1990, 39, 1.

18. F.E. Karasz and W.J. MacKnight in *Multicomponent Polymer Materials*, Eds., D.R. Paul and L.H. Sperling, American Chemical Society, Washington, DC, USA, 1986, 67.

19. D.R. Paul in *Multicomponent Polymer Materials*, Eds., D.R. Paul and L.H. Sperling, American Chemical Society, Washington, DC, USA, 1986, 3.

20. R.J. Roe and D. Rigby, *Advances in Polymer Science*, 1987, 82, 103.

21. W.J. MacKnight and F.E. Karasz, *Comprehensive Polymer Science*, Volume 7, Ed., G. Allen, Pergamon Press, New York, 1989, 111.

22. D.J. Walsh, *Comprehensive Polymer Science*, Volume 7, Ed., G. Allen, Pergamon Press, New York, 1989, 35.

23. M.T. Rätzsch and H. Kehlen, *The Frontiers of Macromolecular Science*, Eds., T. Saegysa, T. Higashimura and A. Abe, Blackwell Science Publishers, New York, NY, USA,1989.

24. M.T. Rätzsch and H. Kehlen, *Progress in Polymer Science*, 1989, 14, 1.

25. K. Kamide, *Thermodynamics of Polymer Solutions, Phase Equilibria and Critical Phenomena*, Elsevier, Amsterdam, 1990.

26. V.P. Privalko and V.V. Novikov, *The Science of Heterogeneous Polymers, Structure and Thermophysical Properties*, John Wiley and Sons, Chichester, 1995.

27. S.I. Kuchanov and S.V. Panyukov, Statistical thermodynamics of Copolymers and their Blends, Chapter 13.4 in *Comprehensive Polymer Science, Second Supplement*, Ed., G. Allen, S.L. Aggarwal and S. Russo, Pergamon Press, New York, 1996, 441.

28. A.A. Askadskii, *Russian Chemical Reviews*, 1999, **64**, 317.

29. M. Muller, *Macromolecular Theory and Simulations*, 1999, **8**, 343.

30. M. Hess, *Korea Polymer Journal*, 1998, **6**, 1.

31. R. Koningsveld and L.A. Kleintjens, *Macromolecules*, 1971, **4**, 637.

32. J. Prigogine, *The Molecular Theory of Solutions*, North-Holland Publishing Company, Amsterdam, The Netherlands, 1959, Chapter 16.

33. R. Simha and T. Somcynsky, *Macromolecules*, 1969, **2**, 341.

34. J. Hijmans, *Physica*, 1961, **27**, 433.

35. D. Patterson, *Macromolecules*, 1969, **2**, 673.

36. D. Patterson and G. Delmas, *Discussions of the Faraday Society*, 1970, 98.

37. B. Rudolf, T. Ougizawa and T. Inoue, *Polymer*, 1998, **39**, 873.

38. B. Rudolf, T. Ougizawa and T. Inoue, *Macromolecular Theory & Simulations*, 1998, **7**, 1, 1.

39. K. Choi, W.H. Jo and S.L. Hsu, *Macromolecules*, 1998, **31**, 1366.

40. P.J. Flory, R.A. Orwoll and A. Vrij, *Journal of the American Chemical Society*, 1964, **86**, 3507.

41. P.J. Flory, R.A. Orwoll and A. Vrij, *Journal of the American Chemical Society*, 1964, **86**, 3515.

42. P.J. Flory, R.A. Orwoll and A. Vrij, *Journal of the American Chemical Society*, 1965, **87**, 1833.

43. P.J. Flory, *Discussions of the Faraday Society*, 1970, **49**, 7.

44. P.J. Flory, *Journal of the American Chemical Society*, 1965, 87, 1833.

45. B.E. Eichinger and P.J. Flory, *Transactions of the Faraday Society*, 1968, 64, 2035

46. I.C. Sanchez and R.H. Lacombe, *Journal of Physical Chemistry*, 1976, 80, 2352.

47. I.C. Sanchez and R.H. Lacombe, *Journal of Physical Chemistry*, 1976, 80, 2568.

48. I.C. Sanchez and R.H. Lacombe, *Journal of Polymer Science, Polymer Letters Edition*, 1977, 15, 2, 71.

49. I.C. Sanchez and R.H. Lacombe, *Macromolecules*, 1978, 11, 6, 1145.

50. I.C. Sanchez, *Journal of Macromolecular Science B*, 1980, 17, 3, 565.

51. J.A.R. Renuncio and J.M. Prausnitz, *Macromolecules* 1976, 9, 898.

52. V. Brandani, *Macromolecules*, 1978, 11, 1293.

53. V. Brandani, *Macromolecules*, 1979, 12, 883.

54. F. Hamada, T.Shiomi, K. Fujisawa and A. Nakajima, *Macromolecules*, 1980, 13, 729.

55. K. Fujisawa, T. Shiomi, F. Hamada and A. Nakajima, *Polymer Bulletin*, 1980, 3, 261.

56. R.L. Scott, *Journal of Chemical Physics*, 1949, 17, 268, 279.

57. H. Tompa, *Transactions of the Faraday Society*, 1949, 45, 1142.

58. H. Tompa, *Transactions of the Faraday Society*, 1950, 46, 970.

59. W.H. Stockmayer, *Journal of Chemical Physics*, 1949, 17, 588.

60. R. Koningsveld in *Polymer Science: A Materials Science Handbook*, Volume 2, Ed., A.D. Jenkins, North-Holland Publishing Company, Amsterdam, The Netherlands, 1972, 1047.

61. R. Koningsveld, *British Polymer Journal*, 1975, 7, 435.

62. R. Koningsveld, L.A. Kleintjens and M.H. Onclin, *Journal of Polymer Science, Part B*, 1980, 18, 363.

63. M. Gordon, H.A.G. Chermin and R. Koningsveld, *Macromolecules*, 1969, 2, 207.

64. R. Koningsveld, H.A.G. Chermin and M. Gordon, *Proceedings of the Royal Society of London, Series A*, 1970, **319**, 331.

65. J.W. Kennedy, M. Gordon and R. Koningsveld, *Journal of Polymer Science*, 1972, **C-39**, 43.

66. Y. Hu, X. Yang, D.T. Wu and J.M. Prausnitz, *Macromolecules*, 1993, **26**, 6817.

67. K. Solc, *Macromolecules*, 1970, **3**, 665.

68. K. Solc, *Macromolecules*, 1983, **16**, 236.

69. K. Solc, *Journal of Polymer Science, Part B*, 1974, **12**, 555.

70. K. Solc, *Journal of Polymer Science, Part B*, 1982, **20**, 1947.

71. R.L. Scott, *Journal of Polymer Science*, 1952, **9**, 423.

72. L. Leibler, *Die Makromoleckulare Chemie - Rapid Communications*, 1981, **2**, 6-7, 393.

73. D.R. Paul and J.W. Barlow, *Polymer*, 1984, **25**, 487.

74. R.P. Kambour, J.T. Bendler and R.C. Bopp, *Macromolecules*, 1983, **16**, 763.

75. G. ten Brinke and F.E. Karasz, *Macromolecules*, 1984, **17**, 815.

76. H.J. Cantow and O. Schulz, *Polymer Bulletin*, 1986, 15, 449.

77. H.J. Cantow and O. Schulz, *Polymer Bulletin*, 1986, 15, 539.

78. P. Munk, P. Hattam and Q. Du, *Journal of Applied Polymer Science, Applied Polymer Symposium*, 1989, **43**, 373.

79. R. Koningsveld, *Die Macromolekulare Chemie - Macromolecular Symposia*, 1994, **78**, 1.

80. R. Koningsveld and W.J. MacKnight, *Polymer International*, 1997, **44**, 356.

81. D. Schwahn, K. Hahn, J. Streib and J. Springer, *Journal of Chemical Physics*, 1990, **93**, 8383.

82. F. Scheffold, E. Eiser, A. Budkowski, U. Steiner, J. Klein and L.J. Fetters, *Journal of Chemical Physics*, 1996, **104**, 8786.

83. J.H. Hildebrand and R.L. Scott, *The Solubility of Non-Electrolytes*, Dover Publishers, New York, 1964.

84. C.M. Hansen, *Journal of Paint Technology*, 1967, **39**, 104.

85. C.M. Hansen, *Journal of Paint Technology*, 1967, **39**, 505.

86. C.M. Hansen, *Industrial Engineering Chemistry: Product Research and Development*, 1969, **8**, 2.

87. D.J. David, Proceedings of Antec '95, Boston, MA, USA, Volume 2, 2748.

88. V.N.S. Pendyala and S.F. Xavier, *Polymer*, 1997, **38**, 3565.

89. D.W. van Krevelen, *Properties of Polymers*, Elsevier, Amsterdam, The Netherlands, 1990.

90. P.A. Small, *Journal of Applied Chemistry*, 1953, **3**, 71.

91. A.F. Barton, *CRC Handbook of Solubility Parameter and Other Cohesive Parameters*, Second Edition, Boca Raton, FL, USA, 1991, Chapter 6, 157.

92. W.P. Hsu, A.S. Myerson and T.K. Kwei, *International Journal of Polymeric Materials*, 1998, **40**, 81.

93. G.J. Pehlert, X. Yang, P.C. Painter and M.M. Coleman, *Polymer*, 1996, **37**, 4763.

94. M.C. Pascu, Gh. Popa, M. Grigoras and C. Vasile, *Synthetic Polymer Journal*, 1998, **2**, 158.

95. E. Espi, M.J. Fernandez and J.J. Iruin, *Polymer Engineering and Science*, 1994, **34**, 1314.

96. A. Wakker and M.A. van Dijk, *Polymer Networks and Blends*, 1992, **2**, 123.

97. D.J. David and T.F. Sincock, *Polymer*, 1992, **33**, 4505.

98. D.R. Paul and J.W. Barlow, *Polymer*, 1984, **25**, 487.

99. C. Silvestre and S. Cimmino, in *Handbook of Polymer Blends and Composites*, Volume 3, Eds., C. Vasile and A.K. Kulshrethsha, Rapra Technology, Shrewsbury, UK, Chapter 11.

100. S. Lee, *Korea Polymer Journal*, 1998, **6**, 145.

101. G.V. Nelson, S.H. Jacobson and D. J. Gordon, *Chemical Design Automation News*, 1992, 7, 39.

102. S.H. Jacobson, D.J. Gordon, G.V. Nelson and A.C. Balazs, *Advanced Materials*, 1992, **3**, 198.

103. A.R. Tiller and B. Gorella, *Polymer*, 1994, **38**, 3261.

104. D. Patterson, *ACS Advances in Chemistry Series*, 1979, **176**, 529.

105. I.C. Sanchez in *Polymer Compatibility and Incompatibility. Principles and Practices*, Ed., K. Solc, Harwood Academic Press, New York, NY, USA, 1982, 59, *Macromolecules*, 1991, **24**, 908

106. T.A. Callaghan and D.R. Paul, *Macromolecules*, 1993, **26**, 2439.

107. T.A. Callaghan and D.R. Paul, *Polymer*, 1993, **34**, 3796.

108. C.K. Kim and D.R. Paul, *Polymer*, 1992, **33**, 1630.

109. C.K. Kim and D.R. Paul, *Polymer*, 1992, 33, 2089.

110. J.J. Gyu and B. Y. Chan, *Journal of Applied Polymer Science*, 1998, **70**, 1143.

111. K.S. Jeon, K. Char and E. Kim, *Polymer Journal*, 2000, **32**, 1.

112. L.P. McMaster, *Macromolecules*, 1973, **6**, 760.

113. O. Olabishi, *Macromolecules*, 1975, **9**, 316.

114. D.J. Walsh and S. Rostami, *Macromolecules*, 1985, **18**, 216.

115. S.C. Chiu, T.K. Kwei and E.M. Pearce, *Journal of Thermal Analysis and Calorimetry*, 2000, **59**, 71.

116. A. Robard, D. Patterson, and G. Delmas, *Macromolecules*, 1977, **10**, 706.

117. J. Cho and I.C. Sanchez, in *Polymer Handbook*, Fourth Edition, Eds., J. Brandrup, E.H. Immergut and E.A. Grulke, John Wiley, New York, NY, USA, 1999, VI/591.

118. H. Tompa, *Polymer Solutions*, Butterworths, London, UK, 1956.

119. I. Zeman and D. Patterson, *Macromolecules*, 1972, **5**, 513.

120. R.J. Kern, *Journal of Polymer Science,* 1956, **21**, 19.

121. Ch. Huglin and A. Dondos, *Die Makromolecular Chemie,* 1968, **126**, 206.

122. M. Bank, J. Leffingwell, and C. Thies, *Macromolecules,* 1971, **4**, 43.

123. D. Patterson and A. Robard, *Macromolecules,* 1978, **11**, 690.

124. G. Riess, J. Kohler, C. Tournut and A. Banderet, *Die Makromolekulare Chemie,* 1967, **101**, 58.

125. G. Riess, J. Periard and A. Banderet in *Colloidal and Morphological Behaviour of Block and Graft Copolymers,* Ed., G.E. Molau, Plenum Press, New York, NY, USA, 1971.

126. G. Riess in *Polymer Blends, Processing, Morphology and Properties,* Eds., E. Martuscelli, R. Palumbo and M. Kryszewski, Plenum Press, New York, NY, USA, 1980, 123.

126a. G. Riess, Proceedings of IUPAC '83, Bucharest, Romania, 1983, 468.

127. A. Dondos, *Die Makromolekulare Chemie,* 1971, **147**, 123.

128. T. Okazawa, *Macromolecules,* 1975, **8**, 371.

129. R. Kuhn and H.J. Cantow, *Die Makromolekulare Chemie,* 1969, **122**, 65.

130. M.G. Prolongo, R.M. Masegosa and A. Horta, *Macromolecules,* 1989, **22**, 4346.

131. L.A. Kleintjens and R. Koningsveld, *Die Makromolekulare Chemie - Macromolecular Symposia,* 1988, **20/21**, 203.

132. J. Kolarik, L. Fambri, A. Pegoretti and A. Penarti, *Polymers for Advanced Technology,* 2000, **11**, 2, 75.

133. C. Huang, M.O. de la Cruz and P.W. Voorhees, *Acta Materialia,* 1999, **47**, 4449.

134. R. Vukovic, V. Kuresevic, F.E. Segudovic, F.E. Karasz and W.J. MacKnight, *Journal of Applied Polymer Science,* 1983, **28**, 1379.

135. R. Vukovic, V. Kuresevic, F.E. Segudovic, F.E. Karasz and W.J. MacKnight, *Journal of Applied Polymer Science,* 1983, **28**, 3079.

136. G. ten Brinke, F.E. Karasz and W.J. MacKnight, *Macromolecules,* 1983, **16**, 1827.

137. G. ten Brinke, F.E. Karasz and W.J. MacKnight, *Macromolecules*, 1992, **25**, 4716.

138. G. ten Brinke, F.E. Karasz and W.J. MacKnight, *Macromolecules*, 1986, **19**, 2274.

139. A.C. Balazs, I.C. Sanchez, I.B. Epstein, F.E. Karasz and W.J. MacKnight, *Macromolecules*, 1985, **18**, 2188.

140. A.C. Balazs, F.E. Karasz, W.J. MacKnight, H. Ueda and I.C. Sanchez, *Macromolecules*, 1985, **18**, 2784.

141. J.V. Hunsel, A.C. Balazs, W.J. MacKnight and R. Koningsveld, *Macromolecules*, 1988, **21**, 1528.

142. A.C. Balazs and M.T. Meuse, *Macromolecules*, 1989, **22**, 4260.

143. T. Shiomi, H. Ishimatsu, T. Eguchi and K. Imai, *Macromolecules*, 1990, **23**, 4970.

144. H.W. Kammer, *Acta Polymerica*, 1986, **37**, 1.

145. H.W. Kammer, *Acta Polymerica*, 1989, **40**, 75.

146. H.W. Kammer, T. Inoue and T. Ougizawa, *Polymer*, 1989, **30**, 888.

147. H.W. Kammer, *Polymer*, 1999, **40**, 5793.

148. D.R. Paul, *Pure & Applied Chemistry*, 1995, **67**, 977.

149. P.P. Gan and D.R. Paul, *Polymer*, 1994, **35**, 3513.

150. L. Leibler, *Macromolecules*, 1980, **13**, 1602.

151. L. Leibler, *Die Makromolekulare Chemie - Macromolecular Symposia*, 1988, **16**, 1.

152. E. Matsen and F.S. Bates *Macromolecules*, 1996, **29**, 2092.

153. G.H. Fredrickson and E. Helfand, *Journal of Chemical Physics*, 1987, **87**, 697.

154. G.H. Fredrickson and E. Helfand, *Journal of Chemical Physics*, 1987, **87**, 124.

155. S.A. Brazovskii, *Soviet Physics JETP - USSR*, 1975, **41**, 85.

156. S.H. Kim, W.H. Jo and J. Kim, *Macromolecules*, 1997, **30**, 3910.

157. S.H. Kim, W.H. Jo and J. Kim, *Macromolecules*, 1996, **29**, 6933.

158. W.H. Jo and S.H. Kim, *Macromolecules*, 1996, **29**, 7204.

159. K. Choi and W.H. Jo, *Macromolecules*, 1997, **30**, 1509.

160. C.C. Lin, S.V. Jonnalagadda, N.P. Balsara, C.C. Han and R. Krishnamoorti, *Macromolecules*, 1996, **29**, 661.

161. A. Dobry and F. Boyer-Kawenoki, *Journal of Polymer Science*, 1947, **2**, 90.

162. S.H. Huang and M. Radosz, *Industrial Engineering Chemistry: Product Research and Development*, 1990, **29**, 2284.

163. S.H. Huang and M. Radosz, *Industrial Engineering Chemistry: Product Research and Development*, 1991, **30**, 1996.

164. S.J. Han, C.J. Gregg and M. Radosz, *Industrial Engineering Chemistry: Product Research and Development*, 1997, **36**, 5520 45.

165. T. Kraska and K.E. Gubbin, *Industrial Engineering Chemistry: Product Research and Development*, 1996, **35**, 4727.

166. B. Folie, *AIChE Journal*, 1996, **42**, 3466.

167. D.Ghonasgi and W.G. Chapman, *Journal of Chemical Physics*, 1994, **100**, 6633.

168. K.S. Schweiser and J.G. Curro, *Advances in Chemical Physics*, 1997, **97**, 1.

169. A. Yethiraj, J.G. Curro, K.S. Schweizer, J.D. McCoy and K.G. Honnell, *Macromolecules*, 1993, **26**, 2655.

170. D.S. Pope, I.C. Sanchez, W.J. Koros and G.K. Fleming, *Macromolecules*, 1991 **24**, 1779.

171. E.F. David and K.S. Schwetizer, *Macromolecules*, 1997, **30**, 5118.

172. S.K. Nath, J.D. McCoy, J.G. Curro and R.S. Saunders, *Journal of Polymer Science, Part B, Polymer Physics*, 1995, **33**, 2306.

173. J.G. Curro, Proceedings of Antec '94, San Francisco, CA, USA, 1994, Volume II, 2083.

174. P.A. Tillman, D.R. Rottach, J.D. McCoy, S.J. Plimpton and J.G. Curro, *Journal of Chemical Physics*, 1997, **107**, 4024.

175. K.F. Freed and J. Dudowicz, *Modern Trends in Polymer Science*, 1995, **3**, 248.

176. K.F. Freed and J. Dudowicz, *Macromolecular Symposia*, 1994, **78**, 29.

177. K.F. Freed and J. Dudowicz, *Macromolecules*, 1998, **31**, 5094.

178. K.F. Freed, J. Dudowicz and K.W. Forman, *Journal of Chemical Physics*, 1998, **108**, 7881.

179. K. W. Foreman and K.F. Freed, *Advances in Chemical Physics*, 1998, **103**, 335.

180. Y. Song, S.M. Lambert and J.M. Prausnitz, *Macromolecules*, 1994, **27**, 441.

181. T. Hino, Y. Song, and J.M. Prausnitz, *Macromolecules*, 1995, **28**, 5717

182. T. Hino, Y. Song, and J.M. Prausnitz, *Macromolecules*, 1995, **28**, 5725.

183. S.I. Kuchanov and S.V. Panyukov, *Journal of Polymer Science, Part B Polymer Physics*, 1998, **36**, 937.

184. P.A. Tillman, D.R. Rottach, J.D. McCoy, S.J. Plimpton and J.G. Curro, *Journal of Chemical Physics*, 1998, **109**, 806.

185. M. Muller and K. Binder, *Macromolecules*, 1995, **28**, 1825

186. M. Muller and K. Binder, *Journal of Computer Aided Material Design*, 1998, **4**, 137.

187. M. Muller and M. Schick, *Journal of Chemical Physics*, 1996, **105**, 8282.

188. I.G. Kokkinos and M.K. Kosmas, *Macromolecules*, 1997, **30**, 577.

189. I. Ya. Erukhimovich, A.R. Khokhlov, T.A. Vilgis, A. Ramzi and F. Boue, *Computational and Theoretical Polymer Science*, 1998, **8**, 133.

3 Phase Behaviour

Cornelia Vasile

3.1 Introduction

In technical applications one often has to handle multicomponent mixtures and the accurate knowledge of their phase separation is essential. In most cases of practical interest, polymers are mixed with many other components, as during the synthesis of copolymers (two monomers, a solvent, a polymer) or in membrane forming systems, numerous additives are also required in the processing of various materials. With all applications a detailed theoretical understanding of the mixing behaviour of such systems is mandatory.

The properties of polymer blends are intimately tied to their morphology, which depends on the miscibility of the components and the mechanism and kinetics of phase separation. For some applications, phase-separated morphologies are required for improving mechanical properties such as impact toughness, while in other instances, a miscible blend is desired, e.g., for improving processability or extending an expensive material with a cheaper one. For either a single phase or two-phase blend, the development of the optimum morphology and properties requires knowledge of the phase diagram and the characteristics of the phase separation process [1-7].

3.2 Phase Diagrams of Binary Polymer Blends and Conditions of Phase Separation

To determine miscibility, isobaric phase diagrams are usually constructed. **Figures 1.1** and **3.1** are represented typical curves for binary polymer systems [1-7]. The coexistence curves, **binodals**, separate the region of complete miscibility from that of partial miscibility. At a given temperature any point situated above the curves with lower critical solution temperature (LCST) and below curves with upper critical solution temperature (UCST) (points C and D) corresponds to the concentration and temperatures of an homogeneous system. Points in the region beneath the curves, e.g., E and F, correspond to a microheterogeneous two-phase system which separates in time into two clear layers in equilibrium with each other. The composition of layers can be found by drawing an isotherm at each temperature. It will intersect the coexistence curve at two points, the abscissas of which correspond to the volume fractions of the second component in each layer. The connecting line is called the **tie line**. Raising the temperature changes the concentrations of the components in both layers, bringing them closer together, but each temperature corresponds to a definite solubility of one component in the other. At the critical miscibility temperature, the composition of both phases becomes the same, the interface between them disappears, and an homogeneous solution results. When critical state is reached, the compositions of the two coexisting phases have became identical and the tie-line has changed into a point, the **critical point**.

In most polymer systems the miscibility of the components improves with decreasing temperature and phase separation occurs by heating and they exhibit a LCST. This temperature corresponds to a minimum on the coexistence curve. There are two types of LCST. The first type of LCST, which is more common one, is characteristic of systems in which strong hydrogen bonds form between the component molecules, promoting mixing of components. By heating these bonds are broken by thermal motion, worsening the miscibility of the components.

The second type of LCST is observed in systems whose components are close in chemical constitution, but differ in the size of their molecules [8].

If for a binary system, the plot of free energy of mixing (ΔG_m) is concave upward throughout the composition range **Figure 1.1a** bottom, then the two components are miscible in all proportions. This is because any point on this curve (Q) has lower free energy than any two phase system of the same overall composition (Q*). Contrary, the lower free energy is obtained by phase separation in the case illustrated in **Figure 1.1b**. In respect to the positions of the mixture with composition they will be metastable against separation into phases in the spinodal region and non stable against separation for composition situated on concave downwards that means under the bimodal curve.

When an uniform mother solution is cooled very slowly, a second phase begins to separate at a definite temperature which is called '**the cloud point temperature**' or the **cloud point**.

In the homogeneous area, the system is stable, the free energy (ΔG_m) is minimal and can only be increased by a demixing process. In the heterogeneous area the overall Gibbs energy of demixed system consisting of the coexistence phases is less than the ΔG_m of the homogeneous system. This area breaks into two domains. One is called metastable, here the beginning of phase separation leads to an increase of ΔG_m. Therefore the phase separation cannot proceeds spontaneously, it is bound to nucleation and growth. The demixed system consists of a continuous matrix-phase and a non-continuous drop-phase.

The second domain located within the two phase area are only touching the homogeneous region at the critical point is called unstable domain, ΔG_m is decreased during the whole demixing process. If the phase separation takes place under these conditions, the well-known **spinodal** pattern with two co-continuous phases is observed.

Binodals and spinodals provide important information because together with the characteristics of the phase state, they give data for analysis of the mechanism of phase separation.

A **binodal curve** indicates the limits of miscibility. They are sometimes (incorrectly) identified with the experimental cloud-point curves. Binodals are computed from the equality of the chemical potential of polymers in two coexisting phases, the general equation for a binodal curve is:

$$\Delta \mu_i = n_i (\partial \Delta G_m / \partial n_i)_{V,T,n_j} \tag{3.1}$$

$$(\mu_A)' - (\mu_{Ao})' = (\mu_A)'' - (\mu_{Ao})'' \quad \text{and} \quad (\mu_B)' - (\mu_{Bo})' = (\mu_B)'' - (\mu_{Bo})'' \tag{3.2}$$

where:

$$\mu_A - \mu_{Ao} = \ln \varphi_A + (1 - r_A/r_B)\varphi_B + \chi_{AB} r_A \varphi_B^2 \tag{3.3}$$

$$\mu_B - \mu_{Bo} = \ln \varphi_B + (1 - r_A/r_B)\varphi_B + \chi_{AB} r_B \varphi_A^2 \tag{3.4}$$

where $\mu_i - \mu_{io}$ are chemical potentials of the i-th component of the mixture, r_A and r_B are the numbers of segments (or polymerisation degree) of polymer A and B, $r_i = V_i / V_r$ (V_i is molar volume of component i) ϕ_A and ϕ_B are volume fractions of components. In the general expression (Equation 3.1), n_i designates various modes of concentration evaluations (volume fraction is the next). χ_{AB} is a free energy interaction parameter calculated from Equations 2.3-2.8 and 2.11 depending on the type of polymer blend studied, practical or theoretical purpose and selected theory for study.

Using a computer it is preferable to use a more direct method in which the ΔG_m function itself is used. The coexisting polymer volume fractions are found by means of numerical iterative construction of the double tangent to $\Delta G (\varphi)$ curves at different χ values [1-4] Normally the calculation of the tie lines connecting the coexisting phases at constant temperature and pressure is carried out by numerically solving the equations describing the equality of the chemical potentials μ_i, which is connecting with the Gibbs energy and the derivatives of the Gibbs energy with respect to the composition variables.

The region of the metastable states, located between the cloud-point curve and the spinodal, is of special significance for polymer systems compared to low-molecular ones. This is because the phase separation in the polymer system proceeds much more slowly than in low-molecular systems. Consequently, the lifetime of the metastable states may often far exceed the duration of a thermodynamic experiment and even the polymer operation time. In such cases, which are typical of concentrated solutions, melts and blends of polymers, a metastable system is expected to exhibit an equilibrium behaviour. A more complicated situation is encountered where more than two phases coexist.

For a complete description of the phase diagram, the spinodals and the critical point must be calculated from the second and third derivatives, respectively:

$$(\frac{\partial^2 \Delta G}{\partial \varphi_A^3})_{P,T} = (\frac{\partial^3 \Delta G}{\partial \varphi_A^3})_{P,T} = 0 \qquad\qquad (3.5)$$

The **spinodal line** is given by the equation containing the second derivatives of ΔG. Only for very simple theoretical equations for ΔG, can the derivatives be easily calculated. Very complicated dependencies of ΔG on the composition result in the case of polymer blends and copolymers. The calculation is more complicated for sheared polymer blends, where ΔG is extended by a term which contains implicit equations and describes the energy stored in the flowing system, where the analytical representation of the derivatives is impossible. The extension of ternary blends needs a fitting function [9].

To calculate the phase diagram on the basis of these equations one must know the concentration (φ_i – volume fraction), temperature T and pressure P.

Several equations useful for spinodal curve evaluation are:

The spinodal condition for binary mixture deduced by lattice fluid theory (LFT) is given by (see also Equation 2.12):

$$\frac{d^2 G}{d\phi_A^2} = \frac{1}{2}(\frac{1}{r_A \phi_A} + \frac{1}{r_B \phi_B}) - \tilde{\rho}(\frac{\Delta P^* v^*}{kT} + \frac{\psi^2 \tilde{T} P^* \beta}{2}) = 0 \qquad\qquad (3.6)$$

where ΔP^* is the binary pair interaction energy (energies), ψ is a dimensionless function, a function of pure component parameter difference and is generally non-zero and β is the isothermal compressibility factor [10-16]. For the other meanings see Chapter 2.

The interaction parameter from Flory-Huggins (FH) model:

$$\chi_{sp} = 0.5[\frac{1}{r_A}\varphi_{sp} + \frac{1}{r_B}(1-\varphi_{sp})] \qquad (3.7)$$

and from LFT for the simplest case of the identical characteristic volumes of components:

$$\chi_{sp} = 0.5[\frac{1}{r_A}\varphi_{sp} + \frac{1}{r_B}(1-\varphi_{sp})]/\rho - T\Psi^2 P^* \beta_A \qquad (3.8)$$

where $\Psi = \rho[\varphi_{sp}/T_A(1-\varphi_{sp}) - (1-\varphi_{sp})/T_B \varphi_{sp} - 1/T (1-\varphi_p) - 1/T\varphi_{sp}] - (1/r_A - 1/r_B)$ (3.9)

subscript $_{sp}$ means 'spinodal'.

The **critical point** is the point where the bimodal and spinodal curves are cotangent:

The corresponding values for critical values of concentration and interaction parameter are:

$$(\psi_A)_{crit} = \frac{1}{[1+(r_B/r_B)^{1/2}]} \text{ and } (\chi_{AB})_{crit} = (1/2)[r_A^{-1/2} + r_B^{-1/2}] \cong 0.5 \qquad (3.10)$$

The FH theory leads to a reasonable prediction of critical values, but there are significant deviations between predictions and experimental measurements of other quantities. A better fitting of calculated and experimental bimodal and spinodal curves has been obtained by assuming a dependence of χ_{AB} on concentration and temperature.

The FH model predicts an improvement in compatibility of the two polymer components during continuous heating above the UCST. However, in the course of such heating the mismatch of not only solubility parameters, but also of the free volume fractions of components becomes so dramatic that segments of more expanded. Low glass transition (T_g) polymers start 'to condense' onto segments of high-T_g components. Since new intersegmental bonds are formed, this process is, of course, exothermic, i.e., a part of the excess enthalpy is lost. Concomitant entropy loss is so severe that the ΔG becomes positive and the system separates into two phases with critical point as LCST. It may be expected that the occurrence of both the UCST and the LCST should be regarded as a universal feature of phase diagram for any combination of linear polymers if other factors like glass transition on cooling or thermal degradation on heating do not interfere.

According to Lipatov and Nesterov, the phase behaviour all the polymer mixtures can be classified in five groups [17]:

1. Systems with specific interactions between components which are miscible blends having only LCST [Examples: oligodimethylsiloxanes/oligomethylphenylsiloxanes; poly(aryl ether ether ketone/poly(aryl ethersulfone); polybenzimidazole/polyimides; 1,4-polybutadiene/*cis*-1,4 polyisoprene; chlorinated polybutadiene/chlorinated polyethylene (PE); poly(2,6-dimethyl-1,4-phenylene oxide/deuterated polystyrene or polystyrene ionomer; brominated poly(2,4-dimethyl-1,4-phenylene oxide) (50-76%Br)/polystyrene (PS) or polymethylmethacrylate (PMMA) or atactic PMMA(aPMMA); poly(hydroxyether of bisphenol A)/poly-ε-caprolactone; polyepichlorohydrine/ polymethacrylate (PMA) or polyethylacrylate (PEA), poly-*n*-propyl acrylate, PMMA, poly(ethylmethacrylate), poly(*n*-butyl methacrylate), poly(*N*-vinyl-2-pyrrolidone), polyethersulfone/poly(ether imide); poly(ethylene oxide)/poly(epichlorohydrin) or poly(D)(-)-(3-hydroxybutyrate), or PMMA; PMMA/poly (α-methyl styrene), or chlorinated polypropylene, or poly(vinylchloroacetate); poly(1,1,1,3,3,3,hexafluoroisopropyl methacrylate)/PMMA or poly methyl ethyl acrylate (PMEA), poly(tetrahydrofurfurylmethacrylate), or poly(2-fluoroethyl methacrylate); poly(1-chloroethylmethacrylate)/PMMA, polyethyl methacrylate (PEMA), poly *n*-propylmethacrylate (Pn-PMA), polyisopropyl methacrylate (PiPMA), poly *n*-butyl-MA (BMA), poly(tetrahydrofurfurylmethacrylate), poly(3-chloroethylmethacrylate)/poly(tetrahydrofurfurylmethacrylate, poly(2-iodoethylmethacrylate)/poly(butylene adipate), or poly(2,2′dimethyl-1,1,3-propylene adipate, or poly(hexamethylene sebacate), poly-ε-caprolactone); poly(2,2′-dichloroethylmethacrylate)/PEMA, or polybutyl methacrylate (PBMA), Poly-*n*-PMA, PiPMA, poly isoamylmethacrylate); poly(2,2,2,-trichloroethylmethacrylate)/PMMA, PEMA, Pn-PMA, PiPMA, poly(tetrahydrofurfurylmethacrylate); polyvinyl acetate (PVAc)/cyclic oligomer bisphenol A polycarbonate, poly(2-bromoethylmethacrylate), poly(2-chloroethylmethacrylate), PMMA; polyvinyl alcohol (PVA), poly(acrylic acid), PVAc, polyvinylchloride (PVC)/polybutyl acrylate (PBAc), PBMA PMMA; chlorinated PVC/poly-ε-caprolactone, PMMA, polyurethane, atactic PMMA; poly(vinyl methyl ether)/poly(2-chlorostyrene), PEA, polypropyl acrylate, PBAc, poly(hydroxyether of bisphenol A); polyvinylidene fluoride/PEA, PEMA; poly(vinylfluoride)/poly(vinyl phenol), poly(*p*-phenylene benzothiazole), polyamide 66 (PA 66); PS/poly(2-chlorostyrene); deuterated PS/poly(vinyl methyl ether; (Li, Mg, Zn, Mn)-sulfonated PS ionomers with sulfonation degree 6.5-9/PA 66; or polypropylene oxide (PPO), PS containing (1,1,1,3,3,3,-hexafluoro-2-hydroxyisopropyl)-α-methyl styrene)/PBMA, biodegradable poly(D,L-lactide)/poly(ε-caprolactone)] [18].

2. Oligomer or polymer mixtures having UCST and both LCST and UCST [Examples: hydrogenated oligocyclobutadiene/isotactic PP or isotactic polybutene; poly(ethylene glycol)/poly(propylene glycol); poly(vinylethylene)/*cis*-1,4-polyisoprene or poly(ethylethylene); poly(vinylidene fluoride)/PMMA].

3. Systems composed of chemically similar (or non-similar) polymers may have both LCST and UCST [Examples: polybutadiene (PB)/deuterated polybutadiene or polyisoprene (M_w = 35 500); protonated polybutadiene/deuterated polybutadiene; brominated poly(2,4-dimethyl-1,4-phenylene oxide (50-76 % Br)/syndiotactic PMMA; PMMA/deuterated PMMA; poly(vinyl methyl ether)/polyisophthalamide (2-10 methylene groups); PS/poly(α-methylstyrene)].

4. Systems of homopolymers with copolymers and mixtures of copolymers with 'miscibility windows' and complicate phase diagrams, frequently they are multicomponent systems so two-, three- or multi-dimensional graphical representation are characteristics – see **Figure 3.1c** and **3.1d**. They represent particular behaviour due to the micro-phase segregation – see Section 3.3.1.

5. 'Sand-glass' (**Figure 3.1c**) type (brominated poly(2,4-dimethyl-1,4-phenylene oxide (50-76 % Br)/syndiotactic PMMA (M_w = 8300 – 55 000) and closed loop gaps of immiscibility (**Figure 3.1b**) (poly(ethylene oxide)/poly(carboxylic acid) or complex (**Figure 3.1d**) (poly(ethylene oxide)/poly(isophthalamide) diagrams, where the UCST occurs at higher temperatures than LCST. Such phase behaviour was found when isotactic polypropylene (iPP) was blended either with hydrogenated oligo(cyclopentadiene) [19] or hydrogenated oligo(styrene-co-indene) [20, 21]; poly(isobutene) and polydimethylsiloxane) (PDMS) [22].

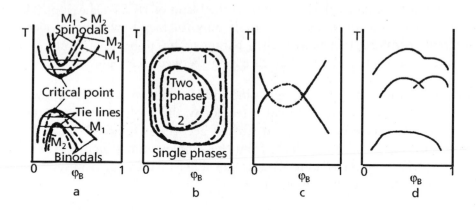

Figure 3.1 Schematic representation of four types of phase diagrams. a) UCST < LCST, the region of complete miscibility lies between curves. The higher is molecular weight of components the narrow will be this region; b) UCST > LCST the region of complete miscibility is found outside the curve; c) No unlimited miscibility region – hourglass type diagram; d) complex phase diagram

Bimodality of phase diagrams can be described within a framework of a new theory by Huggins applied by Koningsveld and co-workers [23] and is predicted by Sanchez-Lacombe theory [10-16]

Only from this summary it is clear that the most common phase diagram type for polymer mixtures is that with LCST, it being characteristic to the majority of miscible polymer blends.

To describe the exothermal process of mixing, the concept of 'complementary dissimilarity' was proposed that assumes that the macromolecules of mixed polymers should have some functional groups which supplement each other and are capable of formation of H-bonds or donor-acceptor complexes [24, 25]. Typical examples are the blends of PVC (and other chlorinated polymers) with polymers containing acceptor groups such as polyacrylates, polymethacrylates, polylactones, etc.

Using a generalised Flory-Huggins parameter, i.e., χ is considered as a function of temperature and volume, Kammer [26] explained the occurrence of an immiscibility loop in polymer blends. This requires both favourable interactions between the constituents and structural dissimilarities of the constituents that lead to a positive noncombinatorial entropy contribution.

With increasing temperature, the interactions are weakened, causing LCST behaviour, while the noncombinatorial entropy contribution becomes more and more dominant, leading eventually again to phase stability of the blend or UCST behaviour. This phase behaviour is experimentally easily accessible only for blends with component having very different polymerisation degree. The χ parameter has to be negative at low and high temperatures and positive in a temperature range between see **Figure 2.1**.

The shape of the phase diagram as well as the width of the 'miscibility window' between the UCST and LCST will be critical affected by the molecular characteristics of components. The phase diagrams constructed according spinodal equation of LFT (Equation 3.6) may change the shape from the simplest one – **Figure 3.1a** to 'closed loops' **Figure 3.1b**, hourglass **Figure 3.1c**, etc depending on the relative mismatch between characteristic parameters of components.

According to the Flory-Huggins theories, solution curves are almost indistinguishable from each other and are concave downward for high density PE (HDPE)/IPP mixture ($\chi = 0$), indicating predicting miscibility over the entire composition range. In the same time it is widely recognised that PE and polypropylene are immiscible especially at room temperature. This immiscibility was explained by the crystallinity differences between the blend components, but it is possible that these blends also exhibit phase separation in the melt, where the crystallinity effects do not exist. Polyolefin blends are currently the focus of intense industrial interest and they continue to receive substantial attention in the literature both from a theoretical and practical perspective. Interest in these materials stems, in part, from their improved mechanical and processing properties and, in part, from the current environmental focus on plastics recycling. The novel polymers of controlled molecular weight, controlled stereoregularity, and comonomer composition, distribution, are being produced from metallocene catalysts, leading to polymers with superior mechanical properties as compared to those produced by conventional catalysts. In this context, blending becomes an invaluable tool for molecularly-engineering resins with desired set of properties and ease of processing. Due to the interest in these blends, especially HDPE/IPP and IPP/ethylene propylene rubber (EPR), the theoretical explanation is necessary to evidence the factors that determine their miscibility/immiscibility (see Section 3.3).

In qualitative terms, the major factor ensuring stability of a single phase state, i.e., a negative value of the excess enthalpy will be favoured both by a negative value of the interaction parameter χ as well as by higher than unity ratios of the characteristic parameters v^*_A/v^*_B and T^*_A/T^*_B.

However, the calculation of phase diagrams for mixtures of more than three components is very difficult. The calculation becomes more complicated when the dependence of interaction parameter on composition and temperature are taken into account. Therefore, a simple method is required with which the phase diagram can be calculated: tie lines and bimodal, spinodal, critical points and their stability. A simple iteration process for the determination of the composition of the coexisting phase was proposed by Horst and Wolf [27, 28].

3.3 Factors Determining and Affecting Phase Behaviour

3.3.1 Structure

The structure of components is the major factor determining phase behaviour. Structural differences determine the compatibility of somehow similar blends. If PVC is miscible with polylactones, they are not miscible neither poly(vinylfluoride) nor with poly(vinylidene fluoride) (PVDF), while poly(vinyl bromide) exhibits an improved compatibility with polylactone, the behaviour being explained by differences in electronegativity. PVDF is miscible with many poly(alkyl acrylates) and poly(alkyl methacrylates).

During the last years many reactive oligomers have been developed, such as oligoesters, oligoethers and oligoesteracrylates, and these are widely used as mixtures in network production by curing. The strong interaction between these oligomers leads to formation of associates in reaction medium that will influence the reaction mechanism and structure of cured polymer. Knowledge of thermodynamic behaviour of such systems is useful in process control and for understanding the morphology and properties of resulting materials. Examples are the PEG/PPG, polyethylene glycol adipate/PPG mixtures. The composition and temperature dependence of χ_{AB} evidences a behaviour with both UCST and LCST. The appearance of the LCST in such systems can result due to the weakening of hydrogen bonds, that is induced by increase in temperature. The minimum evidenced in the dependence of χ_{AB} on temperature and its negative value show that this system demixes on heating.

Besides hydrogen bonds, the miscibility of binary polymer mixtures may depend on ion-ion, ion-dipole and other favourable interactions.

Block Copolymers. Two detailed reviews on phase separation behaviour of heterosystems including block copolymers have been written by Hasegawa and Hashimoto [29] and Kuchanov and Panyukov [30]. Chain connectivity (intramolecular AB junctions) and localisation of such junctions at the interface after phase separation is the fundamental feature by which block copolymers may be distinguished from polymer blends of the same composition. This will lead to a decrease in the excess entropy S_{ex} proportional with the number of junctions. The critical value of interaction parameter will be [31]:

$$\chi_{crit} = S_{ex}/k_B \, \varphi_{crit} \, (1-\varphi_{crit}) \qquad (3.11)$$

where k_B is the Boltzman constant and φ_{crit} is volume fraction at critical point.

The evaluation of χ_{crit} for block copolymers showed that this is invariably larger than that for binary polymer blends of the same composition. In the other words, chain connectivity makes phase separation more difficult in block copolymers than in the corresponding blends, although in other respects both systems exhibit similar characteristics.

An alternative approach in assessing the effect of chain connectivity in phase separated block copolymers involves consideration of their unique morphology. In contrast with polymer blends for which phase separation might ultimately result, at least theoretically, in the formation of two macroscopic phases of pure components. In the case of *non-polar block copolymers with randomly oriented (non-specific) interaction between chain components*, the finite chain length of each component prohibits macroscopic phase separation, hence only microphases ('phase domains') of a limited size may eventually form. The morphology depends on statistical characteristic of copolymer composition as well as upon conditions of observation. According to microscopic examination, the spherical domains (spheres) on a three-dimensional lattice (cubic) is the characteristic morphological feature of components at low volume content (below 30%); as the composition changes a morphological transformation to a system of regularly hexagonally packed cylinders a on two dimensional lattice (amorphous 'superlattice') and finally to interpenetrating continuous alternating lamellar structures around the 'phase inversion' point is observed – **Figure 3.2**. These morphological changes suggest a spinodal decomposition (SD) mechanism of phase separation in melts of block copolymers. Microphase separation transition or order-disorder and order-order between different morphologies transitions can take place.

For *block copolymers whose components specifically interact,* as 'segmented' polyurethanes, microphase separation is caused by self-association of rigid chain fragments. These rigid fragments are separated by 'soft' fragments via chain extenders, so they will be characterised by a fairly broad dispersion. The driving force for microphase separation is essentially the enthalpy gain due to the complete saturation of specific interaction between rigid fragments, rather than the differences between solubility parameters of both types of fragments. The ellipsopids and lamellae are typical morphological forms. The degree of phase separation is lower (larger interphase thickness) the higher the polarity of the soft fragments.

A key parameter controlling such morphologies is the number of segments of the block chains. Also the microdomain morphology of block copolymer systems is determined by the shape of the interface between two phases, i.e., the local curvature of the interface (local aspect) and the topological continuity of the interface in space (global aspect). Curved surfaces can be classified into three groups, namely, elliptic, parabolic and hyperbolic surfaces. The hyperbolic surfaces are the ones that form new, complex, bicontinuous microdomain morphologies. The size of the grain, the coherent region of

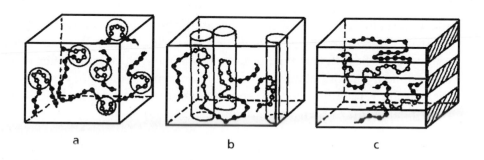

Figure 3.2 Morphology evolution of a non-polar block copolymer as a function of relative content and/or length of the second component: spherical micelles (a), cylindrical micelles (b) and alternating lamellae (c)

the microdomains, depends on the conditions of sample preparation and the segregation power of the constituent polymers in the block copolymers. Three-dimensional continuity of each microphase at the grain boundary is especially important for cylindrical and lamellar morphologies because it is expected to affect significantly the rheological and diffusion properties. The grain boundary is merely a surface with thickness of the order of the repeat distance of the microdomains. Other kinds of morphologies are shown for copolymer/homopolymer and copolymer/copolymer mixtures. Sometimes, microphase separation can be followed by a macrophase separation or inversely microphase separation follows after macrophase separation.

Since microdomain structures of block copolymers have their repeat distances in the order of 10 nm, small angle scattering of X-rays (SAXS) or of neutrons (SANS) and transmission or scanning electron microscopy (TEM and SEM) of ultrafine sections are the major techniques to characterise the structures. Scattering gives information about size and symmetry of the structure, whilst microscopy gives information limited to the local area but related to the structure entity. A 'mesh' structure was shown for a 70/30 wt%/wt% mixture of styrene-butadiene copolymer and PS homopolymer, which consists of alternating parallel sheets of the phase composed of PS block chains and PS in the matrix of PB phase, and the sheets of the major component (PS block or PS homopolymer) are interconnected by catenoidal channels of the same component traversing through the sheets of the minor component (PB).

Therefore, melts and solutions of block copolymers as well as blends containing their macromolecules often form thermodynamically stable mesomorphic states, characterised by an inhomogeneous density distribution on monomer units in space. In the vicinity of critical points, these mesomorphic states consist of blocks localised within the microdomains periodically distributed in space. Microdomain formation attracted much interest both from the theoretical and experimental standpoint [32, 33].

Two extreme limit regimes of the formation of spatially periodic structures are generally considered. Strong Segregation Regime (SSR) and Weak Segregation Regime (WSR). In the first regime (SSR) the different type of units of a binary block copolymer present a strong incompatibility. Microdomains consist mainly of units of the same type, separated by interfaces that are very thin in comparison with their size. Such systems are situated far from the critical points of their phase diagrams. The copolymer chains are strongly extended and gyration radius exceeds that of an unperturbed Gaussian coil.

In the second regime (WSR), the systems is close to critical point and the differences in concentrations of monomer units inside the domains are small, so no sharp interfaces exist between them. Under such a regime the size of macromolecule is only slightly perturbed by volume interactions.

Spatial distribution of concentration of different types of monomer units under certain assumptions of the self-consistent equations was evaluated by Edwards [34] and Helfand [35].

For *interpenetrating polymer networks* (IPN), vapour absorption data and morphological studies revealed that the overwhelming majority of both 'simultaneous' IPN and sequential IPN are phase separated systems. The expression of the interaction parameter contains term depending on network density and crosslink functionality, and in most cases free energy of mixing was found positive [17].

3.3.2 Tacticity

The miscibility of two polymers depends also on the stereoregularity. Isotactic PMMA (iPMMA) is not miscible with PVC in the whole range of compositions whereas syndiotactic PMMA forms miscible mixture when the content of syndiotactic PMMA does not exceed 60%. Polyethylene oxide (PEO) shows miscibility with atactic and syndiotactic PMMA and is not miscible with isotactic PMMA. Other examples are cited in [17]. This is due to the differences in interaction parameters. For example the interaction parameters for the mixtures of PMMA with polyvinylidene fluoride (PVDF) are equal to -0.13; -0.1 and -0.06 for isotactic, atactic and syndiotactic PMMA, respectively.

3.3.3 Branching

The effect of branching can be attributed to exclusion of some part of the branch from the interaction with another polymer. In the polyvinylmethyl ether (PVME) in the mixture with PS the shift of the cloud point curves to the lower temperatures was explained by kinetic effects of phase separation due to a greater number of entanglements of the chains as compared with usual branched macromolecules. Mehta and Honnell [36] used an atomistic skeletal united-atom model to represent HDPE, IPP and EPR and, on the basis of generalised Flory's approach derived an accurate equation-of-state with an compressibility term and concluded that if the phase separation occurs in the melt, it is attributable only to architectural differences. The branching destabilises these mixtures inducing phase separation.

3.3.4 Molecular Weight

With the increase in *molecular weight of components*, the composition range over which the blend is miscible decreases **Figure 3.1a**. Below a critical molecular weight of ~14,000 g/mol of polyolefin, the mixtures exist as single phase over the entire composition range [36].

So they explained practical results obtained with commercial polyolefin blends. This theory can be extended to α-olefin comonomers (1-butene, 1-hexene or 1-octane) in linear low density PE (LLDPE).

In the PS/PS or PEG/PEG systems, the components differing by *average molecular mass*, and consequently by thermal expansion coefficient and having close values of the solubility parameters, there should observed both LCST and UCST and a bimodal concentration dependence of χ_{AB}.

McMaster [37] calculated the phase diagrams using Flory's 'equation-of-state' thermodynamics in a generalised version that accounts for concentration and temperature dependent interaction parameter and polydispersity. By LFM he established that the factors affecting the phase diagrams are mainly: thermal expansion coefficient, molecular weight, interaction parameter, thermal pressure coefficient, etc. **Figure 3.3.**

Figure 3.3 schematically shows some examples of phase diagrams with various gaps of miscibility of binary polymer systems.

McMaster theoretically assessed that when the molecular weight increased, the mutual solubility of the two polymers decreases. Even at molecular weights as low as 50,000, differences in thermal expansion coefficients as small as 10% can lead to virtually complete immiscibility. Small positive values of the interaction energy parameter lead to both UCST and LCST behaviour. The two critical points have already merged to give skewed hourglass shaped bimodal and spinodal curves. The temperature where the LCST and UCST merge is a function of the magnitude difference of the component thermal expansion coefficients. The simultaneously occurrence of an LCST and a UCST should be rare for polymer/polymer systems.

3.3.5 Pressure

In mixtures PEA/PVDF exhibiting LCST, the cloud points decrease with increase of pressure, reach the minimum, and increase again, if the content of the second component does not exceed 30%. At 40% PVDF, only monotonous increase of cloud point takes place. LCST increases with increasing pressure in the entire composition range for the mixtures of copolymers poly(2,6-dimethyl-1,4-phenylene oxide)-poly(styrene-co-fluorostyrene). The UCST of the oligomeric mixtures of styrene (S)/butadiene (B) and B/propylene (P), increases with increasing pressure. The effects can be described by the modified theory of the equation-of-state. As a rule the increase in pressure increases the temperature of phase separation for the systems with LCST. Mixtures with UCST and negative heat of mixing increase the miscibility region along with increase in pressure, whereas for positive heats of mixing this region narrows [38]. Mixtures with LCST and with high molecular mass increases their miscibility with pressure. Both LCST and spinodal temperature increased with pressure [39, 40].

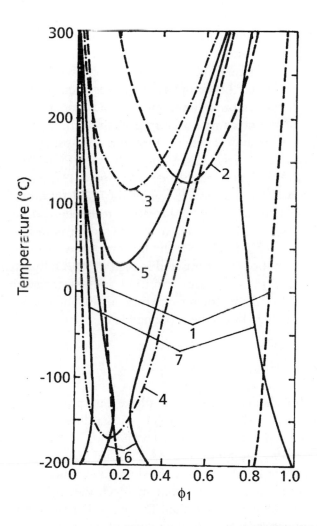

Figure 3.3 Effects of various factors on the shape of the phase diagrams. The curves represent the following effects: the thermal expansion coefficient, molecular weight, and interaction parameter (------------) (1) $M_1 = M_2 = 30\ 000$, $\alpha_2 = 0.640 \times 10^{-3}\,K^{-1}$, relative difference of thermal expansion coefficients 10.3%; (2) $M_1 = M_2 = 50\ 000$, $\alpha_2 = 0.630 \times 10^{-3}\,K^{-1}$, relative difference of thermal expansion coefficients 8.6%; of molecular weight. (—·—·—) (3) $M_1 = 30\ 000$; (4) $M_1 = 80\ 000$; (——) of interaction energy parameter = (5) χ_{AB} – 0.01; (6) χ_{AB} 0.05; (7) χ_{AB} 0.100

3.3.6 Thermal History

The miscibility of polymer components of a mixture depends also on the methods of preparation. Depending on the conditions of evaporation or cooling from the melt blends various aspects can be encountered.

The casting solutions from common solvent is a very useful method for study and obtaining polymer blends. Miscible blends in solvent with great difference in the interaction parameters between each polymer and solvent give narrow region of immiscibility – see **Figure 2.2a**. Such systems may be two-phase mixtures because during evaporation the system passes through the region of immiscibility and increase in viscosity causes the two phase structure of the mixtures to be frozen. In such cases a false UCST can be detected. Rapid evaporation may lead to the formation of a pseudo-homogeneous system with false LCST. Typical cases are PMMA/polycarbonate (PC), PS/PMMA and PMMA/styrene-acrylonitrile (SAN) copolymer blends. The effect of 'apparent dissolution' in the two-phase region is very important for determining the LCST for films dried from solutions and having two-phase structure. Thermal treatment also has an important effect. The pseudo-homogeneous, non-equilibrium systems are more frequently encountered when system undergoes phase separation with a very low rate. Slow phase separation by heating may lead to the overestimation of the miscibility region, especially when LCST is situated near T_g. The effect of low rate of phase separation on the estimation of the cloud point curves was investigated in detail for the mixtures of bisphenol A polycarbonate with PMMA [41]. The annealing allows thermodynamic equilibrium to be reached.

3.3.7 Filler Effect

Incorporation of a filler changes the shape of phase diagram due to the formation of the boundary surface layer near the filler interface, with a different composition than bulk one. As an example incorporation of even 10 wt% filler in PS/PBMA mixture decreases the thermodynamic stability and for the PMMA/PVAc mixture filled with 10 wt% silanised fumed silica, the curves of filled mixtures are situated at much lower temperature than those of unfilled mixtures. Increased filler content affects the asymmetry of the interactions and also the temperature of phase separation. The phase diagrams characterise the systems only when they are phase separated. In the region of incomplete phase separation, e.g., interfacial layer, only changes in the free energy of mixing and thermodynamic interaction parameters can be used to characterise the blends. The difference between free volumes of the components in the surface layer and in the bulk contributes markedly to increases in thermodynamic stability. The thermodynamic explanation consists of the fact that the coordination number depends on the conformational state of macromolecules, and it will be different in the surface layer and in bulk. Therefore the interaction parameter can be considered as a sum of two parameters characterising the surface layer and the bulk:

$$\chi_{AB(filled)} = \chi_{AB}(1-\varphi) + \chi_{AB(interface\ layer)}\varphi \qquad (3.12)$$

where φ is the volume fraction of the mixture in surface layer. The calculations show that the interaction parameter in the surface layer is considerably lower than in the bulk, therefore the interface with solid increases the thermodynamic stability of immiscible polymer pair even in the melt. But not all fillers increase the thermodynamic stability. By addition of filler (kaolin) to the miscible mixture PVAc/EVA copolymer, the interaction parameters increase in the surface layers in the whole range of compositions. This indicates the diminishing thermodynamic stability of the filled mixture. The effect is explained by independent adsorption of both components at the interface decreasing the interaction between component macromolecules [42].

The present theories described well enough the dependence of the phase behaviour on the composition of the blends and the molecular mass of components, but some particularities of such systems have been found that cannot be explained presently.

3.4 Mechanism and Kinetics of Phase Separation [17, 24]

The mechanism and dynamics of phase separation in binary polymer blends have long been the subject for many studies, in an attempt to obtain high performance materials by controlling the morphology.

The mechanism and kinetics of self-organisation in polymer systems under non-equilibrium conditions offers the possibility to manipulate and control the phase morphology in polymer blends to create new materials for various practical applications and open new frontiers in the development of modern technologies.

As has already been shown there are three regions of phase separation of a binary polydispersed blend: miscible, metastable and immiscible. Phase separation takes place in the last two cases but by different mechanisms. In the metastable region between bimodal and spinodal some forms of activation mechanism must trigger the phase separation. In the immiscible region such a triggering is not needed, the phase separation being spontaneous. The metastable mixtures are characterised by a nucleation and growth mechanism of phase separation, with a finite cooling into the metastable region required for nucleation. Due to the fluctuation of density within the metastable region spinodal exhibits diffuse boundary, the need for activation rapidly vanishes as the conditions move from bimodal to spinodal.

Two mechanisms of phase separation are considered: nucleation and growth (NG) and spinodal decomposition (SD). Both take place mainly in three steps: early, intermediate and late step of evolution – **Figure 3.4**.

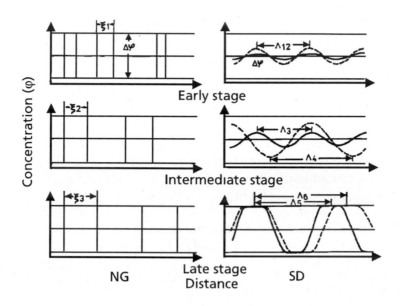

Figure 3.4 A schematic representation of the stages of the phase separation mechanisms NG and SD (concentration patterns or spatial segmental density profile of component A) [9, 16, 43]. λ - is wavelength of concentration fluctuation, ξ is domain size

In the first mechanism, the initial evolution of phase separation takes place by localised fluctuation of concentration (or density) and the mechanism responsible for phase separation is known as nucleation and growth. Into the spinodal region (enclosed by spinodal) the concentration fluctuations are delocalised, spontaneous phase separation takes place known as SD. Due to the large chain dimensions and slow diffusion rates, the mean field theory describes well enough, the phase separation in polymer blends, but other theories have been also developed [24]. A two-step phase separation was described by Hashimoto and others [44, 45].

Under the conditions of gradual temperature decrease and low viscosity, the phase separation begins in a metastable region, where nuclei with increased concentrations of one of the blend components appear in the homogeneous medium. Of these, thermodynamically favourable are only nuclei having sufficiently large dimensions (because the chemical potential is size dependent) whose formation requires a longer time. If the viscosity is sufficiently high, the system will remain spatially homogeneous with decreasing temperature until entering into the region of thermodynamic instability with respect to small fluctuations, characterised by a certain interval of wavelengths of concentration fluctuation which vary with stage of mechanism (λ_1 to λ_6). The process of phase separation under these conditions is conventionally referred to as SD. The mechanisms of phase separation are qualitatively different in the regions of metastability and instability of the phase diagram. Inside each region depending on the position of system in respect to bimodal, spinodal or critical point, qualitative distinctions have been observed in mechanism and kinetics of the phase separation. Near the spinodal the distinction between metastable and immiscible regions becomes diffused and phase separation can occur either by the NG mechanism or by SD.

3.4.1 Nucleation and Growth

The transition of the system from one phase to two-phase state takes place with amplification of fluctuations in the composition with development of the microregions of a new phase. The diffusion flows of components in the metastable region are directed towards diminishing fluctuation. These fluctuations are considered as critical nuclei. If these exceed critical values the second step will be developed. The process is similar to the appearance, growth and coalescence of the micro-regions of the evolved new phase by condensation or crystallisation.

The metastable regions are located on the convex part of the ΔG versus φ curve – **Figure 1.1a**, within the phase separation region, where: $(\delta^2\Delta G/\delta\varphi^2)_{P,T} > 0$ or $((\delta^2\Delta G/\delta\varphi^2)_V > v\beta((\delta^2\Delta G/\delta\varphi\delta v)^2_{P,T}$ where v is the specific volume of the mixture and β is compressibility. When phase separation occurs, the system across the phase diagram from b_1 to b_2, consumes an energy, which is the activation energy required for nucleation. This activation energy depends on the interface energy required for nucleus formation which is the product of interfacial tension coefficient (γ) and surface of nucleus (S). These values characterise the nucleation step. The nuclei or droplets with critical size grow by diffusion of macromolecules into the nucleated domains. The rate of growth can be approximated by the Ostwald ripening equation:

$$dV_d/dt \propto v\varphi V_m D_t/RT \tag{3.13}$$

or

$$d \propto t^{1/n_c} \tag{3.14}$$

where $n_c \approx 3$ is coarsening exponent, d and V_d are the droplet diameter and volume, respectively; φ at points b_1 or b_2 - **Figure 1.1b**, are the equilibrium concentrations. V_m is molar volume of the dispersed phase and D_t is diffusion coefficient. The diffusion step of the droplet growth is followed by the coalescence coarsening determined by the interphase energy balance. The morphology developed at equilibrium via NG mechanism is of droplet/matrix type the composition of the separated domains is constant, with only the size and size distribution of nucleated drops changing in time.

3.4.2 Spinodal Decomposition

The theory of SD of homogeneous system was developed by Cahn and Hillard [46-50], at the initial stage of the process where the concentration fluctuations appear. These fluctuations can be considered as a set of sinusoidal waves with a fixed length (λ). The wavelengths represent the dimensions of the structures formed during the course of SD, when both the composition and size depend on time. For example, in the case of the PS/PVME system, the probability distribution function for concentration *versus* time was determined by digital image analysis and it changes from sharp Gaussian to bimodal [51]. The three stages identified in SD are: diffusion, liquid flow and coalescence accompanied by morphology changes. The coarsening flow mechanism is evident in the last stage, where the process is in part similar with the phenomenon of creation of infinite clusters and could be treated by the percolation theory [52]. The diffusion stage follows the Ostwald equations. The kinetics of the initial stage of SD can be easily estimated by solving the diffusion equation [46-50]:

$$\partial\varphi/\partial t = M(\partial^2 G/\partial\phi^2)\nabla^2\varphi - 2MK\nabla^2\varphi + ... \qquad (3.15)$$

where M is mobility constant and K is the energy gradient term. The coefficient $\nabla^2\varphi$ can be identified with diffusion coefficient D. The solution of Equation 3.15 is:

$$\varphi - \varphi_0 = \sum_q \exp R(q)t[A\cos(qr) + B\sin(qr)] \qquad (3.16)$$

where φ is average uniform concentration, q is the wave number of the sinusoidal composition fluctuation, r is the position variable, A and B are q-dependent parameters and $R(q) = -Mq^2[D + 2Kq^2]$ is the Rayleigh kinetic growth factor or a first order rate constant describing the SD. One has to observe the exponential growth in the intensity of sinusoidal fluctuations of composition during the initial stages of SD. Due to this exponential form, the phase separation process is dominated by the fluctuation with the wave vector q, defined by setting the first derivative of R(q) to zero, so:

$$q_m = (-D/4K)^{1/2} \text{ and } R_m = -MD \, q_m^2/2 \qquad (3.17)$$

where $MD = D^C$ is Cahn-Hillard diffusion constant. Both D and M are temperature dependent. From the examination of these equations it can be seen that outside the spinodal region D>0 and R<0, so that concentration fluctuations are rapidly damped. At the spinodal q=q_m and R_m = D = 0 and a plot of R_m = R(T) or D = D(T) can be used to define the spinodal temperature T_s.

The Cahn-Hillard approach provides a quantitatively correct description of SD, although it over-estimates the difference in the mechanism of phase separation at the spinodal boundary. Infinitesimal changes in concentration or temperature which can change the sign of D will lead to a dramatic change in behaviour. If D>0 the metastable system would not phase separate by SD mechanism, whereas for D<0 the system would spontaneously decompose. In real cases a such a discontinuity is rarely observed, mainly in polymer mixtures where the spinodal is 'diffused' due to the polydispersity and/or thermal fluctuations.

Modern theories of phase separation include the description of processes of NG and SD considering relaxation processes obeying the Markov equation [53, 54], application of the Flory Huggins equation to the concentrated solutions [55], etc. De Gennes used the concept of reptation in a tube formed by labile knots of entanglements [56-59] extending the Flory-Huggins equation by adding a kinetic term and for unsymmetrical molecular weight case. Binder has fundamentally revised the dynamics of concentration fluctuations in polymer blends [60], developing a cluster theory.

The SD was computer simulated using Monte Carlo, Runge-Kutta methods reproducing the observed patterns of phase separation in both regions [61]. Numerical simulation of the region between NG and SD postulated a dynamic spinodal with a transient percolating structure. The simulation also revealed that in the early stage of phase separation the coarsening exponent (n_c) is not necessarily equal with 3, this value corresponds mainly for particle growth (coarsening) in stationary fluid whilst it decreases in the flow presence so the equation is:

$$(d/d_0)^{n_c} = 1 + k_c t \qquad (3.18)$$

where d is average droplet diameter n_c varies from 5/2 to 3/2 depending on the flow type. k_c the coarsening rate constant strongly increases with volume fraction of the dispersed phase.

Knowing the characteristic features of SD, the mathematical modelling of the phase separation in the region of unstable state is important. At a definite composition of the system, there can be found temperature corresponding to the maximum rate of formation of interconnected micro-regions of phase separated. The development of modulated structures by SD corresponds to a short period of development of interacting nuclei. During the later stages of SD, the important role belongs to hydrodynamic initiated coalescence. Two types of morphologies are expected from thermodynamic treatment. The first one originates from the dynamics of phase separation and exists for a short time before phase ripening takes over. The second type of morphology is controlled by the equilibrium thermodynamics where the size and shape of the phase is determined by minimization of the total free energy of the system including that of the interface.

The most useful methods of study of the phase separation are based on time-resolved scattering (light, neutrons, X-ray reflectivity, static secondary ion mass spectroscopy or other irradiation sources, atomic force microscopy (AFM)) techniques, etc. Therefore it is a direct correlation between the scattering variable and thermodynamic quantities such as interaction parameter, concentration fluctuation, etc., usually expressed in terms of the ratios φ/φ_s or T/T_s. This correlation is very important because it establishes that the Cahn-Hillard relationship provide a good and easy method of determining the molecular parameters of polymeric materials.

Linearised theories developed by Cahn and others described the early stages of SD and predict the exponential growth of intensity of light scattering, the appearance and growth of maximum of light scattering at a constant angle with increasing time of decomposition [46-50]. New theories and experimental investigations revealed that the amplitude of composition fluctuations (maximum of light scattering intensity) is shifted with time to smaller angles (intermediate stage) and then amplitude reaches its equilibrium value. Interfacial profile and heterogeneity length are continuously developed [17, 24].

The difference in concentration pattern evolution for NG and SD regions originates in the differences of early stages of separation – **Figure 3.4**. It was demonstrated that at the concentration of the minor phase above 10% to 15%, the SD occurs via rapid growth of regular spaced interpenetrating structures. Neglecting the fine structure of the dispersed phase the phase separation by NG and SD mechanism for low concentration of dispersed phase looks similar.

Time and compositional wavenumber resolved light scattering and nuclear magnetic resonance (NMR) spin-lattice relaxation times during the decomposition time permit discrimination between SD and NG mechanism [62]. NMR experiments on PS/PMVE indicate that the homogeneous phase as well as the phase separated quenched blend contain short scale inhomogeneities. Segment orientation on phase separation has been also treated [63].

A numerical study of the long-term dynamics of phase separation during SD was made by solving nonlinear diffusion equations [64]. A particular formalisms for SD in the mixtures containing liquid crystalline polymers have been developed [65]. The SD mechanism has also been studied in presence of block copolymers [66, 67] and compression [68, 69]

3.5 Phase Separation in Crystalline Polymer Blends

Most thermodynamic theories have been developed for liquid-liquid systems or amorphous polymers and polymer melts that can be assimilated with liquids in respect with distribution of segments. From commercial point of view the semicrystalline polymers are of prime importance, because they constitute ~ 67 wt% of all polymers on the market. For polymer blends, independently on the miscibility in liquid state or during processing step, e.g., extrusion, a phase segregation may be developed at crystallisation temperature. Phase separation and morphology of these kind of blends is discussed in Chapter 11 of this volume.

A kinetic model for SD mechanism was also proposed [70].

3.6 Experimental Data on the Mechanism of Phase Separation

The first data on SD in polymer mixtures were obtained by McMaster [37, 71] and Nishi [72] studying the SAN(28%AN)/PMMA and PS/PVME systems, respectively. According to the electron microscopy data, the domain dimensions formed in the initial stage of SD of SAN/PMMA system were of 50-100 nm and destruction of interconnected structures occurs when the domain size increases. At concentrations far from critical concentrations, the phase separation proceeds according to NG mechanism. For the PS/PVME systems the morphological data showed that both type of phase separation mechanisms are possible. Because the system exhibits LCST, the phase structure becomes finer at elevated temperatures, from few nanometres then diminishes, but the size of micro-regions increases in later stages of the process. The dimensions of micro-regions are small as compared with the period of interconnected structure. NMR data confirm the continuous change of microregion composition with time in accordance with SD mechanism. The diffusion coefficients decreased more three times approaching the spinodal. The PC/PMMA system in the early stage of SD and at low degrees of supercooling ($\Delta T = T-T_s$) obeys Cahn theory while at high ΔT there are departures from this theory [73]. The mechanism and kinetics of phase separation have studied for many systems such as [17]: PB/polyisoprene (PI), PB/styrene butadiene rubber (SBR), PMMA/SAN, PMMA/PVA, polyethersulfone/pole(ethyl oxazoline), PS/PB, etc.

The role of various factors influencing the kinetics of phase separation has been also established as mainly on temperature then on tacticity (PS/isotactic PVME and atactic PVME, PC/iPMMA and aPMMA), addition of a diblock or triblock [74] copolymer (PS/PB/PS-b-PB), presence of a solid filler, shear applied.

In industrial practice other independent variables such as pressure or the imposition of stress affect the phase separation. Use of simple shear stress to generate either NG and

SD morphology at constant temperature and pressure has been experimentally demonstrated. It was shown that the interconnected SD morphologies lead to high modulus and high maximum strain-at-break. Three-dimensional co-continuous morphologies of polymer blends are responsible for outstanding, synergistic performance of blends as chemical resistance and mechanical properties [75]. Compatibilisation by adding solvents or copolymers is another possibility for generating a controlled morphology. It has been found that diblock copolymers affect the kinetics of phase separation more extensively than equilibrium process. Co-continuity can also be obtained by the phase inversion concentration without SD. Between these mechanisms (SD and phase inversion) there are three basic differences: SD originates in a homogeneous, miscible blends via quenching to spinodal region, whereas inversion is a change in morphology in an immiscible blend; SD occurs at any concentration whereas the inversion is usually restricted to a higher range and SD generates fine morphology initially in the nanometer size range, whereas phase inversion leads to much coarser structures of the order of 0.1 to 10 µm. Therefore SD provides finer control of properties in a wider range of concentrations, but is restricted to systems which can be made miscible. Phase inversion is a general phenomenon in immiscible blends taking place usually at high concentrations. A special technique based on SD mechanism was developed to obtain a bilayer structure [76].

3.7 Reaction Induced Phase Separation

A special case of polymer blends are semi- and full IPN and other systems that are formed from two or many polymers during the chemical reactions. In these systems, during their formation, at a definite stage of conversion due to arising thermodynamic incompatibility of constituent network fragments, a process of phase separation proceeds accompanying chemical reactions. A high degree of mechanical entanglements in IPN hinders phase separation processes, which are always incomplete. The regular phase-separated structures in the final polymerised (or cured) materials are interpreted as a result of structure formation by the SD mechanism and is confirmed by light scattering experiments. A statistical theory of microphase separation in IPN was developed by Binder and Frisch [77]. The free energy of mixing strictly depends on the method of IPN formation (sequential or simultaneously). In both cases the expression contains the following contributions: elastic energy and entropy of crosslinking of two networks, the energy of interaction between monomeric units of both networks and free energy of entanglements. The estimation of the limits of stability of one-phase state of various IPN (spinodals) shows that its stability is higher in the region of small amount of one of the networks. The theory of SD in the case of chemical quenching is valid, if curing time is smaller than relaxation time of the system. This leads to the formation of thermodynamic non-equilibrium structures which depend on the prehistory of the system and essentially affect the IPN properties. [78-82]. A comprehensive review on the reaction-induced phase separation was published by Williams, Rozenberg and Pascault [81].

The processes of microphase separation in IPN proceeds according to the SD mechanism, e.g., styrene-divinylbenzene copolymer/PBMA, where a periodic structure appears during curing. A difficulty in the investigation of microphase separation processes by IPN formation is that the composition of the system varies continuously during conversion, so application of the existing theories to describe the phase separation is difficult. To avoid the complications, the reaction conditions should be selected so that the reaction rates are minimal compared with the rate of microphase separation. It has been shown that under process conditions, the reaction kinetics determine the beginning of the phase separation, i.e., the thermodynamic state is determined by the reaction kinetics. It is important to note that phase separation in the reacting system does not affect the kinetic curves, that is the total reaction rate does not change after the onset of phase separation. Such a situation may be observed only when the microregions of phase separation slightly differ in their composition which is typical of SD. The process of microphase separation in semi-interpenetrating network (SIPN) proceeds according to SD mechanism, in spite of the simultaneously proceeding chemical reactions of crosslinking and, in some cases, of polymerisation of monomer, forming linear component of SIPN [82]. Therefore the phase separation proceeds in non-equilibrium conditions. The phase separation begins at a low degree of conversion before gel point and is enhanced by increasing molecular mass and volume fraction of polymer formed. Phase separation begins when the miscibility gap reaches the Φ_{Mc} (initial volume fraction of modifier) at the reaction temperature, T_r – **Figure 3.5** defined as the cloud-point conversion [81, 82].

Most of the modifiers exhibit an UCST behaviour (epoxies cured with diamines and toughened with carboxyl-terminated polybutadiene-acrylonitrile copolymers), i.e., the miscibility increases with increasing temperature. As reaction proceeds the modifier becomes less miscible at T_r. The LCST behaviour is also observed for modifiers as polyethersulfones. A change in temperature shifts the miscibility gap mainly in the pre-gel region, therefore most the phase separation takes place before gelation, although the continuation of the phase separation process after gelation (or even vitrification) of the thermosetting polymer has been observed.

Dusek developed a thermodynamic analysis of the reaction-induced phase separation extended also for equilibrium swelling of networks separating into a highly swollen and a collapsed phase [83-85].

Some chemical groups are continuously transformed into different chemical groups during polymerisation, the interaction parameter between the components may vary throughout the reaction and interaction energy may be a function of conversion. Thus χ may increase

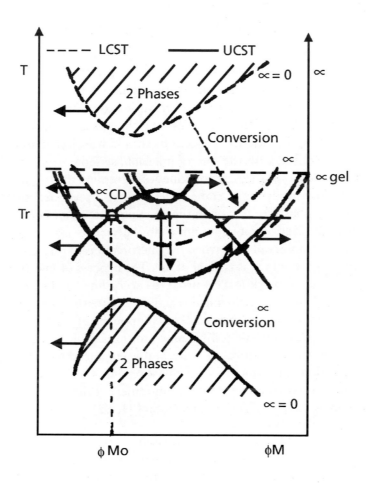

Figure 3.5 Evolution of the miscibility gap with conversion for modified thermosetting polymers showing a UCST (————) and LCST (------------) behaviour. α_{cp}, cloud point conversion, α - conversion [81, 82]

(favouring demixing) remain constant or decrease (favouring mixing) during polymerisation. The dependence of the χ on conversion can be modelled assuming that each segment can have different kinds of interacting sites. Morphology generated during entire process depends on the ratio between the phase separation rate and cure reaction rate. When this ratio tends to zero a bicontinuous structure may be developed, while when it tends to infinity the system evolves along bimodal curve. At the later stages the coarsening of the structure takes place.

Two alternative ways are used to represent the reaction induced phase separation in a polycondensation step: time-temperature-transformation and conversion-temperature transformation diagrams. However a triangular diagram is more useful to describe the miscibility curves in the case of a chain polymerisation, at a constant temperature.

Of course the phase separation mechanism depends on the position of the system on the phase diagram. It is very difficult to plot the phase diagrams for such systems. The phase diagram can be obtained only in a simple case when the IPN is obtained from a linear polymer and a bifunctional polymer that form a three-dimensional network (crosslinked PS /linear PBMA, crosslinked polyurethanes (PU)/linear PBMA). Under crosslinking reaction conditions both composition and thermodynamic compatibility continuously change. Only at very beginning the changes in composition can be considered almost constant and the Cahn's theory may be applied, because the phase separation occurs during 200-300 s.

The morphology and properties of SIPN and IPN are determined by the kinetics of their formation and by processes of microphase separation accompanying the reaction proceeding by NG mechanism or due to SD mechanism, these aspects being interconnected. Three cases are identified:

1. If one network is formed much faster than the other, the initial stage of IPN formation proceeds practically in the liquid state of the components of the second network. The phase separation takes place relatively fast and the evolution of the swollen first network in components of the second network is possible. This case is typical of IPN formation by sequential curing and is characterised by a high degree of phase separation or segregation.

2. The second case corresponds to a vary rapidly development of both networks. Microphase separation is governed by diffusion and it has no time to proceed and the initial structure, one phase, of reaction components is frozen.

3. The third case is the most frequent, the reaction rates of components are comparable (simultaneously IPN) so microphase separation begins simultaneously with the chemical reactions before gel point. The process is stopped after reaching gel-point at a rather high degree of crosslinking. The two-phase structure with incomplete phase separation appears as a result of these two overlapping processes dependent on both reaction rates and kinetics of phase separation. The phase separation in IPN is a two-stage process. The initially 'pure' SD mechanism becomes more complex due to simultaneous processes of phase separation according to the NG mechanism. In some cases one mechanism is replaced by the other. These processes are very complicated thus pure mechanisms of NG or SD can hardly occur in their pure

form. As a result, thermodynamically unstable diffuse micro-regions of incomplete phase separation arise, i.e., the segregation degree is not at its equilibrium value. Phase separation of the oligomer-oligomer system (epoxies and rubber) proceeds according to the NG mechanism due to constantly variation of thermodynamic and kinetic parameters [86]. The derivation of the basic equation for IPN domain diameter is based on a physical model of sequential IPN, according to which polymer B, formed in a swollen network A, constitutes a spherical core and exists in a contracted (deformed state), while polymer A surrounds the core and it is expanded (deformed state) into spherical shell.

If the hardener has a low reactivity, small and large spherical domains are formed, i.e., bimodal structure. Small domains are distributed among large domains. Due to the competition between phase separation and chemical reaction, two-phase morphology can be controlled, as well as physical and mechanical properties and it is possible to keep the spinodal structures at different stages of SD [87]. For example the interconnecting structures generated by SD were frozen in PS/poly(α-chlorostyrene) system by irradiation and in PB/PI system by crosslinking reaction [88, 89]. The period of modulated structure of gel formation varies with the reaction time and conversion degree. It was shown that the phase separation in IPN strongly depends on composition. For IPN based PU, it was found that at small amount of initial monomer in network matrix, more intense interpenetration takes place and therefore smaller segregation degrees are observed. An apparent compatibility due to non-equilibrium frozen state may arise which may be considered as forced compatibility [90, 91]. Forced compatibility should lead to an additional contribution to the free energy of the system. The contribution of elastic energy is small at the onset of phase separation and increases with increase in composition difference between phase separated regions. As a result, the system is stabilised when the thermodynamic driving force for phase separation is balanced by elastic forces from entanglements of network fragments. The stability limit of such a polymer system is characterised by a curve below the chemical spinodal. For the system with UCST, the real spinodal is situated below the chemical spinodal. For some IPN, the real spinodal is so remote from chemical spinodal that the unstable state can not be reached.

The formation of multiphase structures for the given rate of cure, modelling the effect of reaction, was studied by numerically solving of non-linear diffusion equation with a time-dependent free energy term for a series of values characterising the depth of cure. Diffusion equations admit analytical solutions that enable the characterisation of the system above and below the spinodal [92].

Phase separation in chemically reacting polymer blends is a new field of research [93].

3.8 Methods for Determination of Multiphase Behaviour

The methods used to study miscibility of polymers may be classified as I) phase equilibria methods, II) evaluation of interaction parameter and III) indirect compatibility tests (differential scanning calorimetry (DSC), differential thermal analysis (DTA), differential thermo-mechanical analysis (DTM), NMR, infrared (IR) spectroscopy, optical or light microscopy, SEM, TEM, AFM). The applicability of each group of methods depends on domain size which may vary from atomic level (10^{-4} μm) to large dimensions (10^2 μm). The decrease in sensitivity of several groups of methods is AFM > TEM > SEM > optical microscopy and WAXS > SAXS > SANS > light scattering. Dielectric, mechanical, thermal and IR spectroscopy measurements have close limits of detection of 10^{-2} - 10^{-3} (10^{-4}) μm [24].

3.8.1 Phase Equilibria Methods

Phase separation in solution leads to cloud point determination from which cloud point curves are constructed. Cloud point curves should follow spinodal for binary systems. The spinodal can be determined in a series of quenching experiments (heating/cooling rate recommended of 0.1 °C/min). The test is very important for aqueous systems with strong interactions where complexation/decomplexation can be studied (see Chapter 6).

A melt titration method has been also developed.

Phase separation can also be appreciated from film clarity, display of properties which are at least a weighted average of the components, T_g, densitometry, etc.

Light scattering methods are rigorous extended turbidity studies mainly by pulse induced critical scattering. The problem is the time to reach equilibrium in the polymer system, because the temperature is limited by degradation and self-diffusion coefficient of macromolecules is of order of magnitude of 10^{-15} m^2/s. The phase separation is very slow and may never reach completion. By centrifugation an acceleration of the phase can be achieved.

3.8.2 Evaluation of the Interaction Parameters

Intrinsic viscosity measurement, osmotic pressure, vapour sorption, calorimetric studies using Hess law, fluorescence techniques and inverse gas chromatography (IGC), size exclusion chromatography, etc., are the methods used for polymer/polymer/solvent studies for determination of thermodynamic properties, interaction parameter, etc. By studies in solutions, gyration radius and second virial coefficient can be determined.

Melting point depression can be used for interaction parameter determination (see Chapter 2).

IGC [17, 94, 95] which is mainly recommended for polymer-polymer-solvent systems, because the interaction of each polymer with the solvent determines the phase behaviour. For example PS-PVME gave clear a solution when mixed with benzene, toluene or tetrachloroethylene but a cloudy solution with chloroform, dichloromethane or trichloroethylene. The phase diagram determined by IGC is presented in **Figure 2.2b**.

IGC has been somewhat neglected in recent years due to scatter in data and because of solvent (chemical probe) dependence found with B_{ij}. Phase equilibria provide another source of interaction data in polymer solutions. However, they are useful mainly in the vicinity of the temperature of demixing. The most dependable data are from the measurement of the melting point depression by DSC, but this method is applicable only for blends where one of the polymers is crystalline.

3.8.3 Indirect Methods

A large number of methods are available for determining the existence of multiple phases in polymer systems such as: multiple transition temperatures determined by volume - temperature measurements, modulus - temperature measurements, DCS, DTA, phase separation in common solvent, phase contrast optical and electronic microscopy, SAXS, SANS [96-99], positron annihilation lifetime spectroscopy (PALS) technique that gives information on an angstrom scale etc., PALS uses a POSITRONFIT and the MELT program to evaluate the spectra [99, 100].

Thermodynamic interactions between polymer chains are often studied by SANS with one component of a binary blend labelled with deuterium [101]. A surprising effect is the dependence of thermodynamic behaviour on isotopic substitution. SANS in principle, yields the interaction coefficient of polymer blends in a similar procedure to the light scattering of polymer solutions.

3.9 Compatibilisation and Stabilisation Methods

Development of compatibilisation and stabilisation methods for polymer blends is directly related to knowledge of their phase behaviour.

The variability of morphology with the processing conditions, the phase ripening on ageing are two characteristics of heterogeneous blends which limit the penetration of blends into the market. However, from the practical point of view the hetero-phased systems are advantageous; the dispersed phase improves the behaviour of the brittle polymer or has a reinforcing effect. One of the most important problems facing the polymer blend industry is stabilisation of the desired properties, i.e., the elimination of residual stresses and stabilisation of the process-generated morphology and conditions of normal use of the manufactured part. The way to obtain such systems is by stabilisation of the phases by 'compatibilisation' or 'alloying'.

The ultimate method of stabilisation is by chemical or physical crosslinking of the multiphase system, as is done in the rubber industry. Temporary or reversible crosslinking can be achieved by block polymerisation or by introduction of ionic interactions. Another way is to generate various types of IPN. Other methods involve addition of an agent able to modify the interfacial properties of polymer system. In several commercial polymer alloys the stabilisation of the interphase via selective radiation crosslinking, results in reproducibility of performance, processability and recyclability. Three strategies of compatibilisation have been developed:

- Addition of a small quantity of a third component that is either miscible with both phases (co-solvent) or a copolymer with one part which is miscible with one component of the blend and the second part which miscible with another component (0.5-2 wt%); usually block type less frequently a graft one has been used.

- Addition of a large quantity (35 wt%) of a core-shell compatibiliser that behaves like a multipurpose compatibiliser-cum-impact modifier and.

- Reactive compatibilisation by which '*in situ*' the desired quantities of either block and/or graft copolymers) are obtained, e.g., via dynamic vulcanisation.

The compatibiliser successfully used in commercial application is frequently a commercial multicomponent and or a multiphase material. Their use varies from a system to system, not only as a function of compatibiliser efficiency but also in the relation to the overall performance of the final product including, e.g., in weathering. The thermodynamics can again serve a guiding role, calculation indicating that multiblock copolymers should be more efficient than di-block. Block copolymers mainly affect the interfacial tension coefficient and size of dispersed phase with small effect on the shear sensitivity, while by reactive compatibilisation a thick interphase which shows excellent stability under high stress and strain.

Co-solvent can also be used to generate compatible blends. Frequently two immiscible polymers will form a true solution in a common solvent. After the removal (freezing-drying) the interfacial area is so large that even very weak polymer-polymer interactions will sufficiently stabilise such a pseudo-homogeneous system. This method works particularly well in systems with weak hydrogen bonding. Another application of the co-solvent principle is an addition of co-miscible oligomeric plasticisers

Most procedures of compatibilisation are treated in separate chapters of this handbook.

References

1. R. Koningsveld in *Polymer Science*, Ed., A.D. Jenkins, North-Holland Publishing Company, Amsterdam, The Netherlands, 1972.

2. R. Koningsveld and A.J. Staverman, *Journal of Polymer Science, Part A-2*, 1968, **6**, 305.

3. R. Koningsveld and A.J. Staverman, *Journal of Polymer Science, Part A-2*, 1968, **6**, 325.

4. R. Koningsveld and A.J. Staverman, *Journal of Polymer Science, Part A-2*, 1968, **6**, 349.

5. *Polymer Compatibility and Incompatibility, Principles and Practices*, Ed., K. Solc, Harwood Publishers, New York, NY, USA, 1983, pages 1, 25 and 59.

6. *Polymer Blends and Mixtures*, Eds., D.J. Walsh, J.S. Huggins and A. Maconnachie, M. Nijhoff, Boston, MA, USA, 1985, 89.

7. E.J. Beckman, R.S. Porter, R. Koningsveld and L.A. Kleintjens in *Integration of Fundamental Polymer Science and Technology* Eds., P.J. Lemstra and L.A. Kleintjens, Elsevier, London, UK, 1988, 197.

8. P.I. Freeman and J.S. Rowlinson, *Polymer*, 1960, **1**, 20.

9. M. Tokaaki, Y. Satoru, *Journal of Applied Polymer Science*, 1998, **68**, 807.

10. I.C. Sanchez in *Polymer Compatibility and Incompatibility. Principles and Practices*, Ed. K. Solc, Harwood Academic Press, New York, NY, USA, 1982, 59.

11. I.C. Sanchez, *Macromolecules*, 1991, **24**, 4, 908.

12. I.C. Sanchez and R.H. Lacombe, *Journal of Physical Chemistry*, 1976, **80**, 2352.

13. I.C. Sanchez and R.H. Lacombe, *Journal of Physical Chemistry*, 1976, **80**, 2568.

14. I.C. Sanchez and R.H. Lacombe, *Journal of Polymer Science: Polymer Letters Edition*, 1977, **15**, 2, 71.

15. I.C. Sanchez and R.H. Lacombe, *Macromolecules*, 1978, **11**, 6, 1145.

16. I.C. Sanchez and R.H. Lacombe, *Journal of Macromolecular Science B*, 1980, **17**, 3, 565.

17. Y.S. Lipatov and A.E. Nesterov, *Thermodynamics of Polymer Blends*, Technomic Publishing Co. Inc., 1997, Lancaster, PA, USA.

18. J.C. Meredith and E.J. Amis, *Macromolecular Chemistry and Physics*, 2000, **201**, 6, 733.

19. S. Cimmino, E. Di Pace, F.E. Karasz, E. Martuscelli and C. Silvestre, *Polymer*, 1993, **34**, 5, 972.

20. C.H. Lee, H. Saito and T.T. Inoue, *Macromolecules*, 1995, **28**, 8096.

21. C.H. Lee, H. Saito, G. Goizueta and T.T. Inoue, *Macromolecules*, 1996, **29**, 12, 4274

22. H. Wagner and B.A. Wolf, *Macromolecules*, 1993, **26**, 6498.

23. R. Koningsveld and L.A. Kleintjens, *Journal of Polymer Science* 1977, **C-61**, 221.

24. L.A. Utracki, *Polymer Alloys and Blends*, Hanser Publishers, Munich, Germany, 1989.

25. M.M. Coleman, J.E. Graf and P.C. Painter, *Specific Interactions and the Miscibility of Polymer Blends. Practical Guides for Predicting & Designing Miscible Polymer Mixtures*, Technomic Publishing Company Inc., Lancaster, PA, USA, 1991.

26. H.W. Kammer, *Polymer*, 1999, **40**, 21, 5793.

27. R. Horst and B.A. Wolf, *Journal of Chemical Physics*, 1995, **103**, 3782.

28. R. Horst and B.A. Wolf, *Macromolecular Theory and Simulations*, 1995, **4**, 449.

29. H. Hasegawa and T. Hashimoto in *Comprehensive Polymer Science, Second Supplement*, Ed., G. Allen, Pergamon Press, New York, NY, USA, 1996, Chapter 14, 447.

30. S.I. Kuchanov and S.V. Panyukov in *Comprehensive Polymer Science, Second Supplement*, Ed., G. Allen, Pergamon Press, New York, NY, USA,1996, Chapter 13, 441.

31. V.P. Privalko and V.V. Novikov, *The Science of Heterogeneous Polymers, Structure and Thermophysical Properties*, John Wiley and Sons, Chichester, UK, 1995, 106.

32. *Space-Time Organization in Macromolecular Fluids*, Eds., F. Tanaka, M. Doi and T. Ohta, Springer, Heidelberg, Germany, 1989, pages 2, 13, 30 and 155.

33. *Developments in Block Copolymers*, Ed., I. Goodman, Academic, New York, NY, USA, 1982, Volume 1, pages 81 and 99.

34. S.F. Edwards *Journal of Physics A*, 1974, **76**, 332.

35. E. Helfand, *Accounts of Chemical Research*, 1975, 8, 295.

36. S.D. Mehta and K.G. Honnell, Proceedings of ANTEC '97, Toronto, Canada, 1997, Volume 2, 2648.

37. L.P. McMaster, *Macromolecules*, 1973, **6**, 5, 760

38. M. Beiner, G. Fytas, G. Meier, S.K. Kumar, *Physics Review Letters*, 1998, **81**, 594.

39. D.J. Lohse, M. Rabeony, R.T. Garner, S.J. Han, W.W. Graessley and K.B. Migler, *ASC Polymeric Materials Science and Engineering*, 1998, **78**, 87.

40. M. Rabeony, D.J. Lohse, R.T. Garner, S.J. Han, W.W. Graessley and K.B. Migler *Macromolecules*, 1998, **31**, 19, 6511.

41. M. Nishimoto, H. Keskkula and D.R. Paul, *Polymer*, 1991, **32**, 2, 272.

42. A.E. Nesterov and Y.S. Lipatov, *Polymer*, 1999, **40**, 5, 1347.

43. T. Hashimoto, M. Takenaka and H. Jinnai, *Journal of Applied Crystallography*, 1991, **24**, 457.

44. T. Hashimoto, M. Hayashi and H. Jinnai, *Journal of Chemical Physics*, 2000, **112**, 6886.

45. T. Hashimoto, M. Hayashi and H. Jinnai, *Journal of Chemical Physics*, 2000, **112**, 6897.

46. J.W. Cahn and J.E. Hillard, *Journal of Chemical Physics*, 1958, **28**, 258.

47. J.W. Cahn and J.E. Hillard, *Journal of Chemical Physics*, 1959, 30, 1121.

48. J.W. Cahn and J.E. Hillard, *Journal of Chemical Physics*, 1959, **31**, 688.

49. J.W. Cahn and J.E. Hillard, *Journal of Chemical Physics*, 1965, **42**, 93.

50. J.W. Cahn and J.E. Hillard, *Acta Metallurgica*, 1962, **9**, 795.

51. H. Tanaka and T. Nishi, *Physics Review Letters*, 1987, **59**, 692.

52. S. Reich, *Physics Letters*, 1986, **114A**, 90.

53. C. Varea and A. Rodledo, *Journal of Chemical Physics*, 1981, **75**, 5080.

54. M. Metiu, K. Kitahara and J. Ross, *Journal of Chemical Physics*, 1976, **65**, 383.

55. J.J. van Aartsen, *European Polymer Journal*, 1970, **6**, 919.

56. P.G. de Gennes, *Journal of Chemical Physics*, 1971, **55**, 572.

57. P.G. de Gennes, *Journal of Chemical Physics*, 1980, **72**, 4756.

58. P.G. de Gennes, *Journal of Physics Letters* (France), 1977, **38**, 441.

59. P.G. de Gennes, *Scaling Concept in Polymer Physics*, Cornell University Press, New York, NY, USA, 1979.

60. K. Binder, *Journal of Chemical Physics*, 1983, **79**, 638.

61. R. Petschek and H. Metiu, *Journal of Chemical Physics*, 1983, **79**, 3443.

62. N. Parisel, F. Kempkes, C. Cirman, C. Picot and G. Weill, *Polymer*, 1998, **39**, 2, 291.

63. T. Fukuda, K. Fujimoto, Y. Tsujii and T. Miyamoto, *Macromolecules*, 1996, **29**, 9, 3300.

64. E.V. Prostomolotova, I.Y. Erukhimovich and L.I. Manevich, *Polymer Science, Series A*, 1997, **39**, 6, 682.

65. A.J. Liu and G. H. Fredrickson, *Macromolecules*, 1996, **29**, 24, 8000.

66. T. Izumitani and T. Hashimoto, *Macromolecules*, 1994, **27**, 1744.

67. L. Kielhore and M. Muthukumar, *Journal of Chemical Physics*, 1999, **110**, 4079.

68. T. Izumitani and T. Hashimoto, *Polymer*, 1997, **38**, 3409.

69. H. Warkins, G.D. Brown and V.S. Ramachandra Rao, *Macromolecules,* 1999, **32,** 7737.

70. E.V. Prostomolotova and I. Ya. Erukhimovich, *Polymer Science, Series A,* 1998, **40,** 5, 419.

71. L.P. McMaster, *ACS Polymer Preprints,* 1974, **15,** 1, 254.

72. T. Nishi, T.T. Wang and T.K. Kwei, *Macromolecules,* 1975, **8,** 227.

73. T. Kyu and M. Saldanha, *Journal of Polymer Science, Physics Edition,* 1990, **28,** 97.

74. H.J. Liang, *Macromolecules,* 1999, **32,** 8204.

75. T.T. Inoue, T. Ougizawa and K. Miyasaka in *Current Topics in Polymer Science,* Eds., R.M. Ottenbrite, L.A. Utracki, and S. Inoue, Hanser Verlag, Munich, Germany, 1987.

76. P. Muller-Buschbaum, S.A. O'Neill, S. Affrossman and M. Stamm, *Macromolecules,* 1998, **31,** 5003.

77. K. Binder and H.L. Frisch, *Journal of Chemical Physics,* 1984, **81,** 2126.

78. L.H. Sperling in *Interpenetrating Polymer Networks and Related Materials,* Plenum Press, New York, NY, USA, 1981.

79. Y.S. Lipatov, *Journal of Macromolecular Science C,* 1990, **30,** 2, 209.

80. Y.S. Lipatov in *Interpenetrating Polymer Networks,* Eds., D. Klempner, L.H. Sperling and L. Utracki, ACS, Washington, DC, USA, 1994, 125.

81. R.J.J. Williams, B.A. Rozenberg and J.P. Pascault, *Advances in Polymer Science,* 1997, **128,** 97.

82. Y.S. Lipatov, T.T. Alekseeva and V.V. Shilov, *Polymer Networks & Blends,* 1991, **1,** 129.

83. K. Dusek in *Polymer Networks Structural and Mechanical Properties,* Ed. A. Chompff, Plenum New York, 1971, 245.

84. K. Dusek, *Journal of Polymer Science* 1967, **C16,** 1289.

85. K. Dusek, *Chemicke Zvesti,* 1971, **25,** 177.

86. J.K. Yeo, L.H. Sperling and D.A. Thomas in *Polymer Compatibility and Incompatibility*, Ed., K. Solc, Harwood, 1982.

87. H.S-Y. Hsich, *Advances in Polymer Technology*, 1990, **10**, 185.

88. Q. Tran-Cong, T. Nagaki, T. Nakagawa, O. Yano and T. Soen, *Macromolecules*, 1989, **22**, 2720.

89. Q. Tran-Cong, T. Nagaki, T. Nakagawa, O. Yano and T. Soen, *Macromolecules*, 1991, **24**, 1505.

90. I. Sakurada, A. Nakajima and H. Aoki, *Journal of Polymer Science* 1959, **35**, 507.

91. Y.S. Lipatov, L.M. Sergeeva, A.F. Nesterov and L.V. Karabanova, *Reports of the Academy of Science - USSR*, 1975, **220**, 637.

92. S.A. Shaginyan and L. I. Manevich, *Polymer Science, Series A*, 1997, **39**, 908.

93. A.R. Khuklov, Proceedings of Europolymer 2001, Eindhoven, The Netherlands, 2001, PL-18.

94. D.D. Deshpande, D. Patterson, H.P. Schreiber and C.S. Su, *Macromolecules*, 1974, **7**, 530.

95. A. Robard, D. Patterson, and G. Delmas, *Macromolecules*, 1977, **10**, 706.

96. W.W. Maurer, F.S. Bates, T.P. Lodge, K. Almdal, K. Mortensen and G.H. Fredrickson, *Journal of Chemical Physics*, 1998, **108**, 2989.

97. H. Frielinghaus, D. Schwahn, K. Mortensen, L. Willner and K. Almdal, *Journal of Applied Crystallography*, 1997, **30**, 696.

98. J. Klein, T. Kerle and F. Zink, *Macromolecules*, 2000, **33**, 1298.

99. C. Wästlund, *Free Volume Determination in Polymers and Polymer Blends*, Chalmers University of Technology, Göteborg, 1997. [PhD Thesis]

100. C. Wästlund and F.H.J. Maurer, *Macromolecules*, 1997, **30**, 19, 5870.

101. B. Crist, *Macromolecules*, 1998, **31**, 17, 5853.

4 Interface (Interphase) in Demixed Polymer Systems

Cornelia Vasile

4.1 Interface Characteristics

The formation of any heterogeneous system is accompanied by the formation of an interphase or boundary layer, which determines important properties of the systems, mainly the polymeric ones [1]. The interface is characterised by a two-dimensional array of atoms and molecules which are impossible to measure, while the interfacial layer or interphase has a large enough assembly of atoms or molecules to have its own properties such as modulus, strength, heat capacity, density, etc. According to Sharpe [2], interphase is 'a region intermediate for two phases in contact, the composition and/or structure and/or properties of which may be variable across the region and which may differ from the composition and/or structure and/or properties of either of the two contacting phases'. Lipatov [3-6] considers that the interfacial layer and interphase are equivalent. A schematic representation of the interface in a binary mixture is given in **Figure 4.1**.

The main properties of the interphase are thickness (d_{inter}), interfacial tension (γ_{AB}), concentration profile ($\rho_i (x)$), etc.

The boundary or interfacial layers possess an **effective thickness**, beyond which the deviations of local properties from their bulk values become negligible. In polymeric materials the thickness of interfacial layer (or region) may be rather high due to the chain structure and consequently will impart a specific behaviour to the polymers unlike low molecular substances.

Surface (interfacial) tension is the reversible work required to create a unit surface (interfacial) area or excess free energy of a system at constant temperature (T), pressure (P) and composition (n):

$$\gamma = \left(\frac{\partial G}{\partial A} \right)_{T,P,n}$$

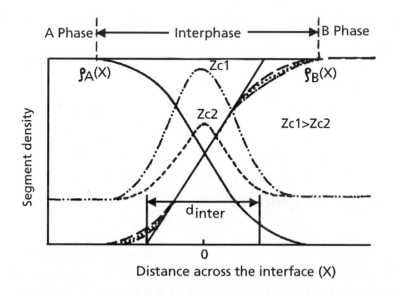

Figure 4.1 Schematic representation of a diffuse interface and its characteristics as interphase thickness, density (concentration) profile across the interface: (——) calculated and approximated by hyperbolic tangent (······) and error function (——). [7, 8] and accumulation of a copolymer with different polymerization degree (Z_{c1} and Z_{c2}) at interface [9]

where γ is the surface (interfacial) tension, G is the Gibbs free energy of the system and A is the surface (interfacial) area. It is a function of the distribution of quasicomponent density.

Surface tension can be separated into two components:

$$\gamma = \gamma^d + \gamma^p$$

where γ^d is dispersive and γ^p is polar component of surface tension. The polar component in its turn can be separated into electron donor and electron acceptor components.

The interfacial tension is related to the surface tension and the polarity of the two contiguous phases by the equation:

$$\gamma_{AB} = \gamma_A + \gamma_B - \frac{4\gamma_A^d \gamma_B^d}{\gamma_A^d + \gamma_B^d} - \frac{4\gamma_A^p \gamma_B^p}{\gamma_A^p + \gamma_B^p} \tag{4.1}$$

as the harmonic-mean of the γ values of individual components, A and B, or as the geometric mean:

$$\gamma_{AB} = \gamma_A + \gamma_B - 2\,(\gamma_A^d \gamma_B^d)^{1/2} - 2\,(\gamma_A^p \gamma_B^p)^{1/2} \qquad (4.2)$$

The harmonic-mean, e.g., Equation 4.1, is much satisfactory as geometric mean, e.g., Equation 4.2 [10, 11], equation valid if reaction does not takes place between polymers.

The surface tension varies with temperature by:

$$\gamma = \gamma_o (1-T/Tc)^{11/9}$$

where γ_o is the surface tension at T = 0 K and Tc is the critical temperature. The temperature coefficient of surface tension is thus given by:

$$-d\gamma/dT = (11/9)\,(\gamma_o/Tc)(1-T/Tc)^{2/9}$$

At ordinary temperatures -dγ/dT for most polymers is practically constant. For polymer pairs that exhibit upper critical solution temperature (UCST), the interfacial tension decreases with increasing temperature and vanishes at the UCST. This to the contrary of polymer pairs having a lower critical solution temperature (LCST), the interfacial tension vanishes at the LCST and increases with increasing temperature in the immiscible region.

The change in entropy of an interfacial layer can by found from a dependence of interfacial tension on temperature and composition of the mixture considered in thermodynamics of surface phenomena.

The surface tension varies also with density ρ according to the Macleod equation:

$$\gamma = \gamma^o \, \rho^m$$

where γ^o and m are constant independent on temperature. γ^o is considered dependent only on the chemical composition and is related to Sugden's parachor (a property defined for a group of atoms).

$$\gamma^o = \left(\left(\sum P_i \right) / M_i \right)$$ where $\sum P_i$ is the parachor of the repeat unit and M_i is molecular weight. The parachor is a group additive quantity and its values are tabulated [12, 13].

m is the Macleod's exponent and for most polymers has a value of 3.0-4.5.

The effect of molecular weight is usually ignored for polymers except for oligomers.

The properties of the interfacial layer are dependent on the conditions of preparation of the mixtures and are controlled by thermodynamic and colloidal chemical factors. The condition $\Delta G < 0$ also describes the process of spontaneous emulsification of one polymer in the medium of another, that leads to the formation of a stable colloid systems. In this case thermodynamic parameters of the system are changed due to dispersion of one polymer in another but also because of formation of an interface between two polymers. By dispersion, parts of the dispersion medium and the dispersed phase move into the interfacial layer (interphase). The interfacial adhesion, structure and properties of the interphase are very important characteristics of the immiscible blends. Interactions between polymers result in the formation of interfaces, i.e., in the formation at least one new polymer phase. It is reasonable to consider the interfacial region as an 'interphase', a third phase in the immiscible blends with its own characteristic properties. The interphase is a region of interdiffusion of the two types of macromolecules.

The segmental mobility within the interphase is lower, as shown by increased glass transition temperature [14]. The relationship between glass transition temperature (T_g) and interphase thickness is non-monotonic, as are the packing density, surface tension and the free mixing energy.

At T_g, surface tension is continuous and the temperature coefficient is discontinuous. The following relationship was established:

$$(d\gamma/dT)_g = (\alpha_g/\alpha_r)(d\gamma/dT)_r$$

where α_g and α_r are isobaric volumetric thermal expansion coefficients in the glassy and rubbery region, respectively.

The concentration profile can be described by various approximations [15]:

- linear gradient: $\varphi = (1/320) \, d_{inter} + 0.5$ (4.3)

- squared sinusoidal function: $\varphi = \sin^2(d_{inter} /320 + \pi/4)$ (4.4)

- hyperbolic tangent: $\varphi = \frac{1}{2}[1 + \tan h(d_{inter} 160)]$ and   (4.5)

- error function (erf): $\varphi = \frac{1}{2}(1 + erf[\sqrt{\pi}d_{inter}/320)$ (4.6)

There are only a few data available which show the equilibrium structure and equilibrium profile at the interface between the two polymers. In that case it is important to study the kinetics of the interphase formation. De Gennes [16] and Harden [17] considered the formation of a diffuse interface for $\chi \leq 1$ and using the scaling concept, obtained the dependence of the interphase width on time $t^{1/2}$ at short times and as $t^{1/4}$ at longer times and that the first dependence is characteristic to miscible mixtures and the second to immiscible mixtures. These studies led to the analysis of the critical fluctuation of concentration near the critical point, spinodal decomposition, dynamics of chain motion, the 'coil-globule transition', etc.

The interfacial tension is a measure of material compatibility and it plays a role to failure because the mixed units in the interphase may contribute to the properties in measurable ways. In a composite material, the interface acts to transmit a stress from one phase to another phase. The efficiency of this stress transfer is dependent on the nature of interface. The interphase region may be a non-negligible fraction in a finely dispersed system.

The formation of the interphase distinguishes the properties of blends from properties of solution. After all, the rheological and mechanical properties of polymer blends are determined by compatibility of the components and by the degree of microphase separation that implies the formation of an interphase. For example, the blends of polycarbonate (PC) with acrylonitrile-butadiene-styrene (ABS) are commercial products that do not use any compatiliser component. The useful properties generated, especially at a copolymer content of ~ 25 wt% are explained by developing a strong interface [18].

The theory of interfacial tension and composition and structure of an interfacial layer in polymer mixtures has been developed by many authors. Helfand, Tagami and Sapse [19-28] used the approximation of mean field theory predicting that the thickness of the interfacial (d_{inter}) layer is ~ 1 nm and depends on χ_{AB}. Roe [29] established that the interfacial tension and thickness of interfacial layer depend on the chain length and on $\Delta\chi = \chi - \chi_{crit}$. Nose [30] showed the change in the polymer coil dimensions at the interface. Sanchez [31-34] used gradient theory and compressible lattice theory to describe the properties of interfacial layer. Other theories [16, 17, 35] have applied the scaling concept to this purpose. New steps in understanding the polymer-polymer interphase have been made by the Wool and co-workers [36-40], who used either methods of molecular dynamics [41] or considered the fractal structure of polymer interfaces [38]. When diffusion occurs at an interface, the concentration profile varies smoothly as a function of the one-dimensional depth. However, when the diffusion process is viewed in the two- or three-dimension diffusion field, it exhibits a rough nature. The random nature of diffusion permits the formation of complex structures with fractal characteristics. The authors analysed gradient percolation using computer simulation for the metal-polymer interface and the fractal structure of a polymer-polymer interface was simulated by reptation diffusion. A definitive statement regarding the validity of the reptation model will be made by development of accurate simulation of dense systems that depends on the current state of computers associated with performance methods for determination of concentration profile [42].

Such studies are important for the following cases: welding of polymers in the melt, lamination of composites, coalescence of powder and pellet resins, drying of latex paints, internal weldlines formed during injection and compression molding, tack of uncured elastomers and a multitude of processing operations where interdiffusion is required to develop full strength at an interface. The fractal analysis compares well with experimental image examination of the interface.

The thickness of the interphase layer depends on the thermodynamic interactions, macromolecular size, concentration and phase conditions. As a result the interphase is not a homogeneous diffused layer, but a complex entity with micro- and macro heterogeneities in orthogonal directions. The interface adhesion can be predicted from values of the interface and surface tensions.

4.2 Theoretical Approaches

4.2.1 Kammer's Theory

Kammer's thermodynamic treatment [43-46] was initially based on the fundamental principles of equilibrium between two phases separated by a boundary layer and later was developed for a non-homogeneous interfacial region, characterised by density and free energy gradients. He assumed that both polymers are in a liquid state and are immiscible and a limited interpenetration occurs in the interphase. The composition (as molar fraction) of interphase (x^s) is considered as depending on temperature. The following equations have been obtained for the interfacial layer characteristic properties:

For the thickness of interphase (d_{inter}):

$$d_{inter} = \frac{(\gamma_A - \gamma_B)V}{RT[\ln \varphi_A^s + r\chi_{AB}(\varphi_B^s)^2]}$$

$$\text{or } d_{inter} = (2\pi/3\sqrt{3}) \frac{b(T^*/T)^{1/2}}{\chi_{AB}^{1/2}[1 - 1.32T/T^*]^{1/2}}$$

where V is geometric average of molar volume of two components, $r = r_1 = r_2$ is the number of segments per chain; γ_i is the surface tension of the components, φ_i^s are the volume fraction of components in the interface layer. For the polyethylene/polystyrene (PE/PS), polyamide (PA)/PE, PS/polymethyl methacrylate (PMMA) systems an interphase thickness of l - 7-10 nm was found [43-46].

For the interfacial tension:

$$\gamma_{AB} = 0.688\ (RT/V)\ \chi_{AB}^{1/2}b(T^*/T)^{1/2}\ (1-T/T^*)$$

where χ_{AB} is calculated from solubility parameters, b is effective length of monomer unit, M is molecular mass of polymer and m is molecular mass of structural unit and T^* is the reduction parameter or characteristic temperature (an adjustable parameter). The effective length (b) of the polymer unit may be calculated from tabulated data [47]. It depends on the root-mean-square end-to-end distance <R> by the relationship:

$$b = m^{1/2}\ (<R>/M^{1/2}).$$

The concentration (as molar fraction) of component in the interphase (x^s) can be obtained from temperature dependence of interfacial tension of phase (1) and (2):

$$x^s \approx \frac{(d\gamma/dT) + (d\gamma/d1)^{(1)}}{1/2\ (d\gamma/dT)^{(2)} + (d\gamma/dT)^{(1)}}$$

It can easily be seen that the direct proportionality of the interfacial layer thickness and b are inversely proportional with χ_{AB}, that means that the interpenetration is realised only on a very small distance which is much less than macromolecular dimensions.

4.2.2 Helfand and Tagami Theory

The configurational statistics model [19-28] is based on a self-consistent field that determines the configurational statistics of the macromolecules in the interfacial region assuming that: (a) the two polymers have the same degree of polymerisation, (b) the segmental density profile (ρ) is the same as for infinitely long macromolecules, (c) $M_w \rightarrow \infty$; (d) the isothermal compressibility is negligible; (e) there is no volume change upon blending; and (f) it was supposed that 'repulsive' interactions are balanced by the entropic effects of the chain intermingling.

The difference between a molecule near a surface and in bulk is that, a polymer molecule tries to get away from a surface because, near the surface it loses conformational entropy (all conformations which will cross the boundary are disallowed). At the interface, the interaction between statistical segments of polymers A and B are expressed by the interaction parameter (Flory's) χ_{AB} [48, 49]. They obtained the following relations for: interfacial thickness (Δd_{inter}) tension coefficient (γ_{AB}) and density profile across the interface (or interfacial composition profile) (ρ) (see **Figure 4.1**):

$$d_{inter} = 2b(6\chi_{AB})^{-1/2}; \tag{4.7}$$

$$\gamma_{AB} = b\rho_o \, k_B T(\chi_{AB}/6)^{1/2}$$

therefore $d_{inter} \, \gamma_{AB}$ = constant.

$$\rho_i/\rho_I^o = y^2(1+y^2) \text{ or } \rho_i/\rho_i^o = \Phi_A(x) = \{1/2 - 1/2 \tan h \, [6 \chi_{AB}^{1/2} (x/d_{inter})]\} \text{ and}$$

$$\Phi_B(x) = \{1/2 + 1/2 \tan h \, [6 \chi_{AB}^{1/2} (x/d_{inter})]\} \text{ with } y = \exp\{6 \chi_{AB}^{1/2} (x/d_{inter})\}$$

where i = A or B, b is a lattice parameter and x/d is the reduced distance across the interface, k_B is Boltzman constant. Therefore interfacial tension can be determined from solubility parameters values.

In binary blends the interphase thickness is inversely proportional to the interfacial tension coefficient; $\chi_{AB} \leq 0.1$ is a quite a large driving force to broaden the interface. The reciprocity between d_{inter} and γ_{AB} is not to be expected to hold in compatibilised blends.

Eliminating the symmetry limitation, interfacial tension is expressed as:

$$\gamma_{AB} = k_B (\rho_o \chi_{AB})^{1/2} \left\{ \frac{\rho_A b_A + \rho_B b_B}{2} + \frac{1}{6} \left[\frac{\rho_A b_A - \rho_B b_B}{\rho_A b_A + \rho_B b_B} \right]^2 \right\}$$

where ρ_o is the density of pure polymer in units of monomer segments per unit volume. The effective length of monomeric units b, is chosen in such a way that mean square distance between chain ends is zb^2, b is Kuhn's statistical segment length, $b' = C_\infty^{1/2} b_o$ with b_o real mean of monomer length. The real mean-square end-to-end distance for long chains is frequently written as: $C_\infty Z b_o^2$ and Z is degree of polymerisation, $C_\bullet$ is Flory's characteristic ratio and k_B is the Boltzman constant. Kuhn's statistical segment length for vinyl polymers has been written as a function of structural characteristics as [47]:

$$b_{vinyl} = 2a[\sin(\theta/2)]/(1+\cos\theta)\sigma^2$$

where (a) is the bond length, θ is the tetrahedral bond angle and σ represents the steric hindrance to internal rotations, $\sigma = <R^2>/<R^2>_f$, $<R^2>$, being the average square end-to-end distance and $<R^2>_f$ is the average square end-to-end distance of the chain in the freely rotating state. The average square end-to-end distance is usually obtained by solution measurements. The b and b_{vinyl} have the same physical significance, but they have different values because each of them is obtained from another type of measurement.

The relations are valid for small values of χ_{AB}.

This theory led to the following conclusions: the surface free energy is proportional to $\chi_{AB}^{1/2}$, the chain-ends of both polymers concentrate at the interface and any low molecular weight third component is repulsed to the interface and the interfacial tension coefficient increases with molecular weight to an asymptotic value:

$$\gamma_{AB} = \gamma_{AB\bullet} - a_o M_n^{-2/3}$$

An effect of this dependence that should take into account is the change in the concentration of polymeric homologues of low molecular weight in the interfacial region because they are much mobile and will lead to decrease in interfacial tension and to promotion of dispersion when the system is heated or subjected to mechanical stress.

The thickness of the interphase is comparable with the dimension of the monomeric unit therefore only limited diffusion of chain segments is possible in the interphase region. The equation for d_{inter} has been corrected to account the number of segments for each in the chains r_A and r_B:

$$d_{inter} = 2b\,(6\chi_{AB})^{-1/2}\,[\,1+ \chi_{AB}^{-1}(\ln 2(1/\,r_A + 1/\,r_B)]$$

Later refinements by Helfand and Sapse [26-28] taking into account interaction energy density (B) gave the relationship:

$$d_{inter} = \sqrt{\frac{2RT}{B_{ij}}}\left[\sqrt{\rho_A/6}(\langle R_A^2\rangle/M_A)^{1/2} + \sqrt{\rho_{Bi}/6}(\langle R_B^2\rangle/M_{Bi})^{1/2}\right]^{1/2}$$

Where $\langle R_B^2\rangle$ is the mean square unperturbed end-to-end chain distance and M_i is the molecular weight. Interfacial tension is expressed as:

$$\gamma_{AB} = \sqrt{\frac{RTB_{ij}}{2}}\left[\sqrt{\rho_A/6}\left(\langle R_A^2\rangle/M_A\right)^{1/2} + \sqrt{\rho_{Bi}/6}\left(\langle R_B^2\rangle/M_{Bi}\right)^{1/2}\right]^{1/2} \text{x}$$

$$\left\{1+\frac{1}{3}\frac{\left(\sqrt{\rho_A/6}\left(\langle R_A^2\rangle/M_A\right)^{1/2} - \sqrt{\rho_{Bi}/6}\left(\langle R_B^2\rangle/M_{Bi}\right)^{1/2}\right)^2}{\left(\sqrt{\rho_A/6}\left(\langle R_A^2\rangle/M_A\right)^{1/2} + \sqrt{\rho_{Bi}/6}\left(\langle R_B^2\rangle/M_{Bi}\right)^{1/2}\right)^2}\right\}$$

Later, Broseta and others [50] using an asymptotic approximation to the square gradient theory have treated the situation with finite molecular weight and due to the entropic gain obtained a broad interface and reduced interfacial tension. This correction can increase interphase thickness with 10%–30%. It was also demonstrated that there is an entropic increase when small chains segregate to the interface because they can lower interfacial tension and increase the interfacial thickness.

A model was proposed to fit neutron reflectivity data that accounts for surface roughness and for a diffuse interface between the polymer layers modelled as either an error function or hyperbolic tangent shaped profile (see **Figure 4.1** and Equations 4.3-4.6), according to theory [51]. The interfacial thickness calculated graphically from the width of a tangent drawn to the inflection of the interfacial profile corresponds with that evaluated by Helfand's theory see **Figure 4.1**. In the case of the polypropylene oxide/styrene-acrylonitrile copolymer) PPO/SAN copolymer, a 10 Å interfacial thickness was found independently on copolymer composition, but various annealing conditions can lead to the increase up to 60-65 Å as copolymer composition decreased from 40 to 25 wt% acrylonitrile (AN). A decrease toward the miscibility limit is recorded for several systems.

The theory correctly predicts an interphase thickness for immiscible blends of about 1-4 nm.

Roe [29] and Joanny and Leibler [52] developed a quasi-crystalline model valid under the conditions of high immiscibility (near the phase separation point) correcting Helfand's relationship. The basic concept is that the nature of the interphase region is substantially dependent on the degree of immiscibility.

The described classical thermodynamic theories treat the interphase as a 'black box' but do not describe how the material is distributed within this 'new phase'. In this sense information is limited even now.

4.2.3 Sanchez-Lacombe Theory

The theory developed by Sanchez and Lacombe [31-33] is based on the gradient theory and the model of compressible liquid lattice where ρ_i is replaced by the segment density of the t-th component. The simplified equations for interfacial tension and interphase thickness (width) are:

$$\gamma_{AB} = 1/2\left(2\tilde{m}_o \Delta\tilde{a}_{1/2} \gamma_{AB}^* \gamma_{AB}^*\right)^{1/2}$$

where $\Delta\tilde{a}_{1/2}$ is the dimensionless free energy which is determined by the expression:

$$\Delta\tilde{a}_{1/2} = \Delta a_{1/2} / (P_{AB}^* \, P_{AB}^*)^{1/2}$$

and the interfacial width is:

$$d_{inter} = (\tilde{m}_o / 2\Delta\tilde{a}_{1/2})(v_{AB}{}^* v_{AB}{}^*)^{1/6}$$

where $\gamma_1{}^* = \varepsilon_{ij}{}^*/(v_i{}^*)^{2/3} = P_i{}^* \delta_{ij}$ is characteristic surface tension of pure components. $\tilde{m}_o = \tau^{1/2} v^{-1/6} \tilde{\rho}_1^2 + 2\tilde{\rho}_1\tilde{\rho}_2 \tau^{-1/2} v^{1/6} \tilde{\rho}_2^{-2}$, $\iota = T^*{}_A/ T^*{}_B$, n= $P^*{}_{A1}/P^*{}_{B2}$, $\tilde{\rho}_A$ and $\tilde{\rho}_B$ are segment densities of the components, calculated from the equation of state at corresponding reduced temperatures $\tilde{T}_A$ and $\tilde{T}_B$.

Other treatments based on the equation-of-state have been proposed [53].

Some values of the interfacial tension and of the interfacial width are given in **Table 4.1**.

Table 4.1 Experimental and calculated values of interfacial tension and of the interfacial width [19-28, 31-34, 54-56]					
	γ (mN/m)			d_{inter} (nm)	
System	Experimental	Calculated Sanchez-Lacombe [31-34, 55, 56]	Calculated Helfand [19-28]	Calculated Sanchez-Lacombe [31-34, 55, 56]	Calculated Helfand [19-28]
PS/PMMA	1.7	1.0	0.3	10.0	16.0
PBMA/PMMA	1.9	2.0	1.5	4.8	2.0
PBMA/PVA	2.9	1.2	3.0	8.0	1.6
PS/PVA	3.7	4.0	1.9	2.3	3.1
PBMA/PDMS	3.8	4.6	3.2	1.6	1.3
PIB/PDMS	4.2	4.6	-	1.5	-
LDPE/PDMS	5.1	5.4	-	1.3	-
PS/PDMS	6.1	6.8	-	1.1	-
PVA/PDMS	7.4	3.8	7.2	2.0	0.8

PBMA: polybutyl methacrylate *PVA: polyvinyl acetate*
PDMS: polydimethylsiloxane *PIB: polyisobutylene*
LDPE: low density polyethylene

4.3 Interface in Systems Containing Copolymers

Copolymers are formed from at least two different monomers arranged in a specific manner. They can be random, alternating, di-block, tri-block, multi-block, star-shaped, graft, comb, etc. Their composition depends on the nature and the amount of comonomers, polymerisation mechanism and conditions (temperature, pressure, etc.). Each group exhibits a particular behaviour. Random graft copolymers [57] have the ability to form micelles, which can be changed by adjusting the copolymer architecture and hydrophilicity/hydrophibicity of backbone and grafts (polybutylene (PB)/PS). Critical micelle concentration (CMc) as appears from spinodal is optimum for a certain number of grafts. Selection of such systems, potentially applicable in drug delivery, is much aided by molecular-thermodynamic calculation, following evolution of a system close to spinodal.

4.3.1 The Interface in Block Copolymers

Theories of block copolymers are complex involving computation of the domain size, the interface thickness, the structure and the order-disorder transition. Helfand and Wasserman [58-60], using narrow interface approximation, found the following expression for interfacial thickness:

$$d_{inter} = 2[(\rho_{oi}b_i^2)/6 + (<R_i^2>/Z_i)/2\chi]^{1/2} \qquad (4.8)$$

where b_i is Kuhn's statistical segment length, Z_i is the degree of polymerisation, ρ_{oi} is the density and $<R_i^2>$ is the radius of gyration of the block i. For identical chains and lattice size $b^2 = b_i^2\rho_i$ and Equation 4.8 is converted to Equation 4.7 for two homopolymers.

4.3.2 Homopolymer/Copolymer Blends

To improve adhesion between phases in incompatible mixtures, compatibilisers are used. Adding polymer A to an A-B block copolymer results in a decrease of the conformational entropy. This loss can be compensated for by the curvature of the interface, which results in a change from the lamellar to discrete cylindrical morphology. The interface in an A/A-B mixture was not theoretically calculated, but most experiments indicate that the addition of a homopolymer to a block copolymer leaves d_{inter} unchanged. The amount of interfacial mixing that occurs in a blend influences the morphology and interfacial adhesion, which ultimately control the physical and mechanical properties of the blend. Interfacial mixing begins when two polymer surfaces are brought into contact and the extent and rate of interdiffusion depend on the interactions between the two polymers.

Several general theories regarding polymer interdiffusion and the techniques used to study polymer-polymer interfacial mixing and to measure diffusion coefficients are described in [61, 62].

Vilgis and Noolandi [9] derived a statistical thermodynamic theory based on the lattice model and using mean-field approximation and reduced equation-of-state variables for a binary blend with an arbitrary block copolymer or solvent (PS, PB and styrene-butadiene block copolymer) and computed the interfacial tension, the width of the interphase, and the concentration profiles of the components – see **Figure 4.1**. The diffusion equations for the density profiles are solved numerically. It has been established that the efficiency of the block copolymer increases with the molecular weight of the blocks and this efficiency is maximum near the CMc. Finite molecular weight and conformational entropy effects were considered, but the excluded volume was not.

The formation of the spherical micelles of a di-block copolymer in a homopolymer was investigated and CMc was expressed as:

$$\Phi_{CMC}^{spherical} = \exp\{1.72(\chi Z_c)^{1/3} f^{1/9}(1.74 f^{-1/3} - 1)^{1/3} - f\chi Z_c\}$$

where f is the fraction of A-polymer in the A-B copolymer and Z_c is the degree of polymerisation of the copolymer. The relationship was later extended to cylindrical and lamellar micelle morphologies.

For blends, compatibilisation by copolymer addition or *in situ* synthesis methods when mostly di-block and graft copolymers are used, it is necessary to know if their chains are auto-miscible or not, and how they modify the interface characteristics. In blends, location of a compatibiliser could be: dissolved in the matrix and dispersed phase, at the interphase, above the CMc it forms micelles in the matrix and dispersed phase and at high concentration it can form mesophases. Therefore the final result depends both on the copolymer structure and its concentration. The microdomains appearing in PS/PMMA blends compatibilised with a diblock copolymer may be useful in lithography because a special pattern can be obtained in very thin films. Presence of copolymer suppresses micellisation.

Interfacial tension of block copolymers formed *in situ* in reactive blending of dissimilar polymers was studies by Weber and co-workers [63].

As it was shown in the Chapter 3, the phase diagrams of a block copolymer exhibit two distinct differences compared to the phase diagram of two homopolymers. Because of the chemical links between blocks, microdomains are present instead of macrophases and the size of microdomain is controlled by the molecular mass of blocks and copolymer composition, concentration and transition temperatures of individual polymers, so the phase diagram will be very complex. Most phase diagrams are characterised by UCST, so the block copolymers form an isotropic melt, while below it they are phase separate, the rigid blocks forming glass or semicrystalline solids. Using various approximations, e.g., narrow interphase approximation, random phase approximation, etc., expressions for domain size and shape and interfacial thickness have been given; theoretical treatments predicting the dependence of these values on degree of polymerisation and magnitude of repulsive interactions between blocks. The same factors also affect the interface in copolymer/homopolymer blends. Blends containing graft copolymers exhibit unusual morphologies. The most common of these were spherical structures (named *onions*) consisting of alternating concentric layers of both components, such structures arising from the incompatibility of the copolymer with the homopolymer.

The ternary systems formed from a copolymer and two homopolymers are of particular interest. From a theoretical (also practical) point of view of compatibilisation, three aspects must be known: copolymer configuration and conformation, composition and concentration, all of these assuring the achievement of desired effect – the morphology corresponding to a certain complex of end-use properties. Addition of a block copolymer either in a single homopolymer or a two-polymer mixture will change its conformation.

The copolymer chain can be located at the interface, dissolved in one of both phases and it can form micelles in one or both polymeric phases. The morphological analyses showed that di-block copolymers have higher interfacial activity than tri-block or grafted copolymers, so it more readily interacts with the two homopolymer phases forming appropriate entanglements that result in a decrease of the interfacial tension coefficient and enhanced interfacial adhesion in the solid state. Analysing the concentration of a block copolymer at the interface, Leibler [64, 65] showed that due to equilibrium adsorption of copolymer at the interface, γ_{AB} should decrease, therefore it is possible that thermodynamically controlled stable droplets can exist, protected by an interfacial film of a copolymer. The size distribution of the droplets is expected to depend on the rigidity and spontaneous radius of curvature of the interfacial film that depends on the molecular structure of copolymer. The various relationships for CMc has been also established [66, 67]. Semi-empirical equations between the interfacial coefficient and compatibiliser concentration were derived that fit the experimental data well, such as:

$$y = y_{CMc} + (y_o - y_{CMc}) \exp\{-k\phi\} \text{ where } y \equiv \gamma_{AB} \text{ or } y \equiv d$$

where k is considered to be an adjustable parameter.

At the surface or interface occurs a selective adsorption of the components of the mixture, which leads to formation of adsorption layers with a different composition when compared with that of the initial mixture. It will appear as an enrichment of surface layer by one component or surface segregation. Surface segregation was shown for a polymer mixture in contact with a solid surface [68], in air, at the interface between phases, and of special importance is copolymer segregation at the interface between homopolymers.

Segregation increases with increasing incompatibility, the higher the interaction parameter, the higher is the surface segregation. Surface segregation in interpenetrating networks (IPN) arises during crosslinking and depend on reaction rates of components.

A curious consequence of the theoretical consideration is the fact that the interfacial tension vanishes to some concentration of block copolymer. To explain this effect it has been suggested that only some of block copolymer macromolecules find it energetically favourable to settle at the interface, while the remainder are randomly dispersed in the bulk phases.

The theoretical treatment was developed both for binary homopolymer/copolymer C systems and ternary homopolymer A/homopolymer B/copolymer C. Upon addition of a homopolymer to block copolymer the interphase thickness either remains constant or it slightly increases.

The interfacial tension coefficient is given by equation:

$$\gamma_{AB/C} = \gamma_{AB} - (\rho\Phi_C RT/Z_C)\, \exp[\chi_{AB}Z_C]$$

where ρ is density and Φ_C is copolymer concentration. The CMc [62, 63] is:

$$\Phi_{+CMc}^{C} = \exp[1.72(\chi Z_c)^{1/3}f^{4/9}(1.74\, f^{-1/3} - 1)^{1/3} - \chi\, Z_c f]$$

where $f = Z_A/Z_B$ and $Z_C = Z_A + Z_B$; Z is degree of polymerisation.

4.3.3 Binary Polymer Blends with a Copolymer as Compatibiliser

Leibler [64, 66] examined the emulsifying effect of the A-B copolymer in an immiscible polymer A with polymer B blend. The block and graft copolymers which have chain fragments of the same chemical nature as both homopolymers, are typically compatibilisers. These fragments (block or grafts) are miscible with the corresponding homopolymers. The factor which determines the choice of a compatibiliser: is its ability to diminish the interfacial tension, because it acts as a surface active substance, and to prevent the phase separation in the solid state, in other words it stabilises the morphology against stress-destruction during formation. The compatibilising effect is obtained if the copolymer is localised at the interface between two immiscible polymers, in other words if it preferentially segregates at the interface. The penetration of blocks into phases makes the interphase region much broader when compared with that of the binary mixture.

Compatibilisation may be accomplished by addition of a small amount of tailored copolymer, e.g., tapered block copolymer A-B or X-Y type with entropic or/and enthalpic contributions, addition of a large amount of a multi-polymeric core-shell or multi-layered type compatibiliser–cum-impact modifier, addition of a co-solvent or low molecular weight compound. In these ways a thermodynamic equilibrium is assumed to occur.

The thermodynamic treatment of this problem was developed by Noolandi [69-77], Shull and Kramer [78, 79], Leibler [64, 65, 80] and Balasz [81] and it is supported by much experimental data. A theory was developed for two immiscible polymers, A and B, dissolved in a solvent in presence of a block copolymer A-B or XY (all three polymers have infinite molecular mass) forming a diluted solution whose concentration is below that corresponding to micelle formation or in absence of solvent [78, 79]. It was supposed that the interaction between solvent and blocks of copolymers are equal to zero. It was found that the decrease of interfacial tension is mainly determined by energetically favourable orientation of blocks at the interface corresponding to the penetration of blocks into the phases of the same chemical nature. The expressions for the decrease in interfacial tension ($\Delta\gamma$) and the amount of copolymer in the interface are different for each situation.

Leibler and Balasz's relations for a di-block copolymer are:

Reduction in interfacial tension:

$$\Delta\gamma = -(kT)(2/\pi)\,(2/3)^{3/2}\,Z_c^{-1/2}\,[\ln(\Phi_+) + \chi Z_c\,f]^{1/2}$$

Where $\Phi_+ = \Phi_c\,/\,[\Phi_B + \Phi_A\,\exp(\chi\,Z_{CA} - Z_{CB})]$

Concentration for saturation of a flat surface:

$$\Phi_{sat} = \exp[(3\pi/8)^{2/3}\,(\chi\,Z_c)^{1/3} - \chi\,Z_c\,f]$$

CMc:

$$\Phi_{+CMc}^{C} = \exp[(3/2)^{4/3}f^{4/9}(\alpha\,f^{-1/3} - 1)^{1/3}\,(\chi\,Z_c)^{1/3} - \chi Z_c f] \text{ with } \alpha = 1.74$$

Tang and Huang [67] and Utracki and Shi [82] gave other semi-empirical expressions for the relation between copolymer concentration and interfacial tension and showed that the radius of the dispersed drop follows the same curve as the interfacial tension coefficient.

With *an arbitrary X-Y block copolymer*, Noolandi and Vilgis obtained [9, 77]:

$$\Delta\gamma = -Z_c^{-1}\,\exp[Z_c\,(\chi_1/2) + \chi_2\Phi]$$

where : $\chi_1 = \chi_{BY} = \chi_{AX} = \chi_{XY} = \chi_{BA}$ and $\chi_2 = \chi_{AY} = \chi_{BX} > \chi_1$, Φ volume fraction in bulk.

For *a comb or graft copolymer* Balasz obtained:

$$\Delta\gamma = -(kT/a^2)(2/\pi)\,(2/3)^{3/2}\,n^{-1}Zc^{-1/2}\,[\ln(\Phi_+) + \chi Z_c\,f]^{3/2}\,(4-3f)^{-1/2}$$

$$\Phi_{sat} = \exp[(3\pi/8)^{2/3}\,(\chi Z_c)^{1/3}\,(4-3f)^{1/3}\,n^{2/3} - \chi\,Z_c\,f]$$

$$\Phi_{+CMc}^{C} = \exp[n^{2/3}(3/2)^{4/3}f^{4/9}\,(\alpha_1\,f^{-1/3} - 4)^{1/3}\,(\chi\,Z_c)^{1/3} - \chi\,Z_cf]$$

or

$$\Phi_{+CMc}^{C} = \exp[n^{2/3}(3/2)^{4/3}(1-f)^{4/9}\,(\alpha_2\,(1-f)^{-1/3} - 1)^{1/3}\,(\chi\,Z_c)^{1/3} - \chi\,Z_cf]\ \text{with}\ \alpha_1 = \alpha+3\ \text{and}\ \alpha_2 = 4\alpha-3.$$

For star copolymers with 4 arms, 2 arms of A and 2 arms of B:

$$\Delta\gamma = -(kT/a^2)(1/n)\,(2/3)^{3/2}\,n^{-1}Z_c^{1/2}\,[\ln(\Phi_+) + \chi Z_c\,f]^{3/2}$$

where n is the number of arms.

These relationships are useful in selecting compatibilisers and evaluation of necessary concentration to achieve the desired degree of dispersion.

The amount of the interfacial agent required to saturate the interface w_{cr}, is related to the molecular weight, the interface area, and the specific cross-sectional area of the copolymer molecule (A) by the relationship proposed by Paul [83]:

$$w_{cr} = 3\ \Phi M/(RAN)$$

where Φ is the volume fraction of the dispersed phase, M is the molecular weight, R is the radius of the dispersed drop, N is the Avogadro number and A is the area occupied by a copolymer molecule.

Vilgis and Noolandi [9] concluded on the basis of this theory that it is possible to design a universal block compatibiliser of XY type operating on the principle of competitive repulsion interactions between homopolymers and copolymer blocks. Several 'universal' compatibilisers are presently on the market. Experimentally it was shown that a di-block copolymer has higher interfacial activity than a tri-block- or a graft- copolymers.

For the system PS/PB it was found that interfacial tension decreases with increasing molecular weight of PS and decreases with increase in polydispersity. Block lengths just above the molecular weight of entanglements are the most practical for the compatibilisation of polymer blends. The efficiency of a block copolymer is limited by the formation of micelles in bulk phase and by the kinetic factors.

Noolandi found that for concentrations below the CMc, the interfacial tension linearly decreases with copolymer concentration, and that the interfacial tension diminishes with increasing molecular mass of the blocks. On the basis of this theory, Thomas and Prud'homme [84] found a relationship between particle size and copolymer concentration.

Correlation of the theoretical results with the experimental data led to the conclusion that the effect of incorporation of a compatibiliser in a mixture of two immiscible polymer mixture causes the decrease both of interfacial tension and the dimensions of the phase-separated domain. The domain dimensions depend not only on the interfacial tension but on the time necessary for phase separation, on the mechanism of coarsening (coalescence) domains, beginning of phase separation and other reasons.

4.4 Experimental Methods for Interfacial Tension Determination

Because of experimental difficulties due to the high viscosity of polymer melts, the first reliable measurements of surface tension were reported in 1965 and in 1969 for interfacial tension of polymer melts by Wu and Roe [85, 86].

Surface (interfacial) tension (γ_i) can be *calculated from the group theories by* additivity, *the parachor method* P_{ij}, *or using solubility parameters* for dispersive, polar and hydrogen interactions these values being tabulated [87, 88]:

$$\gamma_i = \sum_{i=1}^{n} P_{ij} / \sum_{i=1}^{n} v_{ij}^4$$

$$\gamma_{AB} = \gamma_A + \gamma_B - 2 \, \phi \, (\gamma_A \, \gamma_B)^{1/2}$$

Determination of interfacial tension from surface properties [10, 11, 89] is based on Equations 4.1 and 4.2.

The methods for direct determination of interfacial tension coefficient can be classified in two groups [90]:

- Equilibrium methods such as pendant, sessile drop and spinning drop. These methods are time consuming and are limited to low molecular mass resins.

- Dynamic methods such as capillary breakup and deformed drop. The methods from this group are especially recommended for polymer blends, the last one seems to have a great potential.

In the *sessile drop method* the equilibrium shape of the stationary droplet of polymer with higher density immersed in the melt of the second polymer is observed. Usually a small cube of the denser material is first placed on the bottom of quartz container than filled with grains of the less dense one. After melting (usually under a blanket of inert gas) formation of an equilibrium droplet is recorded. The equilibration involves flow of polymer melt under the influence of the interfacial forces. Since the viscosity of most of the commercial resins at processing temperature are much higher than 1 kPa-s and the interfacial tension coefficient is usually much lower than 0.4 Pa-cm, the equilibration time can be of the order of hours or days. From the equilibrium droplet shape, the interfacial tension coefficient can be calculated.

4.4.1 Pendent Drop Method

Pendent drop method is a relatively simple and reliable method [85, 86]. The profile of a pendent drop at equilibrium is governed by the gravitation and the surface tension. The surface (or interfacial) tension is given by:

$$\gamma = g \, \Delta\rho \, d_c^2 / H$$

where γ is the surface (or interfacial) tension, g is the gravitational constant, $\Delta\rho$ the difference in the densities of the two phases (drop and surrounding), d_c is the equatorial diameter of the drop, and H is a correction factor calculated by Adamson [91]. Due to the bigger role of the gravitational forces than in sessile drop method, the equilibrium is sooner reached. The method was computerised and the acquisition of the data allows continuous analysis of the droplet shape.

4.4.2 Capillary Breakup Method

Schematic representation of the interfacial tension measurement is given in **Figure 4.2a** and relationships used are:

$$\alpha \equiv (b-a)/2d_o$$

$$\alpha = \alpha_o \exp(qt) \text{ with } q = \gamma_{12}\Omega \, (\Lambda,\lambda)/\eta_o \, d_o$$

where d_o is initial diameter of the fibre and b and a are maximum and minimum diameter observed during the breakup process, respectively; q depends on interfacial tension, viscosity and Λ, t – time.

- *Capillarity instability method* – the slope of the straight line is a measure of γ_{AB} see **Figure 4.2b**,

- *Deformed drop retraction method (DDRM)*. **Figure 4.2c**. A piece of polymer is inserted between two 1 mm thick films of polymer 2. The sandwich is placed between microscopic glass slides on a hot plate under an optical microscope. After the desired temperature is reached, the central drop (d = 0.2-0.5 mm) is deformed by about 15% by shearing or by elongation. Retraction of the deformed drop is recorded. After retraction is complete, the drop can be deformed again as many times as desired. The method makes possible to verify the relationship $\gamma_{AB} = \gamma_{BA}$. The time dependence (t) of the deformability parameter (D) leads to determination of γ_{AB} according to the dependence established by microrheology or Taylor's theory:

$$D = \frac{L-B}{L+B} = D_o \exp\left(\frac{-80t\gamma_{AB}(\lambda+1)}{(2\lambda+3)(19\lambda+16)\eta_m R_o}\right) = D_o \exp(t/\tau)$$

The drop follows an exponential decay. After a long time the diffusion stops and the CMc plateau is reached.

4.5 Experimental Methods for Determination of Concentration Profile

Advanced techniques of modern spectroscopy have been used to elucidate the interface role in multicomponent systems. An extensive effort has been made in developing methods to verify the interfacial structures of multicomponent materials and improving the analytical techniques for surface characterisation in order to produce better materials.

Currently used instrumental techniques for the analysis of interfacial structures are Fourier transform-infrared spectroscopy (FT-IR), differential scanning calorimetry (DSC), dynamic mechanical spectroscopy (DMS), nuclear magnetic resonance (NMR) [92], transmission and scanning electron microscopy (TEM and SEM), X-ray photoelectron spectroscopy (XPS) and gel permeation chromatography (GPC) [93]. Together with T_g, the T_1 can be considered as a measure of the degree of phase separation. Whereas a large difference in proton T_1 values shows phase separation to a large extent, a small difference indicates less phase separation, because of spin diffusion in a more homogeneous mixture, in addition, the possible formation of entanglements can also decrease the mobility of the mixtures and prevent phase separation from occurring.

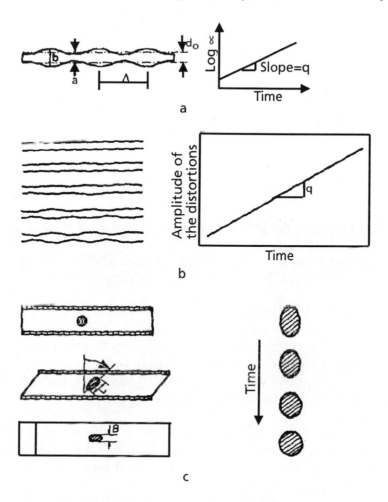

Figure 4.2 a) Schematic representation of the interfacial tension measurement by a) capillary breakup method; b) capillary instability method as amplitude of distortion versus time; c) schematic representation of DDRM method

Diffuse reflectance infrared spectroscopy (DRIFT) was used to characterise directly the structure of blends on molecular level. Agrawal and others [40] obtained results which agreed well with theoretical results. The recent techniques for measuring depth profiles include forward recoil spectroscopy (FRES), FT-IR and attenuated total reflectance (ATR), small angle neutron scattering (SANS), etc. FT-IR and ATR are limited by the depth resolution of about 1000-10 000 Å but provide useful information at large interdiffusion depths. FRES and SANS and specular neutron reflection (SNR) do not measure depth profiles directly and assumptions must be made about the nature of the profile before it can be constructed. SNR has been developed to examine short range interdiffusion in polymers with a resolution of about 5-10 Å. The concentration profile may be also determined by secondary ion mass spectroscopy (SIMS), neutron reflection (determines motion of polymer chains at a penetration depth comparable with radius of gyration). SIMS has ability to measure depth profiles directly (it is useful for tracer studies), its depth resolution is about 100 Å for polymers, but not as good as the neutron reflectivity (SNR) technique with a resolution of about 5 Å [94]. Raman microspectroscopy was also used to determine the depth profile [95].

4.6 Experimental Data on Interfacial Properties of Polymer Blends

Only few measurements of the interphase thickness are available. Generally, values in the range of 2-5 nm without copolymer have been reported – **Tables 4.2** and **4.3**.

Generally the values of surface tension lie between 0.02-0.04 N/m at 140 °C and the values of interfacial tension vary from 0.01-0.015 N/m and both exhibit almost linear variation with temperature. The observed temperature coefficients for polymer melts lie between 5-10 x 10^{-4} N/m/degree for surface tension and are around 20 x 10^{-4} N/m/degree for interfacial tension. Wu [10, 11, 96, 97] reported that molecular weight dependence is significant only in the low molecular range.

The molecules interpenetrate only in a very narrow region (order of magnitude 50 Å). Furthermore, the interfacial region is not formed by whole molecules, but by parts of them. With the increasing temperature, the thickness increases slightly.

An example is commercial blend Xenoy composed of polybutylene terephthalate (PBT), PC and polyphenylene ether/styrene-ethylene-butene-styrene copolymer (PPE/SEBS), a blend with excellent processability, high impact strength and heat deflection temperature and PS/PC/PBT 1/3/7 or 7/3/1. The last blend changes its morphology in function of mixing time: 5 minutes mixing time, PC forms a protective layer around the PS and PBT matrix; after 10 minutes mixing, PC becomes dispersed in the PBT matrix and interlayer thickness decreases and for 15 minutes mixing PBT is dispersed within the PC domains that in turn are dispersed in the PS matrix.

Helfand and Tagami theory predicts the interface to be approximately 35 to 45 Å thick for the system PS/PMMA that was confirmed by neutron reflectivity when a interfacial thickness of 50 ± 10 Å [98] or 42-66 Å [99] and an interfacial tension of 1-1.8 x 10^{-3} N/m.

Mansfield [100] reported an interfacial thickness of 45 Å from neutron reflectivity corresponding to the PC/SAN (25% AN) system and a value of 46 Å was obtained.

Table 4.2 Surface and interface tension of polymer melts [43-46]			
Polymer or blend	T, °C	γ (x 10^3 N/m)	-dγ/dT (x 10^3 N/m, °C)
HDPE	140	28.8	0.057
	250	21.7	0.045
LDPE	140	27.3	0.067
	250	21.1	0.051
PS	140	32.1	0.072
PMMA	140	32.0	0.076
PVAc	140	28.6	0.066
PA	280	37.8	0.101
PET	280	34.5	0.109
HDPE/PS	140	5.9	0.020
HDPE/PMMA	140	9.7	0.018
LDPE/PA	280	17.0	0.041
LDPE/PET	280	9.8	0.015
PS/PMMA	140	1.6	0.013
PS/PVAc	140	3.7	0.0044

HDPE: high density PE
PVAc: polyvinyl acetate
PET: polyethylene terephthalate

These results showed a good agreement between experiment and theory. However, the ellipsometry measurements give a larger thickness suggesting that the Helfand-Tagami theory excludes highly diffuse interfaces.

On the basis of experimental data, some correlations were attempted between the interfacial properties of multiphase polymer systems and sample characteristics, miscibility of components and conditions of interphase formation and agreement with the predictions of various theories. The interphase width increased with annealing time for PMMA/ SAN system from 20 nm to 60 nm after 12 hours at 303 K and remained almost constant for the PMMA/PS system. Therefore this property depends on the compatibility of components, time, temperature, interdiffusion coefficient and is proportional with the

	Table 4.3 Parameters V, b, χ, temperature dependence of interfacial tension and interphase thickness at 140 °C [43-46]				
Polymer pair	V (cm³/mol) (geometric mean)	b (Å)	χ	-dγ/dT (N/m x 10⁻⁵/K)	d_{inter} (Å)
HDPE/PS	60.8	5.9	0.24	0.013	41
HDPE/PMMA	56.0	5.7	0.44	0.017	28
LDPE/PA (280 °C)	67.3	6.9	0.67	0.019	27
LDPE/PET (280 °C)	60.5	7.7	0.66	0.024	37
PS/PMMA	95.6	6.6	0.065	0.006	108
PS/PVAc	89.5	6.7	0.039	0.0025	115

thermodynamic driving force χ - χ_{crit} (difference of interaction parameter at the annealing temperature and at the critical point). The concentration profile determined for PMMA/SAN system can be described by Equations 4.3-4.5 [13].

According to Helfand-Tagami theory the interfacial thickness is very thin for homopolymer pairs polyacrylonitrile (PAN)/polymethylmethacrylate (PMMA), PS/PAN, PMMA/PS. The incorporation of a di-block copolymer (S-co-MMA) in PS/PMMA system broadens the interphase region up to certain limit. At high amounts the block copolymer forms its own structure at the interface. Addition of copolymer slightly increased the interphase thickness from 5 to 8.4 nm in the PS/PMMA system and from 10 to 12 nm for PS/PE with block copolymer of hydrogenated polybutadiene with styrene. To increase the $d_{interphase}$ either tapered di-block copolymers or a mixture of block and random copolymers are used. Tri-block copolymers are less efficient in lowering interfacial tension but they may provide thicker interphase thickness and better adhesion between phases in the solid state. Stabilisation of the morphology by reduction of coalescence can be achieved by selective wetting by the third polymer. Addition of the third immiscible polymer C may stabilise dispersion of an immiscible pair of polymer B in matrix A, if the condition of complete wetting is achieved:

$$\gamma_{AB} > \gamma_{AC} + \gamma_{BC}$$

A partial wetting is expected when:

$$\gamma_{AB} < \gamma_{AC} + \gamma_{BC}$$

126

4.7 Concluding Remarks

Thermodynamic theories have contributed decisively to the understanding of polymer mixtures. A predictive calculation of novel polymer mixtures is presently not yet possible and so experimental studies are of great importance for further theoretical development. Determination of the interaction parameters will result in new theoretical knowledge. Multiphase polymer alloys and compounds are of greatest practical importance. The distribution of the polymers in one another, the stability of the distribution during processing and the interaction energy of the interface are of particular interest. The distribution of the polymers in one another is predictable via thermodynamic parameters, but chemical reactions on the interface, taking place during the mixing process in the melt, are not taken into account. Computer evaluation and simulations led to a faster progress of the thermodynamics of polymer blends helping both in prediction and obtaining tailored properties of the blends.

Though classic, the phase equilibrium of polymer solutions and blends is still an actual subject of research. Academically, the most basic problem in its study is to find analytical expressions for ΔG which allows quantitative prediction of phase relationships of various polymer solutions. Despite many efforts, the expressions derived so far for ΔG of quasi binary solutions remain semi-quantitative.

Non-equilibrium processes of surface segregation have not been yet studied. It is necessary much effort to study the kinetically frozen state and to investigate the kinetics of phase separation. As about 10% of commercial blends are compatibilised by addition rest are reactively compatibilised, the effect of various kinds of compatibilisation methods should theoretically be described in a quantitative manner.

The elaboration of equations describing polymer distribution and interface as functions of determinable parameters of the components will be of decisive importance in the next few years.

It is hoped that, in the not too distant future, a sufficiently broad data base of thermodynamic quantities covering the whole spectrum of mixtures from small molecules to polymer blends will be available to verify theoretical statements.

There is no description of the effect of tri-block and linear multi-block copolymers on the interfacial tension, CMc, interfacial thickness, etc. Limited experimental information is available on their morphology and mechanical performance of blends. The core-shell copolymers are mainly used as impact modifiers. No theoretical information is available on their effect on the interphase properties. Many characteristic parameters of these copolymers are important for the performance of the blends, such as core size, concentration of ingredients, composition of the blends, compounding, etc. The understanding of the surface (or interfaces) of polymers on a molecular level basis is still limited.

The adhesion between polymeric domains in solid state can be demonstrated by the methods developed for composites. The direct measurements of interfacial tension are rare, and mostly were not done in the light of the theoretical development. At this very moment it appears that the theoretical work is ahead of the experimental work. While advances in the theories of interfacial phenomena, which attempt to identify important factors involved and point the way for practical development have been made, there are few quantitative data available which are suitable for drawing comparison between theory and experiment. Theoreticians often look for additional reliable experimental data.

Much less attention has been given to the thermodynamic properties of complex materials like gels, biomacromolecules, micelles, colloids, block copolymers and similar substances that are often called 'soft materials' Such materials abound in nature and technology. Application of chemical thermodynamics to soft materials has been delayed because of experimental difficulties and because, until recently, there were few theoretical models available for describing assemblies of complicated molecules.

Thanks to recent advances in statistical mechanics and molecular physics, and thanks to increasingly fast computers, it is now possible to develop a hard science for 'soft materials'. The creation and understanding of soft materials depends primarily on experimental science. Now statistical mechanics is able to provide guidance toward interpreting experimental results and toward reducing experimental effort. While thermodynamic models are useful for suggesting what experimental work is most likely to lead to a successful result, the hard science for soft materials is still in an early stage but there is a good reason to expect it to grow dramatically in the near future thanks to the creative efforts of dedicated researchers.

References

1. *Microphenomena in Advanced Composites*, Eds., H.D. Wager, G. Maron and B. Harris, Elsevier Applied Science, London, UK, 1993.

2. L. Sharpe in *The Interfacial Interactions in Polymeric Composites*, Ed., G. Akovali, Kluwer Academic Publishers, Dordrecht, The Netherlands, 1993, 1.

3. Y.S. Lipatov, *Polymer Science USSR*, 1978, **20**, 1.

4. Y.S. Lipatov, *Interfacial Phenomena in Polymers*, Naukova Dumka, Kiev, The Ukraine, 1980.

5. Y.S. Lipatov, *Polymer Reinforcement*, ChemTec Publishers, Toronto, Canada, 1995.

6. Y.S. Lipatov, *Colloid Chemistry of Polymers*, Elsevier, Oxford, UK, 1988.

7. L.A. Utracki, *Polymer Alloys and Blends: Thermodynamics and Blends*, Hanser Publishers, Munich, Germany, 1989.

8. Y.S. Lipatov and A.E. Nesterov, *Thermodynamics of Polymer Blends*, Technomic Publishers Co. Inc., Lancaster, PA, USA, 1997, Chapter 6, 363.

9. T.A. Vilgis and J. Noolandi, *Die Makromolekulare Chemie, Macromolecular Symposia*, 1988, **16**, 225.

10. S. Wu in *Polymer Handbook*, Fourth Edition, Eds., J. Brandrup, E.H. Immergut and E.A. Grulke, John Wiley, 1999, VI/521.

11. S. Wu, *Polymer Interfaces and Adhesion*, Marcel Dekker Inc., New York, NY, USA, 1979.

12. D.W. van Krevelen, *Properties of Polymers: Their Estimation and Correlation with Chemical Structure*, Elsevier, Amsterdam, The Netherlands, 1976.

13. D.R. Quayle, *Chemical Reviews*, 1953, **53**, 439.

14. Y.S. Lipatov, *Journal of Applied Polymer Science*, 1978, **22**, 1895.

15. J. Kessler, N. Higashida, T. Inoue, W. Heckmann and F. Seitz, *Macromolecules*, 1993, **26**, 2090.

16. P.G. de Gennes, *Comptes Rendus de l'Academie des Sciences*, 1989, **2**, 300, 13.

17. J.L. Harden, *Journal of Physics (France)*, 1990, **51**, 1777.

18. D.R. Paul, *Macromolecular Symposia*, 1994, 78, 83.

19. E. Helfand and Y. Tagami, *Polymer Letters*, 1971, 9, 741.

20. E. Helfand and Y. Tagami, *Journal of Chemical Physics,* 1971, 57, 1812.

21. E. Helfand and Y. Tagami, *Journal of Chemical Physics*, 1971, **56**, 3592.

22. E. Helfand, *Macromolecules*, 1975, 8, 552.

23. E. Helfand, *Macromolecules*, 1976, 9, 307.

24. E. Helfand, *Journal of Chemical Physics,* 1975, **62**, 999.

25. E. Helfand, *Journal of Chemical Physics,* 1975, **63**, 2192.

26. E. Helfand and A.M. Sapse, *Journal of Chemical Physics,* 1975, **62**, 490.

27. E. Helfand and A.M. Sapse, *Journal of Chemical Physics,* 1975, **62**, 1327.

28. E. Helfand and A.M. Sapse, *Journal of Polymer Science,* 1976, **C-54**, 289.

29. R.J. Roe, *Journal of Chemical Physics,* 1975, **62**, 490.

30. T. Nose, *Polymer Journal,* 1976, **8**, 96.

31. I.C. Sanchez and R.H. Lacombe, *Journal of Macromolecular Science,* 1980, **B17**, 565.

32. I.C. Sanchez and R.H. Lacombe, *Journal of Chemical Physics,* 1976, **80**, 2352.

33. I.C. Sanchez and R.H. Lacombe, *Macromolecules,* 1978, **11**, 1145.

34. C.I. Poser and I.C. Sanchez, *Macromolecules,* 1981, **14**, 361.

35. O.K. Rice, *Journal of Physical Chemistry,* 1979, **83**, 1859.

36. H. Zhang and R.P. Wool, *Macromolecules,* 1989, **22**, 3018.

37. S. J. Whitlow and R.P. Wool, *Macromolecules,* 1991, **24**, 5926.

38. R.P. Wool and J. Long, *Macromolecules,* 1993, **26**, 5227.

39. J.L. Willet and R.P. Wool, *Macromolecules,* 1993, **26**, 5336.

40. G. Agrawal, R.P. Wool, W.D. Dozier, G.P. Felcher, T.P. Russell and J.W. Mays, *Macromolecules,* 1994, **27**, 4407.

41. P.A. Mirau, S.A. Heffner and M. Schilling, *Chemical Physics Letters,* 1999, **313**, 1-2, 139.

42. G.S. Grest, M.D. Lacasse and M. Murat, *Springer Proceedings of Physics,* 1998, **83**, 23.

43. H.W. Kammer, *Wissenschaftliche Zeitschrift der Technischen Universitat, Dresden,* 1975, **24**, 35.

44. H.W. Kammer, *Faserforschung und Textil-technik*, 1977, **28**, 27.

45. H.W. Kammer, *Faserforschung und Textil-technik*, 1978, **29**, 459.

46. H.W. Kammer, *Zeitschrift fur Physikalische Chemie - Leipzig*, 1977, **258**, 1149.

47. A. Ajji and L.A. Utracki, *Progress in Rubber and Plastics Technology*, 1997, **13**, 3, 153.

48. A. Ajji and L.A. Utracki, *Polymer Engineering and Science*, 1996, **36**, 1574.

49. H. Yamakawa, *Modern Theory of Polymer Solutions*, Harper & Row Publishers, New York, NY, USA, 1971, 46.

50. D. Broseta, G.H. Fredrickson, E. Helfand and L. Leibler, *Macromolecules*, 1990, **23**, 132.

51. G.D. Merfeld, A. Karim, B. Majumdar, S.K. Satija and D.R. Paul, *Journal of Polymer Science, Part B, Polymer Physics*, 1998, **36**, 3115.

52. J.F. Joanny and L. Leibler, *Journal of Physics (Paris)*, 1978, **39**, 951.

53. H.S. Lee and J.W. Ho, *Polymer*, 1998, **39**, 2489.

54. S. Wu, *Journal of Macromolecular Science – Reviews in Macromolecular Chemistry*, 1974, **C-10**, 1, 1.

55. I.C. Sanchez, *Polymer Engineering and Science*, 1984, **24**, 79.

56. I.C. Sanchez, *Polymer Engineering and Science*, 1984, **24**, 598.

57. J.M. Prausnitz, *Pure and Applied Chemistry*, 2000, **72**, 1819.

58. E. Helfand and Z.R. Wasserman, *Macromolecules*, 1976, **9**, 879.

59. E. Helfand and Z.R. Wasserman, *Macromolecules*, 1978, **11**, 960.

60. E. Helfand and Z.R. Wasserman, *Macromolecules*, 1980, **13**, 994.

61. E. Jabbari and N.A. Peppas, *Journal of Macromolecular Science C*, 1994, **34**, 205.

62. H.H. Kausch and M. Tirrell, *Annual Review of Materials Science*, 1989, **19**, 341.

63. P. Charoensirisomboon, T. Inoue and M. Weber, *Polymer*, 2000, **41**, 4483.

64. L. Leibler, *Macromolecules*, 1980, **13**, 1602.

65. L. Leibler, *Die Makromolekulare Chemie, Macromolecular Symposia*, 1988, **16**, 1.

66. L. Leibler and L.A. Utracki and Z.H. Shi, *Polymer Engineering and Science*, 1992, **32**, 1824.

67. T. Tang and B. Huang, *Polymer*, 1994, **35**, 281.

68. Y.S. Lipatov, T.S. Khramova, T.T. Todosijchuk and E.G. Gudova, *Journal of Colloid and Interface Science*, 1988, **123**, 143.

69. M.D. Whitmore and J. Noolandi, *Macromolecules*, 1985, **18**, 657.

70. M.D. Whitmore and J. Noolandi, *Macromolecules*, 1985, **18**, 2486.

71. J. Noolandi and K.M. Hong, *Macromolecules* 1980, **13**, 117.

72. J. Noolandi and K.M. Hong, *Macromolecules* 1980, **13**, 964.

73. J. Noolandi and K.M. Hong, *Macromolecules*, 1981, **14**, 727.

74. J. Noolandi and K.M. Hong, *Macromolecules*, 1981, **14**, 736.

75. J. Noolandi and K.M. Hong, *Macromolecules*, 1982, **15**, 482.

76. J. Noolandi and K.M. Hong, *Macromolecules*, 1984, **17**, 1531.

77. T.A. Vilgis and J. Noolandi, *Macromolecules*, 1990, **23**, 2941.

78. K.R. Shull and E.J. Kramer, *Macromolecules*, 1990, 23, 4769.

79. K.R. Shull, E.J. Kramer, G. Hadziioannou and W. Tang, *Macromolecules*, 1990, 23, 4780.

80. L. Leibler, *Macromolecules*, 1982, **15**, 1283.

81. A.C. Balasz, C. Singh, E. Zhulino, S.S. Chern, I. Lystskaya and G. Pickett, *Progress in Surfactant Science*, 1997, **55**, 181.

82. L.A. Utracki and Z.H. Shi, *Polymer Engineering and Science*, 1992, **32**, 1824.

83. D.R. Paul in *Polymer Blends*, Volume 2, Eds., D.R. Paul and S. Newman, Academic Press, New York, NY, USA, 1978, 35.

84. S. Thomas and R.E. Prud'homme, *Polymer*, 1992, **33**, 4260.

85. S. Wu, *Journal of Colloid and Interface Science*, 1969, **31**, 153.

86. R.J. Roe, *Journal of Colloid and Interface Science*, 1969, **31**, 228.

87. J.H. Hildebrand and R.L. Scott, *The Solubility of Non-Electrolytes*, Dover Publishers, New York, NY, USA, 1964, Chapter 3.

88. M.M. Coleman, J.E. Graf and P.C. Painter, *Interactions and the Miscibility of Polymer Blends. Practical Guides for Predicting and Designing Miscible Polymer Mixtures*, Technomic, Lancaster, PA, USA, 1991, Chapter 3.

89. M. Rätzsch and G. Haudel, *Die Makromolekulare Chemie, Macromolecular Symposia,* 1990, **38**, 81.

90. L.A. Utracki, Proceedings of the First International Conference of Polymer Modification, Degradation and Stabilisation, MoDeSt 2000, Palermo, Italy, 2000, Paper No.7/M/1715.

91. S. Adamson, *Physical Chemistry of Surfaces*, Wiley, New York, NY, USA, 1967.

92. K. Boshah and L.K. Molnar, *Macromolecules*, 2000, **33**, 1036.

93. H. Ishida and J. Jang, *Die Makromolekulare Chemie, Macromolecular Symposia,* 1988, **22**, 191.

94. S.J. Whitlow and R.P. Wool *Macromolecules*, 1991, **24**, 5926.

95. S. Hajatdoost, M. Olsthoorm and J. Yarwood, *Applied Spectroscopy,* 1997, **51**, 1784.

96. S. Wu, *Journal of Polymer Science*, 1971, **C-34**, 19.

97. S. Wu, *Journal of Chemical Physics*, 1970, **74**, 632.

98. S.H. Anastasiadis, T.P. Russell, S.K. Satija and C.F. Majkrzak, *Journal of Chemical Physics*, 1990, **92**, 5677.

99. D.W. Schubert and M. Stamm, *Europhysics Letters*, 1996, **35**, 419.

100. T.L. Mansfield, *Polymer Concentration Profiles and Methods of Surface Modification*, University of Massachusetts, 1993. [PhD Dissertation]

5 Water Soluble Polymer Systems - Phase Behaviour and Complex Formation

Georgios Staikos, Georgios Bokias and Gina G. Bumbu

5.1 Introduction

Water soluble polymers have attracted an increasing interest during the last decades, due to their numerous applications in products of everyday life (foodstuffs, cosmetics, drugs, paints) and in various industrial processes (paper making, ceramic processing, textile sizing, water treatment, drilling fluids, enhanced oil recovery). They respond to the desire for replacing organic solvents by water for environmental reasons and are related to biological systems.

They were recognised as a group in its own right in the 1960s after the publication in 1962 of a text edited by Davidson and Sittig concerned with the sources, properties and applications of water soluble polymers, which was revised 6 years later [1]. A more extensive handbook was edited in early 1980s by Davidson [2]. In the same period two monographs have appeared on synthetic water soluble polymers, one by Molyneux, on their properties and behaviour [3, 4] and another by Bekturov and Bakauova on their solution properties [5]. A brief but concise review by McCormick, Bock and Schulz is included in the *Encyclopaedia of Polymer Science and Engineering* [6].

What makes the study of water soluble polymers more exciting is the unique character of water as a solvent [7, 8]. It is a hydrogen-bonded liquid and only polymers with polar groups can interrupt its structure and be dissolved, while the presence of non-polar groups results in the appearance of hydrophobic interactions between the solute molecules [9]. Such interactions, in co-operation with hydrogen bonding, dipole-dipole and Coulombic interactions lead to the spontaneous self-association of certain amphiphilic molecules and play a crucial role in protein structure [10].

5.2 Classification

Water soluble polymers are classified as neutral or charged polymers (polyelectrolytes) depending on their chemical structure and to natural, semisynthetic (chemically modified natural) and synthetic polymers according to their origin (see **Table 5.1**).

Table 5.1 Neutral and charged (polyelectrolytes) water soluble polymers	
A. Neutral	B. Polyelectrolytes
I. Natural	
Proteins: albumins, casein, collagen, gelatins, enzymes, gluten.	Polynucleotides
Polysaccharides	
- storage polysaccharides: starch, glycogen, glucans (laminarin). - seed mucilages: guaran, guar gum, locust bean gum. - extracellular polysaccharides: dextran. - bacterial and fungal polysaccharides: xanthan, scleroglucan, schizophyllan, curdlan, pullulan.	- algal polysaccharides: alginic acid, agar, carragenan, furcellaran. - pectins: pectic and pectinic acids and their derivatives. - plant gums: karaya gum, tragacanth gum; arabic gum, ghatti gum, tamarind gum. - bacterial polysaccharides: gellan. - mucopolysaccharides: hyaluronic acid, chondroitin and chondroitin sulfate, dermatan sulfate, heparin.
II. Semisynthetic	
- starch derivatives: starch acetates, hydroxyethylstarch. - cellulose derivatives: hydroxyethyl cellulose, hydroxypropyl cellulose (HPC), methylcellulose (MC), ethylcellulose, ethylhydroxyethylcellulose.	Carboxymethyl cellulose (CMC), carboxyethyl cellulose, carboxymethylstarch, chitosan (CS).
III. Synthetic	
Poly(ethylene glycol) (PEG), poly(ethylene oxide) (PEO), poly(propylene oxide) (PPO), poly(vinyl methyl ether) (PVME), poly(acrylic acid) (PAA), poly(methacrylic acid) (PMAA), poly(ethylene imine) (PEI), poly(vinyl alcohol) (PVAL), vinyl acetate-vinyl alcohol copolymers (poly(VAC-co-VAL)), poly(*N*-vinyl pyrrolidone) (PVP), polyacrylamide (PAAM), poly(*N,N*-dimethylacrylamide) (PDMAM), poly(*N*-isopropylacrylamide) (PNIPAM).	a) anionic polyelectrolytes: poly(acrylic acid) salts, poly(methacrylic acid) salts, poly(styrenesulfonic acid) (PSSA) and its salts, poly(phosphoric acid) and its salts. b) cationic polyelectrolytes: poly(ethyelene imine) salts, poly(2-vinylpyridine) salts, poly(diallyldimethylammonium chloride) (PDADMAC). c) amphoteric water-soluble polymers: polybetaines, polyampholytes. d) hydrophobically modified polyelectrolytes.
Guaran is a galactomannan, the water soluble fraction (85%) of guar flour *Curdlan is a natural beta glucan* *Pullulan consists either of gluco or maltotriosyl units. It is an extracellular polysaccharide*	

5.3 Water Soluble Polymers in Solution - Phase Behaviour

Polymers are mostly insoluble in water. As an example materials of either natural origin such as rubber, silk or cotton, or synthetically prepared such as plastics or elastomers may be cited. This behaviour is mainly explained by the low polarity of the polymeric substances and the extremely polar character of water. Dissolution of polymers in water is possible only if they contain a polar group such as an ether oxygen (PEO, PVME), an hydroxyl (PVAL), an amine (PEI), a carboxylic (PAA, PMAA), an amide (PAAM, PNIPAM), or an ionic group such as a carboxylate (poly(sodium acrylate)-PNaA), a sulfate (PSSA), or a quaternary ammonium group (PDADMAC).

A special characteristic of the water soluble polymers is a certain amphiphilicity due to the structure of their repeating unit, consisting more or less of a polar and a non-polar part. As a result when such a polymer is dissolved in water, besides the dipole-dipole interactions with the water molecules, hydrophobic hydration, related to the non-polar part of the polymer segments, also occurs. It has a better organisation of the water molecules around the non-polar part, resulting in a negative enthalpy change, due to an increase of the hydrogen bonds formed between them, counterbalanced by a negative entropy change, due to a decrease in their mobility [10]. The solubility of a polymer in water is determined by the balance between these two antagonising factors, i.e., the hydrogen bonding interactions of the water molecules with the polar part and the hydrophobic hydration of the non-polar part. Polymers, in which the polar groups predominate, present an upper critical solution temperature (UCST) behaviour, by separating in two phases from their aqueous solutions as temperature decreases. A good example of such polymers, that are characterised as hydrophilic, is PAA, precipitating out of aqueous 0.2 M HCl on cooling to around 5 °C [3, 11], or phase separating from 0.4 M NaCl solution by decreasing the temperature [12], as shown in **Figure 5.1a** Dissolution of these polymers in water is endothermic, as it also happens with most of the non-polar polymers soluble in non-polar solvents [13, 14].

Another example of such a behaviour is PAAM, for which an ideal Flory (φ) temperature at −8 °C by extrapolating methods has been determined [15]. On the contrary, polymers with a hydrophobic character, present lower critical solution temperature (LCST) behaviour, accompanied by an exothermic dissolution process [3] and precipitate by increasing the temperature. A well known example of such a polymer, with inverse solubility in water, is PNIPAM, exhibiting an LCST at 31 °C [16], with a phase diagram as shown in **Figure 5.1b**. Due to the hydrophobic hydration, dissolution of such polymers in water is entropically disfavoured. On the contrary, their phase separation is entropically driven, in accordance with a postulate that 'at the UCST, phase separation occurs mainly for enthalpic reasons, while at the LCST, phase separation occurs mainly for entropic reasons' [17]. Such an LCST behaviour is exhibited by many of the non-ionic water soluble polymers as shown in **Table 5.2**.

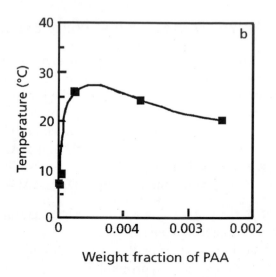

Figure 5.1a Phase separation curve of PAA, of infinite molecular weight, in aqueous 0.4 M NaCl

Redrawn from R. Buscall and T. Corner, European Polymer Journal, 1982, 18, 967, with permission from Elsevier Science

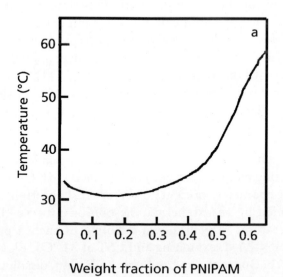

Figure 5.1b Phase diagram for PNIPAM/water

Reprinted with permission from M. Heskins and J.E. Guillet, Journal of Macromolecular Science - Chemistry, 1968, A2, 8, 1441, by courtesy of Marcel Dekker Inc

Table 5.2 Non-ionic water soluble polymers exhibiting LCST behaviour		
Polymer	LCST, (°C)	References
PEO	98	[18]
PPO	−53	[3]
PVME	34.5	[19]
PMAA	50	[20]
PVP	140	[3]
HPC	40-45	[21, 22]

Finally, water soluble polymers can be characterised as hydrophilic if they phase separate from aqueous solutions by decreasing temperature, as hydrophobic if they phase separate by increasing temperature.

5.4 Phase Behaviour of Mixtures of Water Soluble Polymers

By mixing two polymers in a common solvent (water), thermodynamic incompatibility frequently occurs leading to phase separation (at macroscopic or microscopic level). Two kinds of incompatibility could be shown [23, 24]: *segregative*, when each phase is enriched in one of the two components, or *associative*, when one of the separated phases is rich in both polymers, the other being composed almost of pure solvent.

Segregative phase separation is most common with mixtures of two non-ionic polymers or two similarly charged polyelectrolytes or mixtures consisting of a non-ionic polymer and a polyelectrolyte. Mixtures of a polyelectrolyte and a non-polyelectrolyte are less likely to phase separate (preservation of the electrical neutrality requires the charged polymer to segregate with its associated counterions and, as the number of counterions is much greater than the number of charged chains, the entropic disadvantage of segregation is greater) than a mixture of two charged or two uncharged polymers [25-28]. Phase separation can be induced in the first case by addition of extraneous salt [26, 28]. Segregative phase separation has been proposed for separations involving labile biological materials [29].

Associative phase separation, on the other hand, is the result of specific attractive interactions such as electrostatic attraction between two oppositely charged polyelectrolytes or hydrogen-bonding association between a proton donor and a proton acceptor polymer. In such cases, we often speak about the formation of interpolymer complexes, which may be soluble, compact precipitates or liquid coacervates (complex rich phase).

5.5 Interpolymer Complexes

The formation of complexes between water soluble polymeric species was firstly studied in the 1970s [30-32]. Interpolymer complexes (IPC) are divided into two major classes, the hydrogen bonding and the polyelectrolyte complexes. Hydrogen bonding IPC formation takes place between proton-donors, such as PAA or PMAA (weak polyacids) and proton-acceptors, such as PEG or PVP (Lewis polybases). The formation of polyelectrolyte complexes (PEC) is due to attractive Coulombic interactions between a polycation and a polyanion. IPC formation is considerably influenced by hydrophobic interactions. As a result the PMAA/PEO hydrogen bonding complexes are stronger than the PAA/PEO complexes. This difference in stability has been explained in terms of hydrophobic interaction between the α-methyl groups of PMAA and the ethylene backbone of PEO, which might provide additional stability in the complex [33]. Moreover, the composition of the complex of PAA with the polycation $[-N^+(CH_3)_2-(CH_2)_{10}-N^+(CH_3)_2-(CH_2)_{10}-]$, $2X^-$ (10, 10 Ionene) is 1:1 while the composition of the complex of PMAA with 10, 10-Ionene is 9:1 due to the hydrophobicity of the PMAA [34].

Co-operativity is one of the most important characteristics of the polymer-polymer complex formation. Stabilisation of IPC depends on the chain length of the polymers involved. Polyelectrolyte complexes between PMAA, with a molecular weight equal to 4.0×10^4, and the positively charged electrolyte $[-(CH_2)_2-N^+(CH_3)_2-, I^-]_n$, are characterised by stability constants exponentially increasing with the chain length of the cations, n, for an n between 1 and 6 [33]. The hydrogen bonding complex formation between PMAA and PEO starts from a critical PEO chain length of about 45 monomer units [33]. For complex formation between PEO and PAA a critical polyacid chain length equal to 40 monomer units has been estimated to be necessary [35]. It has been also shown that PAA with a degree of neutralisation higher than 12% is not complexable with PEO [36]. Moreover, the acrylic acid copolymer with the anionic monomer 2-acrylamido-2-methylpropanesulfonic acid, a strong acid dissociated even under acidic conditions (pH~3), cannot be complexed by PEO if it contains more than 10% of the non-complexable sulfonate groups [37].

5.6 Hydrogen-Bonding Interpolymer Complexes

Interaction between a proton donor, such as PAA, and a proton acceptor, such as PEG, polymer leads to the formation of a hydrogen-bonding interpolymer complex schematically shown in **Figure 5.2**.

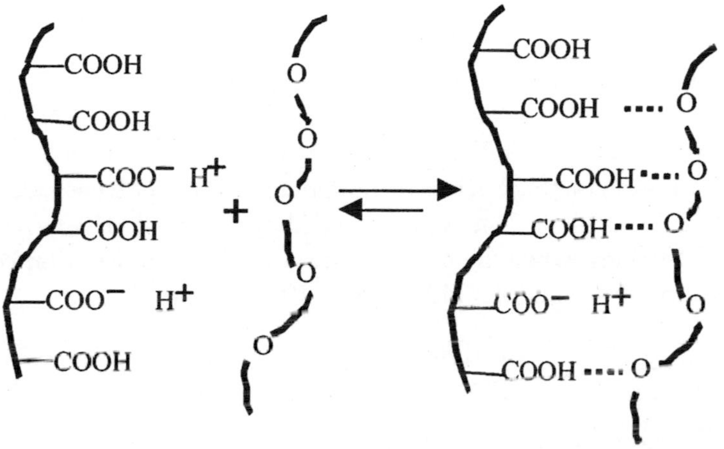

Figure 5.2 A schematic representation of the hydrogen-bonding formation between PAA and PEG

5.6.1 Investigation Methods

Viscometry and potentiometry are the most commonly used methods for the study of hydrogen bonding interpolymer complexes.

- *Viscometry.* Measurement of the specific or of the reduced viscosity of the polymer mixture in a dilute aqueous solution at a constant polyacid concentration as a function of the polybase/polyacid ratio [36, 38] or at a constant total concentration as a function of the polymer mixture composition, and comparison with an ideal value calculated as the weight average of the two constituents [39-41] have been used. Measurement of the intrinsic viscosity of the polymer mixture, based on the isoionic dilution method, and comparison with the ideal value, obtained as the weight average of the intrinsic viscosities of the two constituents, has been also proposed [42, 43]. The viscosity of the mixture usually decreases by addition of the polybase in the polyacid solution and presents a minimum indicating the formation of compact interpolymer complex. Nevertheless an increased viscosity, indicative of a gel-like association, has been also observed for the mixtures PAAM/PAA [40] and PEO/PAA [44], where partially neutralised PAA has been used.

- *Potentiometry.* By titrating a PAA dilute aqueous solution with PEO [36, 45] the degree of complexation of the polyacid can be determined. As a result of the complexation that occurs, the pH of the solution increases until the complete complexation of the polyacid and attainment of the complex stoichiometry [46]. Potentiometry results have been used for the determination of the stability constants of the complexes [47] and for the determination of the thermodynamic parameters of the complexation [48, 33]. Another approach has also been proposed based on the isoionic dilution method. In this method the pH of the polybase solution is adjusted to the same value of that of the weak polyacid by means of HCl. Complexation leads to higher pH values, while the determination of the complexation constant should not depend on dilution effects [42, 43, 49, 50].

- *Conductometry.* The conductivity of a solution of a weak polyacid decreases by titration with a neutral polybase until the IPC stoichiometry is reached [51, 52].

- *Turbidimetry.* Turbidity appears with the formation of insoluble IPC [53, 54].

- *Fluorescence.* Florescence measurements have been used to detect interpolymer complexation from the change of the spectrum of a fluorescence probe covalently bound onto one of the two macromolecular species [55-58] or for the detection of any hydrophobic interactions by using as a fluorescence probe, pyrene free in the solution [59].

5.6.2 Weak Polyacid/Non-Ionic Polybase/Water Ternary Systems

A considerable number of studies of polyacid-polybase pairs forming hydrogen bonding IPC in aqueous solution have appeared in the literature during the last decades and the role of different factors on their formation has been more or less elucidated.

In a study on the complexation of PEO with PAA, PMAA and a styrene/maleic acid copolymer [54] it was shown that, besides the major role of hydrogen bonding, the influence of hydrophobic interaction was considerable, while a definite number of binding sites was necessary for a stable IPC to be formed and cooperative interaction among active sites plays an important role. In the IPC formation between PMAA and PVP, three phase changes were observed by increasing polymer concentration: namely a homogeneous solution, a precipitate and a gel. The interpolymer complex formation was examined in several protic (water, methanol and ethanol) and aprotic (N,N-dimethyl foramide, dimethyl sulfoxide and tetramethyl urea) solvents and the relationship of the complex formation with the dielectric constant of the solvent was determined. The effect of hydrophobic interactions and the role of the α-methyl group of PMAA are also discussed [60]. Moreover, the contribution of hydrophobic interaction in the complexation of PMAA with PVP in aqueous solution has been shown by a calorimetric study [61].

The complexation of PAA with PEO or PVP in dilute aqueous solutions was monitored from the segmental rotational mobility of spin-labelled PAA using the electronic paramagnetic resonance technique. The segmental rotational dynamics revealed that PAA forms strong complexes at low pH with PEO or PVP, whereas no complexes were observed with highly neutralised PAA [62]. A light scattering investigation of the aggregation of the IPC formed between PMAA and PVP showed that the fast complexation in aqueous solution was followed by aggregation, sensitive to pH [63].

The complex formation between vinyl alcohol oligomers (OVA) and PMAA was studied potentiometrically and viscometrically [64] and the influence of the molecular weight of OVA on the characteristics of the complexes formed was elucidated. In another study [65], the influence of the degree of saponification and of the molecular weight of poly(VAC-co-VAL) on its complexation with PAA was investigated. It was shown that a lower degree of saponification leads to the formation of stronger and more compact complexes.

The complex formation between PVP and poly(itaconic acid monomethylester) (PMMI) or the copolymer of maleic anhydride and monomethylitaconate has been studied. The molar composition of the first complex was found to be 1:1, while that of the second was found to equal to 3:2. The complex particles were very compact and their intrinsic viscosity in water was about 5-10 g/cm^3 not depending upon the molecular weight of the polymer components [66]. From diffusion light scattering measurements with the complex formed between PVP and PMMI, it was found that the complex size distributions, two distinct modes, increase when temperature increases. From these results it can be concluded that the driving force for the interpolymer association is hydrogen bonding as much as hydrophobic interactions [67].

Chatterjee and co-workers have presented many studies on hydrogen bonding interactions and complex formation between polycarboxylic acids, or their statistical copolymers with acrylamides (AM), and polybases, such as PVP, PEO and PEI in water [68-70]. They have concluded that the complexation ability of PEI was greater than that of PVP [71]. A comparative study of the interpolymer complex formation between PVAL-*graft*-PAAM and PMAA or PEI showed that the complex formed between the graft copolymer and PMAA was relatively more stable than the complex prepared by mixing stoichiometric portions of the two homopolymers with PMAA. A reverse trend was found with PEI [71].

143

Pyrene-labelled polymers were used to investigate, by excimer fluorescence measurements, the effect of the degree of hydrolysis of PAAM on the PAA/hydrolysed polyacrylamide interactions and the intramolecular interactions between hydrolysed polyacrylamide molecules, at low pH. It was found that stable complexes were formed at a low degree of hydrolysis of PAAM, whereas intramolecular hydrogen-bonding was enhanced with increasing the number of acrylic acid groups on the copolymer chain [72] and a compact structure for the complex was suggested [58]. Moreover, the formation of interpolymer hydrogen-bonding complexes between PAAM and PAA at various ionisation degrees of the latter, ranging from 0% up to 60%, was investigated by solution- and solid-state ^{13}C nuclear magnetic resonance (NMR) [73].

A pair of polymers forming hydrogen bonding complexes, like PAA and poly(N-acryloyl-glycinamide), was used to design interpenetrating network (IPN) hydrogels exhibiting positive temperature-dependent swelling. It was found that hydrogen bonding between the polymers improved the rate and the reversibility of swelling changes in response to temperature fluctuations [74]. The temperature-dependence of the equilibrium swelling of IPN hydrogels constructed with PAA and PDMAM or poly(DMAM-co-AM) was also investigated [75]. IPC between PDMAM and PAA are stable even at 70 °C in aqueous solution, whereas those containing poly(DMAM-co-AM) dissociate with increasing temperature. In these IPN, the transition temperature shifted to higher values with increasing DMAM content. Reversible and pulsative temperature-responsive solute release was achieved by using these IPN hydrogels. The complex formation between PDMAM and PAA was further investigated by measuring the turbidity of the mixture of PDMAM and PAA buffered solutions at various pH values. The complexes were very stable even at 80 °C in pure water (pH 3.2). In buffered solution at pH 3.7 the polymer mixture showed an LCST near 60 °C, although the solutions were transparent till 80 °C at pH 3.8 [76]. The influence of the molecular architecture was recently investigated by studying the characteristic solution properties demonstrated by graft copolymers, PAA-g-PDMAM, in aqueous media [77].

Karibyants and co-workers [78] have examined the effects of the molecular weight of PEO and of temperature on the conformation of the complex formed in aqueous media with a PMAA gel and they have found that a wide range of PEO concentration exists in which the gel may occur in the swollen and in the collapsed state.

Kudaibergenov and co-workers [79] have studied the complex formation between polyvinylether of ethyleneglycol (PVEEG), polyvinylether of diethyleneglycol (PVEDEG) and copolymers of vinylether of ethylene glycol and vinyl butyl ether (poly(VEEG-co-VBE)), with PAA and PMAA in aqueous solution. They showed that PVEEG and PVEDEG did not form complexes with PAA. Nevertheless, introduction of the hydrophobic VBE into the PVEEG chain enhanced the complexation with the polycarboxylic acids. The swelling/deswelling behaviour of composite films, derived from IPC of PVEEG or of poly(VEEG-

co-VBE) with the acrylic acid-VBE copolymer (poly(AA-*co*-VBE)) was studied in water, alcohol and water-alcohol mixtures. Moreover the formation of IPC between poly(AA-*co*-VBE) and polyvinylether of monoethanolamine on a water-butanol interface was studied potentiometrically [80].

Steady-state and time-resolved fluorescence measurements, using pyrene as a free probe, were used to detect hydrophobic interactions in the formation of interpolymer complexes between PAA and PNIPAM, PAAM, PVME or PEG in dilute aqueous solutions. The PNIPAM/PAA and PVME/PAA complexes were found to exhibit a strong hydrophobic character; the PEG/PAA complex showed only a limited hydrophobicity and the PAAM/PAA complex was not at all hydrophobic. This behaviour was related with the LCST behaviour of PEG, PVME, PNIPAM and the UCST behaviour of PAAM [81].

Interpolymer complexation of PAA with PAAM and PNIPAM in dilute aqueous solution was studied. The stoichiometry of the complexes formed was determined [82]. It was shown that hydrogen bonding was the main factor stabilising the PAAM/PAA complex, strengthened by decreasing the temperature, while the much stronger PNIPAM/PAA complex, strengthened by increasing the temperature, was stabilised by hydrophobic interaction [83].

Prevysh and co-workers [84] examined the effect of added salt on the stability of hydrogen bonding IPC consisting of PAA and PEO or PVP. They found that addition of NaCl resulted in interpolymer complex aggregation, on the contrary to what happens in the case of PEC formed between oppositely charged polymers where addition of salt results in the dissolution of the precipitate [85].

Due to the interpolymer complex formation between PMAA and PEG at low pH the diffusional rates of solutes (drugs) in PMAA-*graft*-PEG hydrogels is much lower in acidic than in neutral or basic media [86]. The pH dependent structural changes, attributed to the formation/dissociation of hydrogen bonding IPC, have been studied as a function of the copolymer composition and the PEG graft chain molecular weight. The largest degrees of complexation were observed in gels containing nearly equimolar amounts of monomeric units and the longest PEG grafts [87].

5.6.3 Miscibility Enhancement by Hydrogen Bonding

5.6.3.1 Systems Based on Water Soluble Synthetic Polymers

In many polymer blends hydrogen bonding interactions enhance miscibility in the solid state, while interpolymer complexation is not probably detected in dilute solutions.

However, miscibility is improved upon progressive increase in the density of hydrogen bonding [88].

PVAL and poly(VAC-*co*-VAL) are often used for the preparation of film composites and membranes after mixing with other polymers. It seems that hydrogen bonding interactions play an important role in these systems. The structure of film composites based on PVAL and polyacids like CMC, PAA or PMAA, was studied using IR spectroscopy and dielectric relaxation measurements. The electrophysical properties of the composites were shown to be related to the formation of interpolymer hydrogen-bonding complexes [89]. The specific interactions between poly(4-vinyl pyridine) (P4VP) and poly(VAC-*co*-VAL) were studied by Fourier transform IR spectroscopy (FT-IR). Hydrogen bonding between the two polymers led to significant spectral modification in hydroxyl, carbonyl and pyridine ring spectral bands. Three equilibrium processes compete in the blends, hydroxyl-hydroxyl (self-association), hydroxyl-carbonyl (self-association) and hydroxyl-pyridine (inter-association). The first process was proved to be predominant [90]. Moreover, the miscibility of P4VP and poly(2-vinylpyridine) (P2VP) with PVAL, poly(vinyl acetate) (PVAC) and poly(VAC-co-VAL) was studied by using differential scanning calorimetry (DSC) and FT-IR. The latter method was useful for the study of possible hydrogen bonding interactions [91]. From the study of membranes of mixtures of poly(sodium styrene sulfonate) (PNaSS) or PSSA with PVAL, cast from aqueous solutions, a partial miscibility of the non-ionic polymer with the polyelectrolyte was concluded and it was attributed to interactions between the -OH groups of PVAL and the ionic groups of PNaSS or PSSA [92]. The miscibility of PVAL and PSSA in dilute aqueous solutions, studied by a viscometric method, is favoured by increasing the molecular weight and the number of residual acetate groups of PVAL [93].

The compatibility of PVAL/PVP blends has been studied by means of viscometric and ultrasonic experiments [94]. The results obtained showed that the blends were miscible over the entire composition range. The miscibility was attributed to hydrogen bonding interaction between hydroxyl groups in the PVAL and the carbonyl group of PVP. PVAL/PVP blends cast from aqueous solutions were further studied by DSC, FTIR spectroscopy and X-ray photoelectron spectroscopy (XPS) [95]. A single glass transition temperature (T_g) was observed for all the blends, suggesting that PVAL/PVP blends were miscible within the whole composition range. As a result of the hydrogen bonding interaction between PVAL and PVP the blends showed good miscibility in the bulk, even if, according to XPS results, their surface was enriched with the lower surface energy component, PVAL. So even if hydrogen bonding is the cause for miscibility in the bulk of PVAL/PVP blends, its effect is not strong enough to prevent surface segregation of PVAL. From a DSC study of blends of PVAL with different degrees of hydrolysis (88 and 99%) with PVP with different molecular weights (10,000 and 360,000 g/mol), strong interaction between PVP and PVAL was shown, while the density of interactions was larger in the case of the lower molecular weight PVP and the PVAL with a lower degree of hydrolysis. The crystallinity of the blends was decreased as the molecular weight of PVP was increased [96].

DSC measurements showed that PDMAM was immiscible with PVAC but was miscible with PVAL and poly(VAC-co-VAL). IR spectroscopy studies revealed the existence of specific interactions *via* hydrogen bonding between hydroxyl groups in vinyl alcohol units and the carbonyl group in the tertiary amide, which appeared to be decisive for miscibility [97].

The structure and properties of the blends poly(sodium 2-(3′-thienyl)ethanesulfonate) and poly(2-(3′-thienyl)ethanesulfonic acid) with PVAL, both with a mole ratio 1/1, for improving mechanical properties and processability of the conjugated conducting polymers, have been also investigated. Complex formation has been assessed to hydrogen bonding between the -OH and the $SO_3H(Na)$ groups. The conductivity of both the acid and the salt decreased on blending. Doping the blend with protonic acid increased the conductivity to 0.001 S/cm [98].

Hydrogen bonding interaction has also been used to improve the interfacial bonding between hydroxyapatite and polyactive 70/30, an ethylene glycol-butylene terephthalate biodegradable block copolymer. Hydroxyapatite was first surface-modified by polyacrylic acid or ethylene-co-maleic acid in aqueous solution. The surface-modified hydroxyapatite was then used as filler in composites with polyactive 70/30. The strength and elastic modulus of the composite were significantly improved. This improvement was attributed to hydrogen bonding interactions between the polyacid chains and the polyethylene glycol segments in the polymer matrix [99].

5.6.3.2 Systems Based on Water Soluble Polysaccharides

Miscibility studies suggest that starch is generally thermodynamically incompatible with the hydrocolloids (biopolymers) [100]. Addition of hydrocolloids to starch increases starch viscosity because of the mutual exclusion of the polymers influencing in the same time its gelatinisation and retrogradation [101, 102]. Blends of dextran/amylose [103], amylose/amylopectin [104], amylose/galactomannan [101], potato maltodextrin/locust bean gum [105], potato starch/xanthan [106] present incompatibility, too. Moreover, hydrocolloids interact with amylose accelerating its gelation [106].

From compatibility studies by IR spectroscopy and scanning electron microscopy performed on CS/starch blends [107], it was shown that blends with a starch content lower than 30 wt% are compatible. The crystallisation of starch is inhibited by CS and the recrystallisation of CS was affected by starch, too. In a mixing ratio of CS/starch varying between 8:2 and 7:2 films were obtained with a higher tensile strength of 781 kg cm^{-2}.

5.6.3.3 Systems Based on Water Soluble Polysaccharides/Synthetic Polymers

Rashidova and co-workers [108] studied the thermodynamic compatibility of methyl cellulose and PVP and found that they are compatible only when the ratio of the components is close to 1:1 or close to the pure components. The mixture loses the stability when the ratio is close to 1:2.

A lot of work was done on the CS/PEG blends [109-111]. It was established that CS/PEG aqueous solutions exhibit a pseudoplastic non-Newtonian behaviour described by the Ostwald de Waele model, the rheological behaviour of CS/PEG mixture being determined by CS [112].

Miscibility studies were also performed on CS with a deacetylation degree (DD) of 45-55% in the mixture with PVAL [113] and it was found that the polymers show good miscibility in the blends unlike the other chitosans of DD<40 or >60%. Nishio and co-workers [114] performed miscibility studies of using PAAM, PDMAM, PVP, poly(*N*-vinyl acetamide) (PNVA), poly(acryloyl morpholine) (PACMO) with CS. CS forms a fairly miscible phase with PVP, PAAM and PNVA while CS/PDMAM or PACMO blends exhibit immiscibility (because the T_g of the blends is not dependent on the blend composition). In the CS/PVP blend, it was proved by nuclear magnetic resonance (NMR) measurements that the upper limit of the heterogeneous domain size is < 2.5 nm, so the blend is partially-miscible on the ~2.5 nm scale. For the CS/PAAM mixture the upper limit of the heterogeneous phase is 1.9 nm [114]. They found that the number of hydrogen atoms attached to the nitrogen in the amide group is not an essential factor controlling the miscibility character of the polymer blends with chitosan. Miscibility [115] and compatibility [116] investigations have been performed on the polysaccharides (dextran, pulluan, HPC) and maleic acid copolymers with vinylacetate and styrene mixtures. It was established that the blends with a higher than 85% content in polysaccharide present an evenly fine particle dispersion morphology [117]. In the blends where one of the partners presents LCST, i.e., HPC, hydrophobic interactions could improve the miscibility of the components [118].

5.7 Polyelectrolyte Complexes

PEC represent a special class of chemical compounds, formed as a result of cooperative electrostatic interactions between oppositely charged polyions, as is schematically shown in **Figure 5.3**. The properties of these complexes depend strongly on their composition. Their study has seen a great development since the 1960s, when Michaels [119] prepared well defined 'polysalts' from oppositely charged polyelectrolytes. A work on the synthesis, properties and applications of polyelectrolyte complexes has been completed by Lysacht

[120]. Veis [121] has made a study of polyelectrolyte interactions leading to phase demixing, while compatibility, immiscibility and phase separation, complex coacervation in low and high charge density complexes have been examined.

By mixing aqueous solutions of two oppositely charged polyelectrolytes the following systems shown in **Figure 5.3** can be obtained [122]:

i.   Two-phase systems of supernatant liquid and precipitated complex (water-insoluble stoichiometric polyelectrolyte complexes - SPEC);

ii. Turbid colloidal systems with suspended complex particles (colloidal complexes);

iii. Homogeneous systems containing small complex aggregates (water soluble non stoiciometric polyelectrolyte complexes - NPEC).

A 1:1 stoichiometry has been observed for the polysalt precipitates corresponding to a random charge compensation between the ionic sites of the polyions [122, 123] and with regard to supermolecular order of the polymer chains in a complex precipitate, the 'scrambled-egg model', shown in **Figure 5.4**, has been considered to be closer to reality [122].

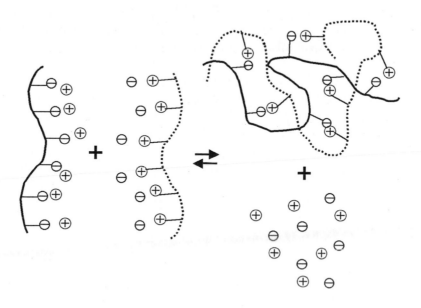

Figure 5.3 A schematic representation of polyanion-polycation interaction and PEC

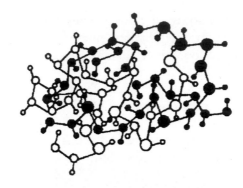

Figure 5.4 The scrambled-egg structure of interpolyelectrolyte complexes

Redrawn from B. Philipp, H. Dautzenberg, K.-J. Linow, J. Koetz and W. Dawydorf, Progress in Polymer Science, 1989, 14, 91. Copyright 1989, with permission from Elsevier Science

Conductometry [124], potentiometry [34] turbidimetry [125], UV spectroscopy [126], viscometry [33], gravimetry [127], light scattering [128], quasi elastic light scattering [130], sedimentation [130], fluorescence [131] and elemental analysis [132], have been mostly used for the study of polyelectrolyte complex formation.

There are two types of polycations, the pendant-type polycations which have charges in the side groups and the integral-type polycations which have charges in the chain backbone. Tsuchida and co-workers [127] observed that when a pendant-type polycation, such as quaternised P4VP is used, an equimolar insoluble complex is formed with PNaSS. On the other hand with an integral type polycation ($-N^+Cl^-(CH_3)_2-(CH_2)_3-N^+Cl^-(CH_3)_2-CH_2-C_6H_4-CH_2- = 3X$) an equimolar insoluble complex is obtained only at [PNaSS]/[3X] < 3, while at a polymer mixture composition [PNaSS]/[3X] > 3, a water soluble complex should be formed with the composition [PNaSS]/[3X]=3.

When a weak polyelectrolyte is involved in a PEC formation, the dissociation state of the weak polyelectrolyte directly affects the composition of the complex obtained. As a result in the complexation of the polycation 3X with PAA, the composition r = [cation]/[carboxylic acid] of the PEC formed, depends on the degree of neutralisation (α) of the polycarboxylic acid. The value of r is 0.2 at $\alpha = 0$, increases with α and becomes unity at $\alpha = 1$ [34].

5.7.1 Water-Insoluble Stoichiometric Polyelectrolyte Complexes

The SPEC are electroneutral, as the charges of the components are mutually neutralised, and they precipitate from solution. They only can be swelled in water and could be used as membranes [133].

The influence of the type of counterions and the concentration and type of added salt on the stoichiometry has been examined in an early conductometric study of the reaction of polystyrene sulfonates with the polyvinyl benzyl trimethylammonium [134]. Furthermore, it has been shown that in addition to coulombic forces, the hydrophobic interaction plays an important role in the complexation [135] as well as the chain length of the polycation and the dissociation degree of the polyacid, when the polyanion is derived from a weak polyacid, such as PAA, PMAA or poly(itaconic acid) [136]. A study on the electrostatic interaction between poly(L-lysine) hydrobromide and poly(L-glutamic acid) at various neutralisation degrees of the glutamic acid, showed that the binding process, involving strong electrostatic interaction between the COO^- and NH_3^+ groups, is strongly cooperative [137].

Fibrous and network-structured complexes were obtained by mixing PMAA and poly(N,N,N',N'-tetramethyl-N-p-xylylene-N'-alkylene diammonium dichloride) solutions. The formation of higher order structures was attributed to coulombic interaction combined with hydrophobic bonding [139]. In a phase separation study of aqueous solutions of the complexes, a homogeneous solution, a coacervate or a precipitate were obtained upon changing such factors as polymer concentration, pH and ionic strength [139]. Three types of water-insoluble PEC formed between poly(vinyl benzyl trimethylammonium chloride) (PVBTAC) and PMAA have been reported. Needle-like structures, radially extended fuzzy spheres and amorphous powders were obtained under different conditions [140].

Adsorption isotherms of water, methanol, acetone and benzene at 25 °C were determined for a series of PEC formed between two oppositely charged polyions (ionene, PVBTAC, PDADMAC, polyaminesulfone, poly(potassium vinyl sulfate) (PKVS), PNaSS and PAA) [141]. Successive differential sorptions have been measured for the system PAA/P4VP + water vapour. The sorption data revealed that the sorption process of water vapour in the complex is controlled not only by a diffusion mechanism but also by a relaxation mechanism of polymer chains [142]. Moreover, the equilibrium sorption of water vapour in complexes of PAA and poly(N-ethyl-4-vinylpyridine bromide) (PNE4VPB), consisting of loosely crosslinked networks, was examined experimentally [143].

Complexes formed between PMAA and P4VP or its methylvinylpyridinium methyl sulfate copolymers have been studied by fluorescence quenching [144], while the fluorescence polarisation method of the motion of 8-anilino-1-naphthalene sulfonic acid covalently attached to PMAA was used to study the formation of PEC in aqueous solution [145].

A study of the complex formation between poly(N,N-dimethylaminoethyl methacrylate) (PDMAEM) and polyphosphates or polyacrylates showed that the degree of conversion was the most important parameter determining solubility, composition, structure and properties of the PEC formed [146]. The interaction of PDMAEM with a polysilicic acid (PSA) sol in dilute aqueous solutions leads to the formation of both soluble and insoluble complexes depending on pH and mixture composition [147]. Similarly the interaction of sols of PSA with quaternised P4VP was studied and the effect of the pH of the medium on the composition of the complexes is reported [148].

A potentiometric study of polyion complexes formed between PMAA and cationic polyorganosiloxanes modified with various apolar groups showed an influence of the chemical structure of the polycation on the complexation in aqueous solution. From a binding study with methyl orange, it was shown that interpolymer complexation was enhanced by an increase of the hydrophobicity in the polycations [149].

High-resolution solid-state ^{13}C NMR spectra obtained for polyanion-polycation complexes of PDADMAC with acrylic acid-acrylamide copolymers (poly(AM-*co*-AA)) showed a stepwise transition to a scrambled-egg structure with decreasing charge density of the polyanion [150, 151]. The influence of the charge density, of the structure of the individual components and of salt addition on the composition of polyanion-polycation complexes were further considered in the complexation of anionic PEO ionomers or of poly(AM-*co*-AA) with cationically modified polyacrylamides, PDADMAC and highly branched PEI [125, 152]. The results obtained showed that a 1:1 stoichiometry is the exception. Of particular note is the gel formation observed when the copolymers of ethylene oxide and sodium carboxyoctylethylene oxide are used.

The enthalpy of the interaction of poly(styrene sulfonate) (PSS) with poly(trimethylammonium-2-ethyl methacrylate) in aqueous solution was determined as a function of the composition of the mixture [153]. Maximum enthalpy was found at a composition of about 1:1 mole. This maximum decreased with increasing mass of the counterion when the alkaline metal salts of PSS were used and no change was observed on the side of the cationic electrolyte. Salts of the alkaline earth metals gave a higher enthalpy.

The thermodynamic parameters of the PEC formation reaction between CMC or partially hydrolysed polyacrylamides of different charge densities with PDADMAC were determined. CMC with low charge density reacted exothermally and the PEC formation reaction was of the enthalpy dominant type, while for CMC with high charge density the exothermic tendency was low and the PEC formation reaction was thus shown to be of the entropy dominant type. The PEC formation reaction with the polyacrylamide systems was slightly endothermic, while they gained large entropy during the PEC formation, indicating that this reaction was of the entropy dominant type [154].

Water-insoluble SPEC redissolve upon salt addition. Trinh and Schnabel [155] studied the redissolution of PEC when ionic strength, I, is increased. The critical ionic strength, $[I]_{crit}$, for redissolution was found to depend on the chemical nature of the polyelectrolytes. Provided that the difference in $[I]_{crit}$ is sufficiently large, isolation of charged polymers can be achieved by selective precipitation. The method was illustrated by separating albumin at pH = 8 from polyphosphate *via* precipitation as PEC with PNE4VPB. PEC of PNE4VPB with PNaSS and/or poly(sodium phosphate) (PNaP) were prepared by precipitation upon mixing aqueous solutions of the corresponding polyelectrolytes. Binary and ternary systems were studied and the chemical composition of the PEC was determined as a function of the mixing ratio. The higher binding ability of PNaSS as compared to that of PNaP was clearly demonstrated [156]. The stability of PEC formed between the weak polyelectrolytes PNE4VPB and PNaMA has been studied by time-resolved light scattering measurements performed in conjunction with the stopped-flow method [157].

A complex formation study between weak polyanions and strong polycations with cationic groups in the main chain showed an influence of the structure of the complementary polymers and of their molecular weight. The greater the structural differences, the higher the end point deviation from stoichiometry [158].

The response of PEC, formed between PNaSS or PNaMA and PDADMAC and its copolymer with AM, to the addition of NaCl, was studied in relation to the anionic group of the polyanion and the charge density of the polycation. The effects of aggregation and macroscopic flocculation were discussed [161]. Schindler and Nordmeier [160] studied the formation and aggregation of PEC formed between PDADMAC and PSSA. The complexes were insoluble precipitates with a stoichiometry of 1:1. The size of the aggregates increases with salt concentration, while a critical salt concentration exists at which this increase becomes extremely abrupt.

PNaSS and PDADMAC have been deposited on spinning silicon wafers. The dependence of polyelectrolyte multilayer thickness was evaluated. Film thickness is approximately proportional to the number of layers and the salt concentration [161].

The complex formation between oppositely charged polyelectrolytes, such as PEI and PDADMAC with maleic acid-propylene copolymers and maleic acid-styrene copolymers, as well as their interaction with silicate powders were investigated [162].

Complex formation between poly (sodium acrylate) PNaA or some copolymers of acrylic acid with itaconic or maleic acid, used as anionic polymers, and cationic polyelectrolytes with quaternary ammonium salt groups in the main chain, was studied [163].

PEC were obtained from polycations with *N,N*-dimethyl-2-hydroxypropylammonium chloride units in the main chain and PNaA or poly(sodium 2-acrylamide, 2-dimethyl propane sulfate) PNaAMPS as polyanions. The influence of the charge density of the cationic polymer and its degree of branching as well as the nature and the molecular weight of the polyanions on complex formation, was examined [164].

Analytical ultracentrifugation was used for the determination of the complex composition and characterisation of the particles of water insoluble polyelectrolyte complexes between PNaSS and poly(DADMAC-*co*-AM) [165]. It was found that the complexes are remarkably polydispersant and form non-stoichiometric particles including the major component in considerable excess. These particles consist of a neutralised, relatively compact core and a surrounding shell of the excess component, which stabilises the particles by electrostatic repulsion. The stoichiometry depended strongly on the ionic strength of the solution and the concentration of the components. Regimes of relative monodispersity were found in water/salt media. PEC with the copolymer containing 50% DADMAC and AM form comparatively monodispersant species in aqueous salt solution with sedimentation coefficients between 4 and 40 Sv. Deviations from the 1:1 stoichiometry increase with increasing amounts of AM in the polycation and with increasing chain length of PNaSS.

Preferential binding of PDADMAC and PNaSS of different molecular weights or PNaSS/PNaMA relative to the ionic strength has been studied [166]. Contrary to theoretical expectations, at extremely low ionic strengths the short chain is preferred in complexation in the systems of low and high molecular weight PNaSS. In competition with PNaMA a slight favouring of PNaSS was observed. These findings were explained by the fact that, under such conditions, complex formation takes place far from the thermodynamic equilibrium, mainly governed by the kinetics of the process. With increasing ionic strength, the binding of PNaSS in comparison to PNaMA and of high molecular weight (HMW) PNaSS in the mixture with low molecular weight (LMW) PNaSS dominates. In contrast to the common assumption that complexation in highly aggregated complexes of strong polyelectrolytes is irreversible, exchange reactions could be observed in long-term experiments. At higher ionic strength, a complex exchange of poly(methacrylate) (PMA) for PSS and of LMW PSS for HMW PSS was found.

Huglin and co-workers [132, 167] have studied the compexation of poly[sodium(2-acrylamido-2-methyl propane sulfonate)] with poly(2-vinyl pyridinium chloride) and poly(4-vinyl pyridinium chloride) (P4VPC) in dilute aqueous solution. Complex

154

composition departed from stoichiometry and depended on the order of mixing. Elemental analysis of precipitates revealed that PEC were deficient in polyanionic units. The influence of charge density on the stoichiometry of the polyelectrolyte complex reaction occurring by mixing the cationic polymer P4VPC with poly(2-acrylamido-2-methyl propane sulphonate-*co*-N,N-dimethylacrylamide) was also studied [168]. Insoluble SPEC were formed by mixing 0.2 M aqueous solutions of PNaP with P4VPC. They were dissolved by strong acids or bases and by weak acids [169].

5.7.1.1 Hydrophobically Modified Polyelectrolytes

Recently, the effect of hydrophobic modification on phase behaviour in aqueous mixtures of oppositely charged polyelectrolytes was studied [170]. If both the oppositely charged polyelectrolytes are hydrophobically modified, the associative phase separation usually observed when mixing oppositely charged polyelectrolytes is effectively prevented in a large mixing region, and there is only a narrow two-phase domain. The observed viscosity enhancement of this mixture was attributed to a formation of mixed aggregates consisting of hydrophobic tails from polyelectrolytes bearing charges of opposite sign [171].

5.7.2 Colloidal Complexes

The phase behaviour of the block polyampholyte poly(methacrylic acid)-b-poly(N-methyl-4-vinylpyridinium chloride) was investigated and interpreted in terms of intra- and inter-polyelectrolyte reactions. It was shown that the polyampholytes and their complexes with a polyelectrolyte could exist in various phase states, depending on the polyelectrolyte concentration, the pH and the ionic strength of the solution [172].

Dautzenberg and co-workers [173] have shown that complex formation between PNaSS and poly(DADMAC-*co*-AM) results in highly aggregated PEC particles, with an aggregation number in the order of 10^3, while in highly dilute solutions it leads to quasi-soluble particles on a colloidal level [122, 174]. Moreover the polyelectrolyte complex formation in diluted aqueous solution was studied on the system PNaSS/PDADMAC using static and dynamic light scattering [175] and it was shown that the polyelectrolyte complex investigated exists as a highly polydisperse system of compact and nearly spherical particles.

The effect of low molecular weight salt on the properties of PEC formed between PDADMAC and copolymers of maleic acid with propylene or α-methylstyrene has been studied [176]. The effect of the ionic strength on the structure of the aggregates formed between PDADMAC and the sodium salt of maleic acid/propene alternating copolymers, with a threefold excess of cationic groups, was studied [177]. The sedimentation coefficient and hydrodynamic size of the aggregates increased with rising salt concentration, while the electrophoretic mobility decreased gradually.

Stable and monodispersive polyion complex micelles were prepared in an aqueous medium through electrostatic interaction between two oppositely charged block copolymers with PEG segments: PEG-*b*-poly(L-lysine) and PEG-*b*-poly(α,β-aspartic acid) [178]. It was shown that the polyion complex micelles were spherical particles in thermodynamic equilibrium without any secondary aggregates.

The formation, structure, and temperature behaviour of complexes between ionically modified PNIPAM copolymers with differing contents of anionic and cationic groups was studied [128]. The level of aggregation of such complexes increases with increasing mixing ratio, while the polymer concentration has only a marginal effect. Nearly stoichiometric complex particles on a 100 nm scale can be prepared. The complexes were highly swollen at 25 °C and collapsed in a temperature range up to 50 °C, in a completely reversible swelling-deswelling process. The results obtained open a new route for the preparation of smart gel-like nanoparticles, which could be used as temperature controlled carrier systems for drugs and enzymes.

5.7.3 Water Soluble Non-Stoichiometric Polyelectrolyte Complexes

The NPEC, containing an excess of one component, have a net charge of the same sign as this excess component, are usually water soluble and can be treated as ordinary soluble polyelectrolytes [179]. They can be obtained by a simple mixing of aqueous solutions of oppositely charged polyelectrolytes taken in non-equivalent ratios. However, a number of specific conditions must be met for the preparation of NPEC [127, 180]. It has been shown that soluble products are formed only if the concentration of one polyelectrolyte, P_1, is higher than the concentration of the other, P_2 [181]. P_1, the polyelectrolyte included in excess in the NPEC is referred as the host polyelectrolyte (HPE). P_2, the polyelectrolyte included in deficient amount is referred as the guest polyelectrolyte (GPE). Moreover, it is also necessary for the solution to contain a certain amount of a low-molecular weight electrolyte, for example 0.002–0.1 M NaCl. In general, as the excess of the HPE is larger, the concentration of the salt required for the formation of a water soluble NPEC is smaller. Regarding the ratio φ = [GPE]/[HPE], it should be lower than a critical value, φ_c. If φ is higher than φ_c, a soluble NPEC of the characteristic composition φ_c and a corresponding amount of an insoluble SPEC coexist in the reaction system. For most of the systems investigated φ_c varies between 0.2 and 0.5 depending on the polyelectrolytes comprising the NPEC [134].

A NPEC can be considered as a 'peculiar' block-copolymer comprised by single and double-strand sequences. The HPE sections free of GPE are single-strand, which are hydrophilic, and the HPE sections occupied by GPE chains are double-strand, which are hydrophobic. One of the most important properties of NPEC is their ability to participate in intermolecular exchange and substitution reactions [182, 183], while a large number of them exhibit an LCST behaviour, suggesting an amphiphilic character [184].

A study of the factors controlling the direction of the macro-substitution reaction of the PNE4VPB polycation with the PNaMA and PNaP polyanions showed that these are the polyanion chain length and the nature of the counterion (potassium, sodium or lithium) [185].

Zezin, Kabanov and co-workers have studied a large variety of water soluble NPEC, such as complexes of PNaA with 5,6-ionene bromide [186], complexes of PMAA and PAA with PNE4VPB [187, 188] and of PNaP or PAA with PDMAEM, ionene bromide and poly(4-vinyl pyridinium bromide) (P4VPB) [189, 190]. The influence of the excess ratio of hydrophobic or hydrophilic sequences of ionised groups not forming salt bonds as well as the state of hydrophilic units on the degree of association of the complexes PDMAEM/PNaP, PAA/ionene bromide, PAA/PDMAEM and PMAA/PNE4VPB) was studied [191, 192]. The effect of the ratio of the degrees of polymerisation of the components on the formation of NPEC between PAA and PDMAEM was investigated [193]. A reaction mechanism based on the concept of a significant amount of defects (loops) in the structure of the water soluble NPEC formed between PNaMA and P4VPB has been proposed [194] and features of the phase transitions in aqueous salt solutions of NPECs formed between PAA and PEI have been presented [195]. Moreover, exchange reactions were studied in three-component mixtures of polyelectrolytes containing one polyanion, PNaMA, and two polycations, PNE4VPB and 5,6-ionene bromide [196].

The kinetics and the mechanism of interpolyelectrolyte exchange was firstly studied for the system of the oppositely charged polyelectrolytes PMAA and PNE4VPB in aqueous salt solutions by a fluorescence quenching technique [197]. In another study the influence of chain length of a competitive polyanion and of the nature of monovalent counterions on the direction of the substitution reaction of polyelectrolyte complexes formed between PNaMA or PNaP and PNE4VPB was investigated [198].

The influence of the nature of the counterions on the formation of NPEC of PAA (blocking (guest) polyelectrolyte) with poly(N-ethyl-4-vinylpyridinium) salts (lyophilising (host) polyelectrolyte) was investigated. It was shown that NPEC were formed only when the degree of binding of the low molecular weight counterion was slight, otherwise SPEC were formed. The stability of the NPEC depended on the degree of binding of the low molecular weight counterions by the lyophilising polyelectrolyte [199].

A study of the influence of salt on the behaviour in dilute aqueous solution of NPEC showed that increase of the ionic strength resulted in a reduction of the complex particles dimensions and of the second virial coefficients of the solutions. At a given salt concentration, phase separation with a stoichiometric complex takes place [201]. The effect of small ions, i.e., bromide, chloride, lithium, sodium, potassium and tetramethylammonium, on the dissociation, ion binding and phase separation of the NPEC formed between PMAA and PNE4VPB was studied by Kabanov and co-workers [201].

A study of the stability of interpolyelectrolyte complexes formed between PMA anions and poly(*N*-ethyl-4-vinylpyridinium) (PNE4VP) cations in NaCl solutions showed that their dissociation begins at a certain critical salt concentration increasing as the chain length of the polycation increases, but being almost independent of the degree of polymerisation of the polyanion [202].

A new family of mechanically reversible gels has been obtained by mixing an aqueous solution of a NPEC of PNaMA and PNE4VPB containing a certain amount of covalent links between the two oppositely charged polyelectrolytes, with an aqueous solution of PKVS. Gelation is attributed to the partial replacement of the electrostatic contacts between the polycation and the PMA anion in the original NPEC with those of polycation and poly(vinyl sulfate) polyanion in the mixture [203]. A study of the behaviour of water soluble NPEC containing lyophilising PMA anions and blocking PNE4VP cations in sodium chloride solutions showed that as the degree of polymerisation of the polycation decreased, the phase separation is prevented, whereas decreasing the charge of the polycation, by decreasing the extent of alkylation, leads to phase separation, explained by a strengthening of hydrophobic interactions in the NPEC particles [130].

Kabanov and co-workers [204] presented results on the formation of NPEC of deoxyribonucleic acid (DNA) with quaternised PVP and they have discussed the applicability of this approach for efficient gene transfer.

DNA polyanions and cationic poly(ethylene oxide)-*b*-polyspermine copolymers form complexes attracting significant attention because of their capability of delivering nucleic acids to target cells. These complexes retain solubility in spite of charge neutralisation due to the presence of the PEO segments [205].

A light scattering study of soluble PEC between PNaSS or PNaMA, used as polyanions, and PDADMAC or a copolymer of DADMAC with AM containing 47 mol% DADMAC, used as polycations, showed that very small amounts of sodium chloride lead to a drastic decrease of the level of aggregation, while a higher ionic strength results in macroscopic flocculation [206].

The formation of PEC between PMA or dextrans functionalised with anionic groups, both polyanions being labelled with fluorescent pyrenyl groups, and PNE4VP as a quencher in comparison with complex formation with highly sulphated or sulphonated polyanion competitors, like heparin and polystyrene sulfonate, has been studied by means of the fluorescence quenching technique. The data obtained showed that both the chemical nature of the negatively charged groups and the charge density of the polyanion chains control the equilibrium of competitive interpolyelectrolyte reactions [207].

The complexation of polypropyleneimine dendrimers (Astramol), of five generations, with linear PNaA, PAA, PNaSS or native DNA has been studied in salt-free solutions as a function of pH. A pH controlled interpolyelectrolyte reaction resulting in the formation of the corresponding PEC occurred on mixing the dendrimer with the polyanion solutions, while all protonated amine groups of the dendrimer could form ion pairs with the carboxylate or sulfonate groups of the polyanions [208].

A mathematical model describing water soluble NPEC has been recently presented [209]. It is a mixture of the Madelung's theory for ionic crystals and of the Manning's counterion condensation theory. The predictions are compared with experimental results obtained with non stoichiometric complexes formed between ionene and PAA or PMAA. The central parameters are the degree of complexation and the degree of counterion binding.

In a recent study Osada and co-workers [210] showed that the complexes formed between the zwitterionic polymer, poly[3-dimethyl(methacryloyloxyethyl) ammonium propane sulfonate] (PDMAPS) and the anionic polymer poly(2-acrylamido-2-methyl propane sulfonic acid) (PAMPSA), or the cationic polymers poly-3-acryloylamino propyl trimethyl ammonium chloride and the ionene bromides are soluble and exhibit an UCST behaviour, dependent on the molar ratio. The PDMAPS/PAMPSA complex exhibits a dramatic increase in viscosity due to a network structure through electrostatic interaction.

5.8 Polymer-Protein Complexes

Protein complexation with polymers in aqueous solution may be driven by hydrogen bonding, coulombic attractions and hydrophobic interaction [211]. Kokufuta and co-workers [212, 213] have demonstrated that pepsin forms a water soluble complex with PEG at pH 3.0, presumably through the hydrogen bonding of the protein-COOH groups with the PEG ether groups. Dubin and co-workers [214] have proposed the use of a cationic polyelectrolyte, such as PDADMAC, for the separation of globular proteins according to their surface charge density as a result of protein-polyelectrolyte complex formation through coulombic attractions. Audebert and co-workers [215] have considered that hydrophobic interactions play an important role in the association of hydrophobically modified polysodium acrylate with lysozyme or bovine serum albumin (BSA).

Protein-polyelectrolyte complexation and the structure of the complexes formed are influenced by different factors such as: salt addition generally decreases the degree of binding, while increasing pH, above the protein isoelectric point (IEP), promotes the formation of protein-polycation complexes, and decreasing pH, below the IEP, promotes the formation of protein-polyanion complexes [216]. Depending on pH, ionic strength, and the protein-polyelectrolyte stoichiometry, these interactions may result in soluble complexes, complex coacervation, or precipitation [217]. Soluble complexes have been of particular interest as they appear to be the precursors of more extensive aggregates and knowledge of their structure is a prerequisite for understanding of the higher-order systems, because they can be easily studied with many experimental methods.

The effect of the ionic strength and pH on the complex formation between BSA and PNE4VPB [218] or copolymers of P4VP with 4-vinyl-N-cetylpiridinium bromide has been studied [219]. The binding of haemoglobin with dextran sulfate and diethylaminoethyl dextran has been studied. The results indicate that a complex is formed, either soluble or insoluble depending on pH. Binding appears to be stronger for the haemoglobin-dextran sulfate compounds [220].

A study of the complexation between potassium poly(vinyl alcohol) sulfate (KPVAS) and papain, human serum albumin, lysozyme, ribonuclease, trypsin and pepsin showed that in a salt free solution at pH = 2 electrically neutral protein-polyelectrolyte complexes (aggregates) with a uniform size were formed from all the proteins, other than pepsin [221].

A systematic study of protein binding to a homologous series of amphiphilic polyelectrolytes has considerably contributed to the quantitative evaluation of the hydrophobic interactions between them [211]. A minimum alkyl side chain length of 3-4 carbons is required for significant hydrophobic interactions. As the length of the alkyl side chains increases, the tendency of the copolymers to form both intrapolymer micelles and to bind with proteins increase. Amphiphilic polymers of low molecular weight (M_w < 34,000), random copolymers of acrylic acid, N-octylacrylamide and N-isopropylacrylamide, have been used for the stabilisation of aqueous dispersions of three integral highly hydrophobic membrane proteins of nanometric dimensions [222]. Hydrophobically modified poly(sodium acrylate)s in alkaline media interact strongly with BSA in dilute solution whereas hydrophilic precursors do not. Complexation is favoured by the hydrophobicity of the polymers at both high and low ionic strengths while coulombic repulsion plays a minor role compared with hydrophobically driven attractions [223]. Moreover, reversible gelation in hydrophobic polyelectrolyte/protein mixtures has been recently studied [224]. These reversible gels exhibited similarities with chemically cross-linked macromolecules.

A complexation study of human serum albumin, hemoglobin and bovin trypsin with PDADMAC and KPVS in salt-free conditions, showed that complexes with a 1:1 charge stoichiometry are formed and that changes in the conformations of the protein molecules caused by the complexation are not so large as to lose their biochemical capacity [225].

The interaction between three globular proteins of substantially different IEP (BSA, bovine pancreas ribonuclease, and chicken egg lysozyme) and a number of synthetic cationic and anionic polyelectrolytes of different charge densities was studied in dilute aqueous solution. For each polyion-protein pair, there is a well-defined critical pH at which binding starts by the formation of a soluble complex, while a further pH change produces phase separation (complex coacervation) [129, 226-231]. In a study of the interaction of PAA with egg white proteins (EWP) and lysozyme the following results were obtained: for EWP precipitation, larger molecular weight PAA had a higher efficiency in protein removal, while lysozyme removal was largely independent of PAA molecular weight, indicating that the high positive charge density of this protein is sufficient to ensure strong electrostatic binding. Particle size was also affected. Larger molecular weight PAA gave larger precipitates. For turbidimetric titrations a larger molecular weight PAA gave higher critical pH values [232]. Coacervation induced by pH in complexes of BSA and PDADMAC has been studied. The state of macromolecular assembly of complexes formed prior to and during the pH-induced coacervation could be characterised by specific pH values at which recognisable transitions took place [233].

Patrickios and co-workers [234] studied the interaction of dilute solutions of synthetic low-molecular weight block acrylic polyampholytes with proteins by turbidimetric titration. As in the homopolyelectrolyte-protein systems, the onset of interaction is manifested by a large increase in turbidity at a certain pH lying between the polyampholyte's self-aggregation pH and the protein's isoelectric point. Increasing salt concentration suppresses the protein-polyampholyte and polyampholyte-polyampholyte interactions, suggesting that the main driving force in these phenomena is electrostatic.

Addition of very small amounts of PNaSS in α-gelatin solutions resulted in a massive increase in their viscosity attributed to the formation of a complex between the two negatively charged polymers [235].

The interactions between soluble collagen from calfskin and PAA were studied. In the 2.5-4.0 pH range PEC between the oppositely charged polymers are formed and precipitate. Blending of collagen with PAA allows the possibility of producing bioartificial materials with good hydrolytic stability and better physicochemical and mechanical properties [236].

5.9 Three-Component Interpolymer Complexes

A new type of three-component IPC of two like-charged polyelectrolytes *via* a low molecular weight oppositely charged mediator has appeared recently [237]. The mediator is either a bifunctional or a monofunctional compound incorporating a hydrophobic substituent. The synthesis of the ternary IPC based on PAA, PNaP and a low molecular weight base, 4,4′-dipyridyl [238] or piperazine [239], acting as mediators, has been presented and the involvement of aromatic and aliphatic amines in the formation of such ternary complexes has been investigated.

References

1. *Water Soluble Resins*, 2nd Edition, Eds., R.L. Davidson and M. Sittig, Van Norstrand Reinhold, New York, NY, USA, 1968.

2. *Handbook of Water Soluble Gums and Resins*, Ed., R.L. Davidson, McGraw-Hill, New York, NY, USA, 1980.

3. P. Molyneux, *Water Soluble Synthetic Polymers: Properties and Behaviour*, Volume I, CRC Press, Boca Raton, FL, USA, 1983.

4. P. Molyneux, *Water Soluble Synthetic Polymers: Properties and Behaviour*, Volume II, CRC Press, Boca Raton, FL, USA, 1984.

5. E.A. Bekturov and Z.K. Bakauova, *Synthetic Water Soluble Polymers in Solution*, Huthig and Wepf Verlag, Basel, Switzerland, 1986.

6. C.L. McCormick, J. Bock and D.N. Schulz in *Encyclopedia of Polymer Science and Engineering*, Eds., H.F. Mark, N.M. Bikales, C.G. Overberger and G. Menges, John Wiley and Sons, New York, NY, USA, 1989, Volume 17, 730.

7. F. Franks, *Water*, The Royal Society of Chemistry, Cambridge, UK, 1983.

8. *Water: A Comprehensive Treatise*, Volume 4, Ed., F. Franks, Plenum Press, New York, NY, USA, 1975.

9. C. Tanford, *The Hydrophobic Effect: Formation of Micelles and Biological Membranes*, 2nd Edition, John Wiley and Sons, New York, NY, USA, 1980.

10. *Intermolecular and Surface Forces*, Ed., J. Israelachvili, Academic Press, New York, NY, USA, 1991.

11. A. Silberberg, J. Eliassaf and A. Katchalsky, *Journal of Polymer Science,* 1957, **23**, 259.

12. R. Buscall and T. Corner, *European Polymer Journal,* 1982, **18**, 967.

13. P.J. Flory, *Principles of Polymer Chemistry*, Cornell University Press, Ithaca, NY, USA, 1953.

14. P-G. de Gennes, *Scaling Concepts in Polymer Physics*, Cornell University Press, Ithaca, NY, USA, 1979.

15. A. Kanda, M. Duval, D. Sarazin and J. Francois, *Polymer,* 1985, **26**, 406.

16. M. Heskins and J.E. Guillet, *Journal of Macromolecular Science A,* 1968, **2**, 8, 1441.

17. J. Dayantis, *Macromolecules,* 1982, **15**, 4, 1107.

18. S. Saeki, N. Kuwahara, M. Nakata and M. Kaneko, *Polymer,* 1976, **17**, 685.

19. R.A. Horne, J.P. Almeida, A.F. Day and N-T. Yu, *Journal of Colloid and Interface Science,* 1971, **35**, 77.

20. J. Eliassaf and A. Silberberg, *Polymer,* 1962, **3**, 555.

21. E.K. Just and T.G. Majewicz in *Encyclopedia of Polymer Science and Engineering*, Eds. H.F. Mark, N.M. Bikales, C.G. Overberger and G. Menges, John Wiley & Sons, New York, 1985, 3, 248.

22. *Hydroxypropylcellulose, Chemical and Physical Properties*, HERCULES Incorp., KLUCEL Firma Papers, 1979.

23. E. Dickinson in *Gums and Stabilisers for the Food Industry 4*, Eds., G.O. Phillips, D.J. Wedlock and P.A. Williams, IRL Press, Oxford, UK, 1987, 249.

24. L. Piculell and B. Lindman, *Advances in Colloid and Interface Science*, 1992, **41**, 149.

25. L. Piculell, K. Bergfeldt and S. Nilsson, *Biopolymer Mixtures*, Eds., S.E. Harding, S.E. Hill and J.R. Mitchell, Nottingham University Press, Nottingham, UK, 1995, 13.

26. L. Piculell, S. Nilsson, L. Falck and F. Tjerneld, *Polymer Communications*, 1991, **32**, 158.

27. I. Iliopoulos, D. Frugier and R. Audebert, *Polymer Preprints,* 1989, **30**, 371.

28. M.B. Perrau, I. Iliopoulos and R. Audebert, *Polymer,* 1989, **30**, 2112.

29. P-A. Albertsson, *Partition of Cell Particles and Macromolecules: Separation and Purification of Biomolecules, Cell Organelles, Membranes and Cells in Aqueous Polymer Two-Phase Systems and their Use in Biochemical Analysis and Biotechnology,* 3rd Edition, Wiley, New York, NY, USA, 1986.

30. V.A. Kabanov and I.M. Papisov, *Polymer Science USSR,* 1979, **21**, 261.

31. E.A. Bekturov and L.A. Bimendina, *Advances in Polymer Science,* 1981, **41**, 99.

32. E. Tsuchida and K. Abe, *Advances in Polymer Science,* 1982, **45**, 1.

33. E. Tsuchida, Y. Osada and H. Ohno, *Journal of Macromolecular Science B,* 1980, **17**, 4, 683.

34. E. Tsuchida, *Journal of Macromolecular Science - Pure and Applied Chemistry,* 1994, **A31**, 1, 1.

35. I. Iliopoulos and R. Audebert, *Journal of Polymer Science: Polymer Physics Edition,* 1988, **26**, 10, 2093.

36. I. Iliopoulos and R. Audebert, *Polymer Bulletin,* 1985, **13**, 171.

37. I. Iliopoulos and R. Audebert, *Macromolecules,* 1991, **24**, 2566.

38. V.Y. Baranovsky, S. Shenkov and G. Borisov, *European Polymer Journal,* 1993, **29**, 8, 1137.

39. H. Ohno, K. Abe and E. Tsuchida, *Die Makromolekulare Chemie,* 1978, **179**, 3, 755.

40. G. Staikos and K. Tsitsilianis, *Journal of Applied Polymer Science,* 1991, **42**, 3, 867.

41. G. Staikos, G. Bokias and C. Tsitsilianis, *Journal of Applied Polymer Science,* 1993, **48**, 2, 215.

42. G. Bokias, G Staikos, I. Iliopoulos and R. Audebert, *Macromolecules,* 1994, **27**, 2, 427.

43. G. Bokias and G. Staikos, *Recent Research Developments in Macromolecular Research,* 1999, **4**, 247.

44. I. Iliopoulos, J.L. Halary and R. Audebert, *Journal of Polymer Science: Polymer Chemistry Edition*, 1988, **26**, 1, 275.

45. I. Iliopoulos and R. Audebert, *European Polymer Journal*, 1988, **24**, 2, 171.

46. S. Shenkov and V.Y. Baranovsky, *Journal of Polymer Science: Polymer Chemistry Edition*, 1994, **32**, 1385.

47. Y. Osada and M. Sato, *Journal of Polymer Science: Polymer Letters Edition*, 1976, **14**, 129.

48. Y. Osada, *Journal of Polymer Science: Polymer Chemistry Edition*, 1979, **17**, 3485.

49. G. Staikos, G. Bokias and K. Karayanni, *Polymer International*, 1996, **41**, 3, 345.

50. G. Staikos, K. Karayanni and Y. Mylonas, *Macromolecular Chemistry and Physics B*, 1997, **198**, 9, 2905.

51. L.A. Bimendina, V.V. Roganov and E.A. Bekturov, *Journal of Polymer Science: Polymer Symposia Edition*, 1974, **44**, 65.

52. S.K. Chatterjee, A. Malhotra and D. Yadov, *Journal of Polymer Science, Polymer Chemistry Edition*, 1984, **22**, 12, 3697.

53. D.V. Subotic, J. Ferguson and B.C.H. Warren, *European Polymer Journal*, 1991, **27**, 1, 61.

54. T. Ikawa, K. Abe, K. Honda and E. Tsuchida, *Journal of Polymer Science: Polymer Chemistry Edition*, 1975, **13**, 7, 1505.

55. B. Bednar, Z. Li, Y. Huang, L-C.P. Chang and H. Morawetz, *Macromolecules*, 1985, **18**, 10, 1829.

56. H.T. Oyama, W.T. Tang and C.W. Frank, *Macromolecules*, 1987, **20**, 8, 1839.

57. Y. Wang and H. Morawetz, *Macromolecules*, 1989, **22**, 1, 164.

58. K. Sivadasan, P. Somasundaran and N.J. Turro, *Colloid and Polymer Science*, 1991, **269**, 2, 131.

59. M. Koussathana, P. Lianos and G. Staikos, *Macromolecules*, 1997, **30**, 25, 7798.

60. H. Ohno, K. Abe and E. Tsuchida, *Die Makromolekulare Chemie*, 1978, **179**, 3, 755.

61. K. Abe, H. Ohno, A. Nii and E. Tsuchida, *Die Makromolekulare Chemie*, 1978, **179**, 8, 2043.

62. J. Pilar and J. Labsky, *Macromolecules*, 1994, **27**, 14, 3977.

63. A. Usaitis, S.L. Maunu and H. Tenhu, *European Polymer Journal*, 1997, **33**, 2, 219.

64. H. Horiuchi and K. Morisava, *Kobunshi Ronbunshu*, 1980, **37**, 1, 1980.

65. G. Staikos and G. Bokias, *Die Makromolekulare Chemie*, 1991, **192**, 2649.

66. L.A. Bimendina, E.A. Bekturov, G.S. Tleubaeva and V.A. Frolova, *Journal of Polymer Science: Polymer Symposia Edition*, 1979, **66**, 9.

67. A. Leiva, L. Gargallo and D. Radic, *Polymer International*, 1994, **34**, 4, 393.

68. S.K. Chatterjee and A. Malhotra, *Angewandte Makromolekulare Chemie*, 1984, **126**, 153.

69. S.K. Chatterjee, A. Malhotra and D. Yadav, *Journal of Polymer Science, Polymer Chemistry Edition*, 1984, **22**, 12, 3697.

70. S.K. Chatterjee, D. Yadav, A.M. Khan and K.R. Sethi, *Angewandte Makromolekulare Chemie*, 1989, **171**, 191.

71. S.K. Chatterjee and N. Misra, *Macromolecular Chemistry and Physics*, 1996, **197**, 12, 4193.

72. K. Sivadasan and P. Somasundaran, *Journal of Polymer Science: Polymer Chemistry Edition*, 1991, **29**, 6, 911.

73. F.O. Garces, K. Sivadasan, P. Somasundaran and N.J. Turro, *Macromolecules*, 1994, **27**, 1, 272.

74. H. Sasase, T. Aoki, H. Katono, K. Sanui, N. Ogata, R. Ohta, T. Kondo, T. Okano and Y. Sakurai, *Makromolekular Chemie: Rapid Communications*, 1992, **13**, 12, 577.

75. T. Aoki, M. Kawashima, H. Katano, K. Sanui, N. Ogata, T. Okano and Y. Sakurai, *Macromolecules*, 1994, **27**, 4, 947.

76. T. Aoki, K. Sanui, N. Ogata, N. Maruyama, H. Ohshima, A. Kikuchi, Y. Sakurai and T. Okano, *Kobunshi Ronbunshu*, 1998, **55**, 4, 225.

77. T. Shibanuma, T. Aoki, K. Sanui, N. Ogata, A. Kikuchi, Y. Sakurai and T. Okano, *Macromolecules*, 2000, **33**, 2, 444.

78. N.S. Karibyants, O.E. Filipova and S.G. Starodubtsev, *Polymer Science Series B*, 1995, **37**, 7-8, 385.

79. S.E. Kudaibergenov, Z.S. Nurkeeva and G.A. Mun, *Macromolecular Chemistry and Physics*, 1995, **196**, 7, 2203.

80. S.E. Kudaibergenov, Z.S. Nurkeeva, G.A. Mun, V.V. Khutoryanskiy, and A.D. Gazizov, *Polymer Advances for Technologies*, 2000, **11**, 1, 15.

81. M. Koussathana, P. Lianos and G. Staikos, *Macromolecules*, 1997, **30**, 25, 7798.

82. G. Staikos, G. Bokias and K. Karayanni, *Polymer International*, 1996, **41**, 3, 345.

83. G. Staikos, K. Karayanni and Y. Mylonas, *Macromolecular Chemistry and Physics*, 1997, **198**, 9, 2905.

84. V.A. Prevysh, B.C. Wang and R.J. Spontak, *Colloid and Polymer Science*, 1996, **274**, 6, 532.

85. C.K. Trinh and W. Schnabel, *Macromolecular Chemistry and Physics*, 1997, **198**, 5, 1319.

86. A.M. Lowman and N.A. Peppas, *Journal of Biomaterials Science: Polymer Edition*, 1999, **10**, 9, 999.

87. A.M. Lowman and N.A. Peppas, *Polymer*, 2000, **41**, 1, 73.

88. M. Jiang, M. Li, M. Xiang and H. Zhou, *Advances in Polymer Science*, 1999, **146**, 122.

89. T.G. Lazareva, I.A. Il'yushchenco and I.F. Alimov, *Polymer Science Series A*, 1994, **36**, 9, 1231.

90. L.C. Cesteros, J.R. Isasi and I. Katime, *Macromolecules*, 1993, **26**, 26, 7256.

91. L.C. Cesteros, J.R. Isasi and I. Katime, *Journal of Polymer Science: Polymer Physics Edition*, 1994, **32**, 2, 223.

92. M.O. David, and T.Q. Nguyen, *European Polymer Journal,* 1994, **30**, 9, 1013.

93. C.O. M'Bareck, M. Metayer, D. Langevin and S. Roudesli, *Journal of Applied Polymer Science,* 1996, **62**, 161.

94. R.V. Rao and P. Iatha, *Journal of Materials Science Letters,* 1999, **18**, 457.

95. L. Li, C-M. Chan and L-T. Weng, *Polymer,* 1998, **39**, 11, 2355.

96. S.N. Cassu and M.I. Felisberti, *Polymer,* 1997, **38**, 15, 3907.

97. L.G. Parada, L.C. Cesteros, E. Meaurio and E. Katime, *Polymer,* 1998, **39**, 5, 1019.

98. S-A. Chen and M-Y. Hua, *Macromolecules,* 1996, **29**, 14, 4915.

99. Q. Liu, J.R. De Wijn, D. Bakker and C.A. Van Blitterswijk, *Journal of Materials Science: Materials in Medicine,* 1996, **7**, 9, 551.

100. W-M. Kulicke, D. Eidam, F. Kath, M. Kix and A.H. Kull, *Starch,* 1996, **48**, 3, 105-114.

101. M. Alloncle, J. Lefebvre, G. Llamas and J.L. Doublier, *Cereal Chemistry,* 1989, **66**, 90.

102. M. Yoshimura, T. Takaya and K. Nishinari, *Carbohydrate Polymers,* 1998, **35**, 71.

103. M.T. Kalichevsky, P.D. Orford and S.G. Ring, *Carbohydrate Polymers,* 1986, **6**, 145.

104. M.T. Kalichevsky and S.G. Ring, *Carbohydrate Research,* 1987, **162**, 323.

105. P. Annable, M.G. Fitton, B. Harris, G.O. Phillips and P.A. Williams, *Food Hydrocolloids,* 1994, 8,351.

106. B. Conde-Petit, A. Pfirter and F. Escher, *Food Hydrocolloids* 1997, **11**, 393.

107. D. Yumin, X. Zuyong and R. Lu, *Journal of Natural Science,* 1997, **2**, 2, 220.

108. S.S. Rashidova, N.L. Voropaeva, T.D. Kalantarova and A.A. Tager, *Vysokomolekulyarnye Soedineniya Seriya A and Seriya B,* 1997, **39**, 3, 556.

109. W.H. Jiang and S.J. Han, *European Polymer Journal,* 1999, **35**, 11, 2079.

110. A. Nikolova, N. Manolova and I. Rashkov, *Polymer Bulletin,* 1998, **41**, 1, 115.

111. J. Piekielna and M. Mucha, *Inzynieria Chemiczna Procesowa*, 1998, **19**, 1, 145.

112. A. Nikolova, N. Manolova and I. Rashkov, *Polymer Bulletin*, 1998, **41**, 1, 115.

113. Y. Miyashita, M. Sato, N. Kimura, Y. Nishio and H. Suzuki, *Kobunshi Ronbunshu*, 1996, **53**, 409

114. Y. Nishio, T. Koide, Y. Miyashita, N. Kimura and H. Suzuki, *Journal of Polymer Science: Polymer Physics Edition*, **37**, 13, 1999, 1533.

115. G.G Bumbu, C.G Chitanu, A. Carpov, H. Darie and C. Vasile, Proceedings of the 4th International Conference on Hydrocolloids, 1998, Osaka, Japan, p.171.

116. A. Carpov, G.G. Bumbu, G.G. Chitanu, H. Darie and C. Vasile, Proceedings of Fifth Mediterranean School on Science and Technology of Advanced Polymer Based Materials, Capri Naples, Italy, 1997, 513.

117. G.G. Bumbu, C. Vasile, G.C. Chitanu and A. Carpov, *Polymer Degradation and Stability*, 2001, **72**, 1, 99.

118. G.G. Bumbu, G. Staikos, C. Vasile and G.C. Chitanu, unpublished data

119. A.S. Michaels and R.G. Miekka, *Journal of Physical Chemistry*, 1961, **65**, 10, 1765.

120. M.J. Lysacht in *Ionic Polymers*, Ed., L. Holliday, Applied Science, London, UK, 1975, 281.

121. A. Veis in *Developments in Ionic Polymers*, Eds., A.D. Wilson and H.J. Prosser, Applied Science Publishers Ltd., Barking, UK, 1983, 293.

122. B. Philipp, H. Dautzenberg, K-J. Linow, J. Kötz and W. Dawydoff, *Progress in Polymer Science*, 1989, **14**, 91.

123. A.S. Michaels, *Industrial Engineering and Chemistry*, 1965, **57**, 10, 32.

124. A.S. Michaels, L. Mir and N.S. Schneider, *Journal of Physical Chemistry*, 1965, **69**, 5, 1447.

125. J. Kötz, K-J. Linow, B. Philipp, L.P. Hu and O. Vogl, *Polymer*, 1986, **27**, 10, 1574.

126. N. Karibyants and H. Dautzenberg, *Langmuir*, 1998, **14**, 4427.

127. E. Tsuchida, Y. Osada and K. Sanada, *Journal of Polymer Science: Polymer Chemistry Edition*, 1972, **10**, 3397.

128. H. Dautzenberg, Y. Gao and M. Hahn, *Langmuir*, 2000, **16**, 23, 9070.

129. J. Xia, P.L. Dubin, Y. Morishima, T. Sato and B.B. Muhoberac, *Biopolymers*, 1994, **35**, 411.

130. V.A. Izumrudov and S.K. Lim, *Polymer Science Series A*, 1998, **40**, 3, 276.

131. Y. Itoh, K. Negishi, E. Iizuka, K. Abe and M. Kaneko, *Polymer*, 1992, **33**, 14, 3016.

132. M.B. Huglin, L. Webster and I.D. Robb, *Polymer*, 1996, **37**, 7, 1211.

133. V.A. Kabanov and A.B. Zezin, *Pure and Applied Chemistry*, 1984, **56**, 3, 343.

134. A.S. Michaels, L. Mir and N.S. Schneider, *Journal of Physical Chemistry*, 1965, **69**, 5, 1447.

135. E. Tsuchida and Y. Osada, *Die Makromolekulare Chemie*, 1974, **175**, 2, 593.

136. E. Tsuchida, Y. Osada and K. Abe, *Die Makromolekulare Chemie*, 1974, **175**, 2, 583.

137. A. Domard and M. Rinaudo, *Macromolecules*, 1980, **13**, 4, 898.

138. E. Tsuchida, *Die Makromolekulare Chemie*, 1974, **175**, 2, 603.

139. K. Abe, H. Ohno and E. Tsuchida, *Die Makromolekulare Chemie*, 1977, **178**, 8, 2285.

140. J.C. Salamone, S. Poulin, A.C. Watterson and A.P. Olson, *Polymer*, 1979, **20**, 5, 611.

141. Y. Kurokawa, N. Shirakawa, M. Terada and N. Yui, *Journal of Applied Polymer Science*, 1980, **25**, 8, 1645.

142. Y. Hirai and T. Nakajima, *Journal of Applied Polymer Science*, 1988, **35**, 1325.

143. Y. Hirai and T. Nakajima, *Journal of Macromolecular Science*, 1989, **26**, 10, 1465.

144. N.R. Pavlova, Y. E. Kirsh and V.A. Kabanov, *Polymer Science USSR*, 1979, **21**, 9, 2276.

145. H. Ohno and E. Tsuchida, *Die Makromolekulare Chemie, Rapid Communications*, 1980, **1**, 9, 585.

146. A.V. Kharenco, E.A. Neverova, R.I. Kalyuzhnaya, A.B. Zezin and V.A. Kabanov, *Polymer Science USSR*, 1981, **23**, 9, 2268.

147. E.N. Yermakova, P.V. Nuss, V.A. Kasaikin, A.B. Zezin and V.A. Kabanov, *Polymer Science USSR*, 1983, **25**, 7, 1605.

148. L.N. Yermakova, Y. G. Frolov, V.A. Kasaikin, A.B. Zezin and V.A. Kabanov, *Polymer Science USSR*, 1981, **23**, 10, 2529.

149. K. Nozuyama, Y. Veno, T. Kato and Y. Sekine, *Die Makromolekulare Chemie*, 1986, **187**, 5, 1159.

150. J. Kötz, J. Kunze, K.J. Linow and B. Philipp, *Polymer Bulletin*, 1986, **15**, 3, 247.

151. J. Kötz, E. Ebert, J. Kunze, B. Philipp, J. Lindberg and K. Soljamo, *Die Makromolekulare Chemie*, 1990, **191**, 3, 651.

152. H. Dautzenberg, J. Kötz, K.J. Linow, B. Philipp and G. Rother, *Polymer Preprints*, 1991, **32**, 1, 594.

153. W. Oppermann and T. Schulz, *Die Makromolekulare Chemie, Macromolecular Symposia*, 1990, **39**, 293.

154. S. Kanbayashi and T. Arai, *Konbunshi Ronbunshu*, 1992, **49**, 5, 407.

155. C.K. Trinh and W. Schnabel, *Angewandte Makromolekulare Chemie*, 1993, **212**, 167.

156. C.K. Trinh and W. Schnabel, *Angewandte Makromolekulare Chemie*, 1994, **221**, 127.

157. C.K. Trinh and W. Schnabel, *Macromolecular Chemistry and Physics*, 1997, **198**, 1319.

158. S. Dragan, M. Cristea, C. Luca and B.C. Simionescu, *Journal of Polymer Science: Polymer Chemistry Edition*, 1996, **34**, 3485.

159. H. Dautzenberg and N. Karibyants, *Macromolecular Chemistry and Physics*, 1999, **200**, 1, 118.

160. T. Schindler and E. Nordmeier, *Polymer Journal*, 1994, **26**, 10, 1124.

161. S.T. Dubas and J.B. Schlenoff, *Macromolecules*, 1999, **32**, 24, 8153.

162. G. Kramer, H-M. Buchhammer and K. Lunkvitz, *Journal of Applied Polymer Science*, 1997, **65**, 1, 41.

163. S. Dragan, M. Cristea, C. Luca and B.C. Simionescu, *Journal of Polymer Science: Polymer Chemistry Edition*, 1996, **34**, 17, 3485.

164. M. Cristea, S. Dragan, D. Dragan and B.C. Simionescu, *Die Makromolekulare Chemie, Macromolecular Symposia*, 1998, **126**, 143.

165. N. Karibyants, H. Dautzenberg and H. Coelfen, *Macromolecules*, 1997, **30**, 25, 7803.

166. N. Karibyants and H. Dautzenberg, *Langmuir*, 1998, **14**, 16, 4427.

167. L. Webster, M.B. Huglin and I.D. Robb, *Polymer*, 1997, **38**, 6, 1373.

168. L. Webster and M.B. Huglin, *European Polymer Journal*, 1997, **33**, 7, 1173.

169. N. Acar, M.B. Huglin and T. Tulun, *Polymer*, 1999, **40**, 23, 6429.

170. K. Thuresson, S. Nilsson and B. Lindman, *Langmuir*, 1996, **12**, 2, 530.

171. M. Tsianou, A-L. Kjoninksen, K. Thuresson and B. Nystrom, *Macromolecules*, 1999, **32**, 9, 2974.

172. E.A. Bekturov, S.E. Kudaibergenov, R.E. Khamzamulina, V.A. Florova, D.E. Nurgalieva, R.C. Schulz and J. Zoeller, *Die Makromolekulare Chemie, Rapid Communications*, 1992, **13**, 4, 225.

173. H. Dautzenberg and G. Rother, *Journal of Polymer Science: Physics Edition Part B*, 1988, **76**, 353.

174. H. Dautzenberg, J. Hartmann, S. Grunewald and F. Brand, *Berichte Bunsengesellschaft fur Physik Chemalische*, 1996, **100**, 6, 1024.

175. H. Dautzenberg, G. Rother and J. Hartmann in *Macro-Ion Characterisation. From Dilute Solutions to Complex Fluids*, Ed., K.S. Schmitz, ACS Symposium Series No.548, Washington, DC, USA, 1994, 210.

176. H-M. Buchhammer, K. Lunkvitz and D.V. Pergushov, *Die Makromolekulare Chemie, Macromolecular Symposia*, 1998, **126**, 157.

177. D.V. Pergushov, H-M. Buchammer and K. Lunkvitz, *Colloid and Polymer Science*, 1999, **277**, 2-3, 101.

178. A. Harada and K. Kataoka, *Macromolecules*, 1995, **28**, 15, 5294.

179. E. Tsuchida and K. Abe, *Developments in Ionic Polymers-2*, Eds., A.D. Wilson and H.J. Prosser, Elsevier Applied Science Publishers Ltd, Barking, UK, 1986, 191.

180. V.A. Kasaikin, O.A. Kharenco, A.V. Kharenco, A.B. Zezin and V.A. Kabanov, *Vysokomolekulyarnya Soedineiya*, 1979, **B21**, 84.

181. O.A. Kharenco, A.V. Kharenco, R.I. Kalyuzhnaya, V.A. Izumrudov, V.A. Kasaikin, A.B. Zezin and V.A. Kabanov, *Vysokomolekulyarnya Soedineiya*, 1979, **A21**, 2719.

182. V.A. Kabanov, A.B. Zezin, V.A. Izumrudov, T.K. Bronich and K.N. Bakeev, *Die Makromolekulare Chemie, Supplement*, 1985, **13**, 137.

183. K.N. Bakeev, V.A. Izumrudov, S.I. Kuchanov, A.B. Zezin and V.A. Kabanov, *Macromolecules*, 1992, **25**, 4249.

184. Z.G. Gulyaeva, M.F. Zansokhova, I.V. Chernov, V.B. Rogacheva, A.B. Zezin and V.A. Kabanov, *Polymer Science, Series A*, 1997, **39**, 2, 213.

185. V.A. Izumrudov, T.K. Bronich, O.S. Saburova, A.B. Zezin and V.A. Kabanov, *Die Makromolekulare Chemie, Rapid Communications*, 1988, **9**, 1, 7.

186. Z.G. Gulyaeva, O.A. Poletaeva, A.A. Kalechev, V.A. Kasaikin and A.B. Zezin, *Polymer Science USSR*, 1976, **18**, 12, 3204.

187. V.A. Izumrudov and A.B. Zezin, *Polymer Science USSR*, 1976, **18**, 11, 2840.

188. V.A. Izumrudov, V.A. Kasaikin, L.N. Yermakova and A.B. Zezin, *Polymer Science USSR*, 1978, **20**, 2, 452.

189. O.A. Kharenco, A.V. Kharenco, R.I. Kalyuzhnaya, V.A. Izumrudov, V.A. Kasaikin, A.B. Zezin and V.A. Kabanov, *Polymer Science USSR*, 1979, **21**, 12, 3002.

190. O.A. Kharenco, A.V. Kharenco, V.A. Kasaikin, A.B. Zezin and V.A. Kabanov, *Polymer Science USSR*, 1979, **21**, 12, 3009.

191. O.A. Kharenco, V.A. Izumrudov, A.V. Kharenco, V.A. Kasaikin, A.B. Zezin and V.A. Kabanov, *Polymer Science USSR*, 1980, **22**, 1, 274.

192. V.A. Kabanov and A.B. Zezin, *Makromolekulare Chemie Supplement*, 1984, **185**, 6, 259.

193. A.B. Zezin, V.A. Kasaikin, N.M. Kabanov, O.A. Kharenco and V.A. Kabanov, *Polymer Science USSR*, 1984, **26**, 7, 1702.

194. V.A. Izumrudov, A.B. Zezin and V.A. Kabanov, *Polymer Science USSR*, 1983, **25**, 9, 2296.

195. V.B. Rogacheva, S.V. Ryzkhikov, A.B. Zezin and V.A. Kabanov, *Polymer Science USSR*, 1984, **26**, 8, 1872.

196. V.A. Izumrudov, A.P. Savitskii, A.B. Zezin and V.A. Kabanov, *Polymer Science USSR*, 1984, **26**, 8, 1930.

197. V.A. Izumrudov, A.P. Savitskii, K.N. Bakeev, A.B. Zezin and V.A. Kabanov, *Die Makromolekulare Chemie, Rapid Communications*, 1984, 5, 709.

198. V.A. Izumrudov, T.K. Bronich, O.S. Saburova, A.B. Zezin and V.A. Kabanov, *Die Makromolekulare Chemie, Rapid Communications*, 1988, 9, 7.

199. N.K. Nefedov, T.G. Yermakova, V.A. Kasaikin, A.B. Zezin and V.A. Lopyrev, *Polymer Science USSR*, 1985, **27**, 7, 1677.

200. V.A. Izumrudov, O.A. Kharenco, A.V. Kharenco, Z.G. Gulyaeva, V.A. Kasaikin, A.B. Zezin and V.A. Kabanov, *Polymer Science USSR*, 1980, **22**, 3, 767.

201. D.V. Pergushov, V. A. Izumrudov, and V. A. Kabanov, *Polymer Science, Series A*, 1993, 35, 7, 940.

202. D.V. Pergushov, V.A. Izumrudov, A.B. Zezin and V.A. Kabanov, *Polymer Science, Series A*, 1995, 37, 10, 1081.

203. V.A. Kabanov, A.B. Zezin, V.A. Izumrudov T. Bronich, N.M. Kabanov and O.V. Listova, *Die Makromolekulare Chemie, Macromolecular Symposia*, 1990, **39**, 155.

204. V.A. Kabanov, A.V. Kabanov and I.N. Astafyeva, *Polymer Preprints*, 1991, **32**, 1, 592.

205. A.V. Kabanov, S.V. Vinogradove, Y.G. Suldaltseva and V.Y. Alakhov, *Bioconjugate Chemistry*, 1995, **6**, 639.

206. H. Dautzenberg, *Macromolecules*, 1997, **30**, 25, 7810.

207. V.A. Izumrudov, F. Chaubet, A-S. Clairbois and J. Jozenfonvicz, *Macromolecular Chemistry and Physics,* 1999, **200**, 7, 1753.

208. V.A. Kabanov, A.B. Zezin, V.B. Rogacheva, Z. G. Gulyaeva, M.F. Zansochova, J.G.H. Joosten and J. Brackman, *Macromolecules*, 1999, **32**, 6, 1904.

209. E. Nordmeier and P. Beyer, *Journal of Polymer Science: Polymer Physics Edition*, 1999, **37**, 4, 335.

210. L. Chen, Y. Honma, T. Mizutani, D-J. Liaw, J.P. Gong and Y. Osada, *Polymer*, 2000, **41**,1, 141.

211. J.Y. Gao and P.L. Dubin, *Biopolymers*, 1999, **49**, 185.

212. E. Kokufuta and H. Nishimura, *Polymer Bulletin*, 1991, **26**, 277.

213. J. Xia, P.L. Dubin and E. Kokufuta, *Macromolecules*, 1993, **26**, 24, 6688.

214. P.L. Dubin, T.D. Ross, I. Sharma and B. Yegerlehner in *Ordered Media in Chemical Separations*, Eds., W.L. Hinze and D.W. Armstrong, ACS Symposium Series No.342, Washington, DC, USA, 1987, 162.

215. F. Petit, R. Audebert and I. Iliopoulos, *Journal of Colloid and Polymer Science*, 1995, **273**, 777.

216. J. Xia and P.L. Dubin in *Macromolecular Complexes in Chemistry and Biology*, Eds., P.L. Dubin, J. Bock, R.M. Davis, D.N. Schulz and C. Thies, Springer-Verlang, Berlin, Germany, 1994, 247.

217. J.Y. Gao, P.L. Dubin and B.B. Muhoberac, *Journal of Physical Chemistry*, 1998, **102**, 28, 5529.

218. V.A. Kabanov and M.I. Mustafaev, *Polymer Science USSR*, 1981, **23**, 2, 280.

219. V.A. Kabanov, M.I. Mustafaev and V.V. Goncharov, *Polymer Science USSR*, 1981, **23**, 2, 287.

220. T.Q. Nguyen, *Die Makromolekulare Chemie*, 1986, **187**, 2567.

221. A. Tsuboi, T. Izumi, M. Hirata, J. Xia, P.L. Dubin and E. Kokufuta, *Langmuir*, 1996, **12**, 26, 6295.

222. C. Tribet, R. Audebert and J-L. Popot, *Langmuir*, 1997, **13**, 21, 5570.

223. C. Tribet, I. Porcar, P.A. Bonnefont and R. Audebert, *Journal of Physical Chemistry*, 1998, **102**, 7, 1327.

224. R. Borrega, C. Tribet and R. Audebert, *Macromolecules*, 1999, **32**, 23, 7798.

225. E. Kokufuta in *Macromolecular Complexes in Chemistry and Biology*, Eds., P.L. Dubin, J. Bock, R. Davis, D.N. Schulz, C. Thies and D. Schulz, Springer-Verlag, Berlin, Germany, 1994, **18**, 301.

226. J.M. Park, B.B. Muhoberac, P.L. Dubin and J. Xia, *Macromolecules*, 1992, **25**, 1, 290.

227. J. Xia, P.L. Dubin, Y. Kim, B.B. Muhoberac and V.J. Klimkowski, *Journal of Physical Chemistry*, 1993, **97**, 17, 4528.

228. J.Xia, P.L. Dubin and H. Dautzenberg, *Langmuir*, 1993, **9**, 8, 2015.

229. L.S. Ahmed, J. Xia and P.L. Dubin, *Journal of Macromolecular Science - Pure and Applied Chemistry*, 1994, **A31**, 1, 17.

230. J. Xia, P.L. Dubin, L.S. Ahmed and E. Kokufuta in *Macro-Ion Characterisation, From Dilute Solutions to Complex Fluids*, Ed. K.S. Schmitz, ACS Symposium Series No.548, Washington, DC, USA, 1994, 225.

231. Y. Li, K.W. Mattison, P.L. Dubin, H.A. Havel and S.L. Edwards, *Biopolymers*, 1996, **38**, 527.

232. J-Y. Shieh and C.E. Glatz in *Macromolecular Complexes in Chemistry and Biology*, Eds., P.L. Dubin and J. Bock, R. Davis, D.N. Schulz, C. Thies and D. Schulz, Springer-Verlag, Berlin, 1994, **16**, 273.

233. K. Kaibata, T. Okazaki, H.B. Bohidar and P.L. Dubin, *Biomacromolecules*, 2000, **1**, 1, 100.

234. C.S. Patrickios, C.J. Jang, W.R. Herter and T.A. Hatton in *Macro-Ion Characterisation, From Dilute Solutions to Complex Fluids*, Ed., K.S. Schmitz, ACS Symposium Series No. 548, Washington, DC, USA, 1994, 257.

235. J.H.E. Hone, A.M. Howe and T. Cosgrove, *Macromolecules*, 2000, **33**, 4,1199.

236. N. Barbani, L. Lazzeri, C. Cristallini, M.G. Cascone, G. Polacco and G. Pizzirani, *Journal of Applied Polymer Science*, 1999, **72**, 971.

237. O.V. Kargina, O.V. Prazdnichnaya, *Polymer Science, Series A*, 1993, **35**, 4, 561.

238. O.V. Kargina, O.V. Prazdnichnaya, I.D. Yurgens and E.Y. Badina, *Polymer Science, Series A*, 1996, **38**, 8, 937.

239. O.V. Kargina, E.Y. Badina, O.V. Prazdnichnaya and I.D. Yurgens, *Polymer Science, Series B*, 1999, **41**, 11-12, 331.

6 Water Soluble Polymer Systems - Applications of Interpolymer Complexes and Blends

Georgios Staikos, Georgios Bokias and Gina G. Bumbu

6.1 Introduction

Water soluble polymers, being environmentally friendly substances, show good potential for further development in industrial, agricultural and medical applications, either as themselves or in blends. Blends based on water soluble polymers have been prepared to improve certain of their properties used in various applications or to obtain new properties.

This chapter is not an exhaustive presentation of all water soluble polymer systems studied in the literature. An outline of the specific processing and characteristics of such blends in their main field of applications is presented.

6.2 Applications of Interpolymer Complexes

Interpolymer complexes (IPC) have been considered as promising functional materials for various applications. In particular materials consisting of polyelectrolyte complexes (PEC) have been used in the preparation of membranes with controlled permeabilities as well as in drug delivering systems, as enzyme carriers, as effective binders for dispersed systems, especially for soils. Moreover, the process of PEC formation is used for flocculation and water purification, isolation and fractionation of biopolymers, etc. PEC are likely to be a source of new materials and by combining unique physico-chemical properties with high biocompatibility, are very important for biomedical applications. Their applications have been surveyed by Philipp and co-workers [1].

An early review of the interaction of synthetic polyelectrolytes with physiologically active components, enzymes and hormones, in relation to the problem of producing systems for medical purposes has been published by Samsonov [2]. Later, Kabanov [3] presented a very brief review of the literature on the properties and potential biomedical applications of PEC. Also, the prospects of using PEC of deoxyribonucleic acid (DNA) with polycations to deliver genetic material to the cell have been reviewed [4].

Water soluble non-stoichiometric polyelectrolyte complexes (NPEC) have been considered as promising materials for biotechnological and biomedical applications [5]. Promising results on controlling the enzyme catalytic activity have been reported, through the covalent binding of the enzyme molecule onto a polyelectrolyte chain able to form a soluble PEC with an other polyelectrolyte [6]. The ability of NPEC to experience phase separation under slight changes in pH or composition of the mixture of polyions allow classification of NPEC as 'smart' or 'intelligent' polymers, which attract considerable attention due to their promising practical applications mainly in biotechnology and medicine [7].

NPEC have been investigated as potential drug carrier systems for parenteral administration. Complexes between the polycationic quaternised polyvinyl imidazole and an excess of a higher molecular weight partially sulphonated dextran were designed to inherit the biocompatible properties of dextran. Platelet aggregation studies showed that toxic aggregatory effects normally induced by the polycations were eliminated *in vitro* when they were a component of a soluble PEC. However this complexation was not sufficient to prevent un-wanted interactions *in vivo*, leading to removal of the PEC from the circulation [8]. In another study complexing of the DNA plasmid with poly(*N*-ethyl-4-vinyl pyridinium) polycations results in the formation of a membrane active polycomplex that is much more active in transformation of competent cells than pure plasmid [9]. A PEC, comprised of poly(diallyldimethylammonium chloride) (PDADMAC) and sodium cellulose sulfate, has been used for the microencapsulation of gonadotropin releasing hormone [10].

The interactions of proteins with polyelectrolytes are important in natural biological systems as well as in biotechnological applications. Protein-polyelectrolyte complex coacervation provides an alternative to conventional protein separation and purification methods [11, 12]. The efficiency of separation was examined using the cationic polyelectrolyte PDADMAC and the proteins bovine serum albumin, β-lactoglobulin, γ-globulin and ribonuclease A [13]. The existence of soluble protein-polyelectrolyte complexes makes polyelectrolytes potentially important enzyme carriers, in which the enzyme activity could be controlled [14]. Immobilisation of other active substances on PEC has been also studied. Complexation of flavin-containing polycations to poly(methacrylic acid) (PMAA) influences favourably the catalytic activity of the immobilised flavin units [15]. Drug delivery could be another use of such systems [16, 17].

PEC have been used for the preparation of membranes [18, 19] of high water flux and good selectivity. Combinations of anionic polyelectrolytes like PMAA or polystyrene sulfonic acid (PSSA) with cationic ones like poly(vinyl benzyl trimethylammoniun chloride) have been used. Moreover, the interfacial reaction between aqueous solutions of an ionic and a cationic polyelectrolyte has been proposed for the microencapsulation of biologically active substances [20]. PEC formed between sodium cellulose sulfate and PDADMAC have been used for encapsulation of artificial tissues [21]. A membrane consisted of poly(maleic acid-co-styrene)–polyethylene glycol (PEG) IPC has been used for the selective permeation of carbon dioxide and oxygen [22]. Composite membranes containing a separating layer of PEC consisting of polyacrylic acid (PAA) and a polycation were developed for the separation of water-ethanol by pervaporation. Among the polycations, ionenes with quaternary ammonium groups in the backbone chain were effective in giving membranes of higher permselectivities [23].

PEC formed between a pipcridinium cationic polymer and various polyanions, such as polysodium acrylate (PNaA), polysodium styrene sulphonate, dextran sulfate and carboxymethylcellulose (CMC), in aqueous solution have been used for binding of azo dye anions (methyl orange and its homologues) [24].

An automatic colourimetric titration method has been described for the determination of polyelectrolytes in aqueous solution, based on the formation of PEC between cationic and anionic polymers. Examples are given for polyethylene imine (PEI), polyvinyl sulfate and polyacrylate [25]. The number of basic groups of human carboxyhaemoglobin was evaluated by colloid titration with potassium polyvinyl alcohol sulfate [26].

A method, for imparting controllable hydrophilicity to the surface of hydrophobic polymers, including polyethylene, polypropylene, saturated polyesters and silicon polymers, using synthetic polyelectrolytes and their complexes, has been described. It is based on chemical modification of the surface by radiochemical grafting of PAA, PMAA or polyvinylpyridine, followed by formation of PEC. Surfaces with controllable moisture absorption, controllable charge and enhanced blood compatibility were obtained [27]. The blood compatibility of PEC formed between cellulose derivatives in aqueous solution has been also examined by both *in vivo* and *in vitro* tests [28, 29]. The reaction between oppositely charged polyelectrolytes, leading to the formation of a PEC, was used to modify the surface of organic and inorganic polymers. In this manner, it was possible to prepare a strongly cationic surface charge on materials like cellulose, polyester, diatomaceous earth and clay. The reaction was successfully used to modify filter sheets, free of asbestos, usually used in the food and pharmaceutical industries. Filters with a high adsorption capacity for bacteria and pyrogens were obtained in an aqueous medium [30].

Interactions between branched PEI and PNaA with kaolin have been investigated for structuring phenomena in pigment systems [31].

Complexation of polyaniline (PAn) with PSSA has been used to render PAn soluble in water. It occurs as a result of the formation of a NPEC, with a structure resembling to that of a block copolymer. The complexation was executed in N-methyl-2-pyrrolidone

181

in the presence of lithium chloride. The precipitate, obtained with addition of acetone, were soluble in water [32].

Stable dispersions of PEC nanoparticles can be used to remove dissolved organic molecules from aqueous solution *via* hydrophobic or electrostatic interactions. The sorption capability of such macromolecular assemblies increases with increasing molar mass and hydrophobicity of the macromolecules used and new materials for removing organic pollutants from wastewater can be designed [33].

6.3 Applications of Polysaccharide-Based Systems

Polysaccharides are cyclolinear or branched polymers, with structural units composed of one or many various sugars or sugar derivatives. Most of them are obtained by extraction from renewable resources like plants, (i.e., starch and gums from plants seeds, pectin from fruits and algin and carrageenan from algae [34]), and the animal kingdom, (i.e., chitin from shrimp, crab or lobster). In plants, polysaccharides are mainly found as polysaccharide mixtures while in animal sources they are generally in a mixture with proteins. Other polysaccharides are manufactured by microbial synthesis (dextran, curdlan, pullulan, xanthan) or by chemical modification of natural products (cellulose derivatives).

Polysaccharides show large differences in solubility and in their solution or gel properties due to a great variation in their chemical (primary) structure that determines the conformation of the macromolecules adopted both in aqueous systems and in the solid state.

6.3.1 Gelation Behaviour

6.3.1.1 General Aspects of Gelation

A particular feature of the polysaccharides is their ability to form gels in certain conditions (concentration, temperature, ionic strength). The gelation phenomenon of polysaccharides is of a major interest for their food and pharmacological applications.

A gel structure occurs when the permanent chain-chain interactions like hydrogen bonding, dipole-dipole and ionic interactions as well as the interactions with the solvent are strong enough to obtain an ordered polymer structure. A single polysaccharide chain could participate in several ordered regions (junction zones), to form a three-dimensional network, or gel structure.

Physical gels can be obtained by:

- Reduction of the water activity. This causes an increase in interchain binding compared to the chain-solvent association. Reduction of the water activity could be achieved by addition of a low molecular weight, hydrophilic molecule, binding water in competition with the polymer.

- Lowering the pH of the acid polysaccharides in order to suppress ionisation.

- Freeze-thaw cycles. By freezing the polysaccharide solution, the ice formation progressively raises the effective polymer concentration promoting the association [35]. Many interchain junctions formed in this way redissolve on thawing; where the barrier to spontaneous association in solution is kinetic rather than thermodynamic, the junctions may persist.

When one or both polymers of a mixture can gel, the phase separation phenomenon is of a particular interest. If only one of the polymers can gel, the gelation will occur when the concentration of this polymer in the continuous phase is above the minimum gelling concentration [36], **Figure 6.1**. Segregative phase separation will prevent gelation of the mixture if the non-gelling polymer forms the continuous phase and the concentration of

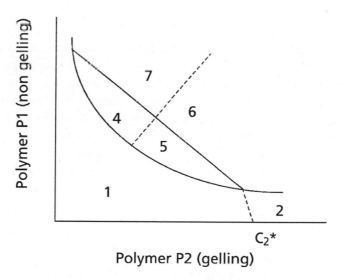

Figure 6.1 Phase diagram for a mixture of a gelling (P2) and a non-gelling polymer (P1). Gelation occurs in the zones 2 and 6. Zone 3 would be the image of zone 2 for a mixture of two gelling polymers. C_2^* is the minimum gelling concentration

the gelling agent in this phase is lower than the minimum gelling concentration, C_2^*. Gelation of the overall mixture is therefore possible only in areas 6 and 2 (**Figure 6.1**) [37].

Charged polymeric cosolutes inhibit gelation of charged gelling agents and uncharged cosolutes inhibit the gelation of uncharged gelling agents [36].

When both polymers form thermoreversible gels, clear mixtures could be obtained beyond the gelation temperature in a large range of composition, but at low temperatures, macroscopic phase separation is prevented by gelation, i.e., mixtures of lime hide gelatin and iota-carrageenan [38].

The gels obtained from mixtures of two different biopolymers often have a biphasic structure induced by segregative interactions (thermodynamic incompatibility) between the constituent polymers [39]. The ultrastructure of the resulting co-gels can vary widely depending on the time-temperature course of gelation in relation to the rate of the segregation and network formation [40].

6.3.1.2 Co-Gelling Behaviour of Some Water-Soluble Polysaccharides

Galactomannans exhibit different interactive properties with other polysaccharides, the content of D-galactosyl units playing a major role in the co-gelling behaviour. The most effective galactomannans in the co-gelling interactions with agaran and κ-carrageenan are those in which the mannan backbone is less substituted. It could be presumed that the regions of the galactomannan chain, which are sparsely substituted or unsubstituted by D-galactosyl units, are primarily involved in the crosslinking [41]. For example agarose and galactomannans gel together. Although the gelation of the two polysaccharides occurs in the same time on cooling, on reheating, the order-disorder transition of the galactomannan takes place first. Considering the mixed agaran-galactomannan gelling systems it was shown that locust bean gum (LBG) is more interactive than galactomannan [35]. Agar in mixture with LBG and gelatin acts as emulsifier and stabiliser and gives sherbets and ices with superior texture, syneresis (gradual phase separation), and flavour stability. The optimum stabiliser concentrations are agar 0.12%, LBG, 0.07% and gelatin, 0.2%.

Co-gelling behaviour of carrageenan with galactomannans depends on sulfate ester content of carrageenan. κ-Carrageennan, which has one sulfate half-ester per disaccharide unit exhibits co-gels with galactomannans. κ-Carrageenan and LBG are known to form thermoreversible and synergistic gels when mixed together [42-44]. Mechanical properties of these materials often change during food processing and storage [45, 46]. ι-Carrageenan which has two half-esters per disaccharide unit does not present such interactions. Furcellan (a polysaccharide similar to carrageenan) that contains one sulfate half-ester group for every two disaccharide repeating units shows a greater ability to interact with galactomanans than κ-carrageenan does. Substituents occurring on the outside surfaces of agaran and κ-carrageenan decrease the binding of galactomannans.

Galactomannans and xanthan co-gel, too. Mixtures of xanthan and LBG form firm, true, thermoreversible gels at low total polysaccharide concentration [47] by heating and cooling their mixed aqueous solutions in 0.1 M NaCl [48]. The gels are thermally stable, do not flow and do not recover from mechanical damage. These gels exhibit sharp melting and gel setting over a narrow temperature range. Gel setting increases by increasing the total polysaccharide concentration and shows little dependence on the relative levels of concentrations of the two polysaccharides. The synergistic interaction between xanthan chains and unsubstituted regions of galactomannan chains leads to aggregated subunits that may stick together at rest or under shearing, forming shear sensitive 'superaggregates' and inducing thixotropic properties even at very low concentrations.

At high concentration, on cooling, the xanthan/LBG mixtures form strong thermoreversible gels [49, 50]. In a semi-dilute solution, xanthan chains display parallel packaging association, particularly at high ionic strengths [51, 52]. The morphology of xanthan/LBG mixture consists of superstrands of xanthan helices [53], with a high tendency to form bundles in the presence of LBG. At high cooling rate, xanthan chains are rapidly trapped by the formation of stable xanthan/LBG junction zones while at low cooling rates, it forms unstable xanthan-xanthan bundles, that appear like a light (less dense) network which could then slowly be reorganised. A xanthan gum/LBG blend heated at 140 °C forms a true gel. Structural features of the gels were independent of the thermal treatment and of the LBG fraction. A xanthan/LBG network was formed from xanthan supramolecular strands. Addition of LBG did not influence the xanthan supramolecular structure [53]. The rheological properties of the gel depend on the mannose:galactose ratio. Decreasing the mannose:galactose ratio from 3.35 (LBG) to 3.0 (tara gum-galactomannan), the temperature of gelation is not significantly affected, but the gel strength decreases.

Xanthan interacts with konjac mannan (KM) more strongly than with galactomannans forming mixed gels [47]. They form gels at a total polysaccharide concentration of 0.02%, the lowest gelling concentration observed for a carbohydrate system. A gel of 0.2% xanthan with 0.25% KM melts at around 63 °C while a gel of xanthan/galactomannan melts at 41 °C. By deacetylating xanthan, its gelation behaviour with LBG is improved, while gel melting and setting temperatures of its mixture with KM are reduced by 20 °C.

Carrageenan and gellan, gel in the presence of high levels of hydroxypropylmethylcellylose (HPMC) forming continuous networks, whereas the gelation of agarose is inhibited.

Agarose and carrageenan, gel in the presence of a high level of arabic gum (14%) at the highest level inclusion. The gelation of gellan is inhibited at low levels of arabic gum. Higher molecular weight of gellan compared with carrageenan could be an explanation for the inhibition of its gelation [36, 37].

Gelation properties of starch are enhanced by the presence of another polysaccharide with higher hot paste viscosities [54]. Addition of a hydrocolloid (LBG, guar gum, xanthan) strongly influences the gelatinisation and retrogradation of starch and can lead to a considerable viscosity increase [55]. Alloncle [56] described the starch/hydrocolloid system as a suspension of swollen starch particles dispersed in a solution of hydrocolloid. Phase volumes in starch-biopolymer composites are dictated solely by the swelling behaviour of the starch component with no evidence of any significant competition from the 'solvent avidity' of the biopolymer [57]. The degree of starch gelatinisation affects the optical clarity of the films and this could be important for some applications [58].

Studies of the gelatinisation behaviour of the corn starch in the presence of xanthan, guar, LBG and sodium alginate showed that the effect was more pronounced with ionic polysaccharides [54]. The presence of another polysaccharide in the starch-based blends has little effect on the gelatinisation temperature or the gelatinisation enthalpy of starch [59-61]. Ferrero and co-workers [61] attributed the increases in the midpoint temperature of the gelatinisation process, T_{gel}, and the final temperature of the gelation process, T_c, to the reduced availability of water required for the disruption of the crystalline regions within the granule. For the starch/KM system [60] and sweet potato starch in presence of various cellulose derivatives [62] T_{gel} and T_c increase by increasing the total polysaccharide concentration.

Sudhakar and co-workers [63] studied the influence of the inorganic salts on the gelatinisation or 'pasting' behaviour of starch/xanthan systems and the pasting characteristics of corn starch and waxy maize with guar gum and LBG. The increase in the storage moduli of corn starch in the presence of guar, LBG or xanthan was attributed to phase separation due to incompatibility of starch and added polysaccharides [56, 64].

The morphology and compatibility of different starch-based blends of theoretical and industrial interest were studied. For example, Yoshimura and co-workers [65] prepared corn starch/KM blends by mixing their powders and then dispersing the mixture in distilled water heated at 95 °C for 15 minutes and held at 95-98 °C for 30 minutes. They established that corn starch is thermodynamically incompatible with KM. The microstructure of the mixtures is, corn starch continuous phase under most conditions. Corn starch forms a weak gel even if the KM phase volume appears to dominate the microstructure. KM behaves as a concentrated polymer solution which is trapped by a weak gel solution of corn starch [65].

Alginate/starch gels have also been prepared. An effective diffusivity at the centre of the gel of 4.6 x 10^{-10} m^2/s was reported. Similar values of 7.8 x 10^{-10} m^2/s and 3.3 x 10^{-10} m^2/s were reported for diffusion of glucose (25 °C) into carrageenan (2%) [66] and sugar (61%) in an agar gel/milk bilayer system [67]. Mixtures of alginate and high-methoxyl pectin give firm gels at low pH and low solids (20% sucrose or less). Gelation is not observed above a pH of 3.8. Alginate/pectin gels are thermoreversible; the melting point increases as the pH of gel formation is reduced [68]. This kind of low pH gelling system can be used for low calorie or diet jams and jellies, salad dressings and low pH fruit-filled gels. A similar system using alginate/whey protein complex and fish protein can be utilised to produce fish analogs from fish mince.

Proteins can co-gel as well with polysaccharides. For example, films of starch/gelatin were prepared by mixing their aqueous solutions, at 80 °C, at a total polymer concentration of c_{tot} = 5.0 %wt and then casting them from the gel state at 20 °C. The films have a crystalline structure and a greater thermodynamic stability [69]. The gelation temperature and gel strength of the gelatin/starch gel were dominated by the gelatin component, with no indication of network formation by the starch.

Because of the interactions between casein micelles and carrageenan (iota), carrageenan is used in the food industry to make milk gels and to stabilise milk fat emulsions, ice creams and chocolate milk. These kinds of gel have been intensely studied. Dalgleish and Morris [70] established that λ, κ and ι-carrageenans may be absorbed on casein micelles and their charge density influences their absorption. Snoeren [71] reported that κ-carrageenan interacts with κ-casein, mostly at the periphery of the micelles, but not with αS1-β-caseins. Langendorff [72] demonstrated by electophoretic mobility determinations that ι-carrageenan chains are absorbed progressively on casein micelles at 25 °C while at 60 °C only λ-carrageenan is absorbed. Coil-helix transition of ι-carrageenan is essential to its electrostatic absorption on casein micelles, this transition resulting in an increased charge density of the polymer chain [71, 72]. When ι-carrageenan is added to milk, two modes of structure could occur: a network of ι-carrageenan chains reinforced by the presence of casein micelles or another network, much weaker, involving ι-carrageenan chains in helical conformation adsorbed on casein micelles.

6.3.2 Edible Films and Packaging Materials

6.3.2.1 General Considerations

One of the main directions of the applications of the water soluble polysaccharide-based blends is in the food packaging industry as edible films (see **Table 6.1**). Food packaging

Table 6.1 Blends used for edible films and food packaging

Blend	Mixing process	Characteristics	Observations	Ref.
Soluble gelatin/starch or hydroxypropyl starch	Casting of mixed solution of 5% aqueous gelatin and 2% aqueous soluble potato starch solutions at 60 °C or 20 °C.	Films prepared at low temperature have higher tensile strength (σ) and percentage elongation at break (ε) than those prepared at high temperature.	Plasticiser: glycerol or sugars. Conditioning the films at relative humidity higher than 60% increases their ε and causes a drop in σ of about 20%.	[89]
Sodium caseinate/starch	Casting of mixed solution followed by evaporation at high temperature.	σ = 7.5-8.0 GPa [76]; Increasing the amount of sodium caseinate in the blend, the WVTR decreases from 7.3 to 5.4 x 10^{-11} g/m s Pa [92].	Plasticiser: polyols.	[88, 99]
100 starch (corn, wheat, potato)/ (1-7) cellulose (CMC, EC, MC, acetyl fibre)/(6-18) water-soluble resin (polyethylene glycol, PVAL, alkyd resin).		Biodegradable.	Blend contains: plasticiser, water (50-70 parts), hygroscopic agent (2-6 parts), lubricant (0.5-2 parts), foaming agent (0.3-2 parts), foaming aid (0.3-2 parts). Materials may contain additives as: antioxidant, fungicide, brightener, and coupling agent.	[100]

Table 6.1 continued...

Blend	Mixing process	Characteristics	Observations	Ref.
(56-86) Pectin/(44-14) starch	*Laboratory scale:* casting from 5-8 wt%, relatively aqueous solutions, dry at room temperature. *Industrial:* extrusion on a twin screw extruder mass flow rate for blends: 27-102 g/min; water: solid feed rates = 0.43-2.85 cm³/g. Six different extruder temperature profiles were used. The screw speed: 350 rpm for profiles I-III and 350 and 450 rpm for profile IV.	Wide range of good mechanical properties [101-103]. $T_g \sim -50\ °C$, flexible films at room temperature. Storage modulus = 103 MPa at room temperature. Expected elongation at break $\sim 10\%$-20%. Moisture content 30%-75%. Excellent oxygen barrier properties [104].	Plasticiser: glycerol. Films are used for: water soluble pouches for detergents and insecticides, flushable liners and bags and medical delivery systems and devices. Edible bags for soup and noodle ingredients.	[58]
PVAL/CS	Solution mixing under vigorous stirring and heating; plasticiser is added into the solution for 10-15 minutes until it is dispersed.	Up to 20% plasticiser in the blends does not show a phase separation. σ of unplasticised b ends is ~0.55 GPa for plasticised blends with 5% sucrose and 5% water, σ ~305 MPa.	Pressure (20-121 MPa) does not affect the diffusion coefficients. Chitosan does not affect the crystallinity of PVAL [105].	[79]

material have to provide optimum properties (appropriate gas and water barrier properties in order to prevent the destruction of food by microbial or insect attack) so that the supplied product remains in a satisfactory condition for its anticipated shelf life. Consumer demand for high-quality, long shelf life, ready-to-eat foods has initiated the development of mildly preserved products that keep their natural and fresh appearance as long as possible [73, 74]. Edible and biodegradable polymer films for food packaging offer alternative packaging options, advantageous to the synthetic 'recalcitrant' packaging polymers, being environmentally non-polluting [75, 76], lowering in the same time the moisture, aroma and lipid migration among food components [77] and improving the textural properties and coherent structure [78].

The value of polysaccharides as edible films is well recognised [79, 80] but apart from some special applications [81, 82], polysaccharide- and/or protein-based edible films have not been used extensively in the food industry, yet [78]. The main drawbacks of edible films are related to their poor moisture barrier performance (apart from wax), colour, appearance and their mechanical and rheological properties [83-85]. The main focus is to improve their water vapor transmission (WVT) barrier properties [79].

Although most edible films are not effective as WVT barriers, they are quite efficient as gas barriers. Due to their potential application as films, the theoretical or empirical calculation of gas permeation was the subject of several investigations [86, 87]. A methodology to evaluate the gas permeability for amorphous synthetic polymers was proposed by Salame [87].

Gas diffusivity of the films containing hydrogen-bonding groups increases with increasing the water content [88]. Changes in the polarity of the hydrated matrix as a result of water absorption may further complicate the solubility and diffusivity of gases and thereby affect the permeability values. For example for polyol-plasticised starch/methylcellulose blends an increase in gas diffusion rate of at least 3-4 orders of magnitude was noticed compared to non-plasticised blends [89].

Natural polymers are characterised by extensive water clustering and plasticising of their matrix properties, which is regarded as an important shortcoming for packaging applications.

The biodegradable polymers have proved to be inferior to non-biodegradable packaging materials in terms of both functional and cost effectiveness due to their hydrophilic nature and their inadequate mechanical strength. Polymers such as polyhydroxy butyrate (PHB), modified starches, chitosan (CS) and pullulan have food packaging applications or are used as blend components. PHB, polyvinyl alcohol (PVAL) and several modified starches, i.e., hydroxy propylate, ethylated, seem to be good candidates as they are already being produced on an industrial scale by several European and Japanese companies (Boehringer, The Netherlands and Nippon Starch, Japan) [79].

Due to their processing difficulties, the edible films and coatings based on polysaccharide (or protein) blends are still at an experimental level. The most extensively and successfully polymers used in this field are starch and gelatin [79].

6.3.2.2 Starch-Based Edible Polymer Blends

Polysaccharide blends based on gelatinised (corn starch) or soluble starch and methylcellulose (MC) or microcrystalline cellulose were produced either by extrusion and subsequent hot pressing or by casting [88]. Polyols are generally used as plastifiers for polysaccharide or polysaccharide based-blends. Their plasticising ability for starch/ MC blends is: glycerol < sorbitol < xylose < sucrose. The presence of about 30% total plasticiser content in the network of starch/MC blends increases considerably the mobility of the gases throughout the rigid matrix. When a plasticiser, either water or polyol is incorporated in a blend, the percentage elongation, water vapour transmission rate (WVTR) and gas permeability increased considerably, whereas the thermal and mechanical properties showed a substantial decrease due to plasticising of the matrix [79].

The presence of a single glass transition (T_g) for these blends could be due to both the close proximity of the T_g of the individual polymeric components and their similar behaviour in the presence of plasticisers [89].

As a processing method, extrusion is often preferred to casting because the throughput of the process is faster and it requires less energy for the removal of water. When casting is used, starch has to be pre-gelatinised whereas in the extrusion process starch granules are gelatinised *in situ* during the process. All of these factors reduce the production costs of manufactured films, i.e., pectin-starch-glycerol, by extrusion rather than by casting [90] (see **Table 6.1**).

6.3.2.3 Chitosan-Containing Blends

CS/PVAL films plasticised and non-plasticised with sorbitol, sucrose or water [79] were prepared and studied for food packaging applications. For these blends, a low sorption of CO_2 was found. The low levels of sorbed CO_2 could be assigned to the high level of hydrogen bonding in both plasticised and non-plasticised blends. Carbon dioxide diffusivity was studied because of its strong effective bacteriostatic action toward preventing microbial growth. The low CO_2 transport for CS/PVAL blends could be interesting for the food packaging companies, i.e., in the bakery industry.

6.3.2.4 Polysaccharide-Protein Blends

Proteins have good film forming properties and may be used in coating formulations for fruit and vegetables as they are effective as gas barriers for O_2 and CO_2, but their WVTR is high [74, 91]. For example, casein and casein derivatives with different molecular weights (19,000-23,900) have been extensively used in the food industry (dairy, meat and confectionery) as well as in medical and pharmaceutical applications [92, 93]. Gelatin coatings with or without polyols, i.e., glycerol, carrying antioxidants were effective in reducing rancidness when they were applied by spraying or dipping to cut up turkey meat or to smoke-cured chicken [94].

A relatively low number of protein-polysaccharide systems were investigated for their potential use as edible film applications, i.e., gelatin/starch blends. In gelatin/starch blends as well as in the plasticised starches the efficiency of polyols or glycerol as plasticisers depends on the amylose/amylopectin ratio [95]. Incorporation of polyols in conjunction with the presence of water in protein or protein/starch blends results in a broadening of the T_g transition step. When the polyol content exceeded a threshold of 22%, double peaks in the glass transition region were recorded [96]. This behaviour indicates the incompatibility of the components of the blend.

For the charged polysaccharides, the compatibility with proteins decreases by increasing the salt concentration [97, 98].

6.3.3 Pharmaceutical and Medical Applications

6.3.3.1 Cellulose Derivative-Based Blends

Cellulose derivative based blends are used for the formulation of delay or sustained drug release. Cellulose ether-ester films [106] are flexible and transparent, exhibit moderate strength, resistance to oil and fat migration. They act as moderate barriers to moisture and oxygen [78, 107, 108].

Sakellariou and Rowe [109, 110] studied the compatibility of HPMC and PVAL without or with plasticiser (glycerol) because these polymers are extensively used as film coatings of solid pharmaceutical formulations. They have found that the polymers are totally incompatible due to the significant discrepancies in the polar and hydrogen bonding characteristics [111]. The plasticisers are often added to depress the T_g and the modulus of the polymers by eliminating the internal stresses that are responsible for coating cracking and bridging of the intagliations (engravings) [112].

Film coatings for pharmaceutical purpose made of ethylcellulose (EC) with HPMC [113, 114] and of HPMC with sodium CMC [115] blends were prepared and studied. The performance of the final product depends strongly on film thickness, rate of wetting, solvent system, temperature profile and air flow conditions of the coating process. The interactions and the morphology in the blend are of major importance for the drug release profile [116].

In EC/HPMC film coatings, the drugs are released through pores created in the film coating by the dissolution of the dispersed phase (HPMC). Sakellariou [117] reported on the retention/leaching of water soluble (HPMC, HPC and PEG 6000) and pH-dependent-soluble cellulose derivatives from the blends with 20% and 80%w/w ethylcellulose content, used in delayed or sustained release film. A small amount of HPMC is retained in the blends with high (80%w/w) EC content [116].

Influence of the polymer viscosity on the drug release from the hydrophilic erodable matrix of HPMC and sodium CMC was investigated by Bonferoni and co-workers [115]. They found that when the erosion occurred there were obtained different release profiles depending on polymer viscosity.

Hydrogels suitable for drug release were obtained from blends based on CMC with agar–agar. These blends form a strong structural network of the initial polysaccharides and at temperatures lower than 36 °C appears to be a non-thixotropic component [118].

Administration of the drug through buccal and sublingual sites [119] are advantageous compared with other routes [120] due to the rapid onset of action, high level of the drug in blood, avoidance of the first pass effect and the exposure of the drug to the gastrointestinal tract. Therefore efforts are being focused on researching these kinds of tablets. In Japan are now commercially available tablets named Aftach [121, 122] for the treatment of aphthous ulcers. They have HPC and carboxyvinyl polymer as the major excipients and have appropriate oral mucosal adhesion properties and controlled drug release features.

6.3.3.2 Chitosan-Based Blends

The CS complexes are well known through their application for the immobilisation of enzymes, the micro-encapsulation of cells and for controlled drug release [123, 124]. For example albumin was microencapsulated within a CS-alginate complex membrane in the presence of calcium chloride [125]. Collagen (C)/CS composites, membranes and hydrogels were studied for their potential applications in biomedical and pharmaceutical field [126, 127].

Many formulations for coatings with applications in colonic drug delivery, with a bimodal drug release profile [128] were proposed based on pectin/CS blends [129, 130]. This could be explained by the great degree of protection against premature drug release in the gastrointestinal tract of pectin/CS compression-coated tablets when compared to tablets using pectin alone.

For example Macleod and co-workers [131] prepared a coating from a mixture of pectin/CS/HPMC/0.1 M HCl with glycerol as plasticiser. They found that a ratio of pectin/CS (3:1) is favourable in terms of swelling and permeability due to the maximum interaction between the NH_3^+ groups from chitosan and the COO^- groups from pectin. The permeability of the coating for a model drug is higher in the presence of pectinolytic enzyme than in buffer alone [132]. A tablet coated with a film containing pectin/CS/HPMC in a ratio 3:1:1 and subsequently coated with a methacrylic acid copolymer would deliver the majority of the drug dose into the colon; the release of the drug will start at a low rate in conditions that simulated the small intestine without pectinolytic enzyme and accelerated in the conditions simulated the colon (that implies the presence of pectinolytic enzyme) [131].

Fernández-Hervás and Fell [129] have prepared a coating for tablets from pectin/CS by compressing around the core mixture. They used normal concave punches at a core/coat ratio of 1:5. The chosen weight ratio of pectin/CS was 10/1 because at this ratio the maximum interaction between pectin and CS occurs. The coat thickness depends on the type of the drug to be released, i.e., 1.59 mm for indomethacin and 1.48 mm for paracetamol. They concluded that the coat is successful to protect and then release relatively insoluble compounds. Soluble materials showed premature release in conditions analogous to those in the upper gastrointestinal tract. These coatings are capable of retarding the release of the tablet core materials until they reach the colon, an environment rich in bacterial enzymes which degrade the coating and allow the release of drug [130].

Polymeric matrices for controlled drug release systems were prepared based on CS and different polyacids (PAA and PSSA). They were obtained by radical polymerisation of acrylic acid monomer [132] or sodium styrenesulfonate [133] in the presence of CS or CS hydrochloride, respectively. CS/PAA systems have a crystalline structure, the polycomplexes being quite ordered. CS/PSS systems have a 'scrambled' egg structure. CS/PAA have a negative influence on mitocondrial activity [132].

Formulations for membranes or microspheres containing CS/alginate (or sodium alginate) were proposed for protein sustained release. For example a CS/alginate complex membrane made by ionotropic gelation with tripolyphosphate [134, 135] in the presence of $CaCl_2$ was envisaged for controlled release of albumin [125]. Liu and co-workers [136] have prepared microspheres with narrow size distribution (diameters ranging from 20-100 μm) from a solution of sodium alginate in distilled water, sprayed into 0.5% $CaCl_2$ solution with magnetic stirring and transferred to a chitosan hydrochloride solution (0.1-2 w/v%). A porous structure was obtained by lyophilisation. The microspheres could be used for immune activating growth factor interleukin-2 (Il-2) and bovine serum albumin-fluorescein isothiocyanate release, in tumour immunotherapy [136] and they have been approved for oral use like 'generally recognised as safe' materials. The sustained release of proteins from these microspheres is of longer duration than release from another kind of microspheres based on alginate, i.e., alginate/polylysine or alginate/$CaCl_2$ microspheres.

Miyazaki and co-workers [137] prepared from CS/alginate mixtures mucosal adhesive tablets for delivery of diltiazem, a calcium channel blocker useful in cardiovascular diseases such as angina and systemic hypertension. Tablets were made by compression of weighed dry powder of CS (deacetylation degree = 80%), sodium alginate (mixing ratio CS/alginate varied from 1:4 to 4:1) and the drug.

Bioadhesion studies *in vitro* indicate an adhesive force of 80.2-100.6 g/cm^{-2} depending on CS:alginate mixing ratio and alginate average molecular weight. The tablets adhered within a few seconds and remained in place for at least 1 hour. Diltiazem is rapidly released from the bioadhesive tablets, the release rate being modified by changing the mixing ratio of the polysaccharides and the average molecular weight of alginate. Tablets with low alginate content or a high molecular weight alginate have a slower release rate. Tablets with a CS:alginate composition of 4:1 were intact in the dissolution vessel after 6 hours. Diltiazem administered sublingually shows a significant improvement of bioavailability (60.6% for tablets consisting of CS/alginate (1:4) and 64.1% for the tablets with a 1:1 ratio of the polysaccharides) compared to that achieved by oral administration (30.4% bioavailability).

CS/protein blends have been investigated for drug encapsulation too. Remuñán-López and Bodmeier [138] studied the CS/gelatin type B coacervates for encapsulation of some drugs like piroxicam, clofibrate, sulfamethoxazole. The coacervation was carried out by solution mixing, at 40 °C, for 4 hours, at a pH of 5.5 (optimum pH: 5.25-5.5) and a total polymer concentration of ~3.25% w/w. Spherical and isolated coacervate CS:gelatin droplets were obtained at a ratio of 1:20 till the total polymer concentration reaches 3.15% w/w. At higher concentrations they become a gel. The amount of coacervate formed reaches the optimum at a ratio CS:gelatin of 1:10, and is suppressed at a ratio of 1:50.

For the controlled release of propanolol hydrochloride, it was proposed that a C/CS membrane be obtained by a solvent evaporation technique. For this purpose a 0.5% w/v solution of chitosan and collagen in 0.5 M acetic acid were mixed in different ratios (C/CS of 1:3, 1:1 or 3:1), poured on plate and dried at 4 °C. The membranes obtained were neutralised by immersion in 1% aqueous NaOH and dried at 4 °C [139]. By varying the C/CS ratio in the membrane, its swelling potential and permeability to the drug could be regulated. Partition of propanolol hydrochloride into the composite membranes is much greater when compared to that of the drug into either collagen or chitosan membranes and this is attributed to the lower polarity of composite membranes due to charge neutralisation and thereby better interaction with the drug molecules. From this membrane the drug is released following a zero-order kinetic law [139].

6.3.3.3 Applications of Hydrogels

Properties of pure synthetic polymers and those of pure biological polymers alone are often inadequate to produce materials with good chemical, mechanical, thermal and biological performance characteristics. So researchers have tried to prepare blends of synthetic polymers with biological macromolecules to obtain *bioartificial polymeric materials* [140]. They are produced in different forms such as, films, sponges and hydrogels and were evaluated as biomaterials for dialysis membranes [141], wound dressings [142] and drug delivery systems [142, 143].

Biodegradable hydrogels have a wide application in the improvement of existing dosage forms and the development of better drug delivery systems [144]. Hydrogels could be obtained by the repeated freezing and thawing cycles of aqueous polymer solutions [145]. Hydrogels are appropriate for drug delivery systems because of their high water content, high permeability to small molecules and adequate mechanical properties.

Cascone and co-workers [147-149] have prepared hydrogels based on PVAL and different polysaccharides, such as CS, dextran, hyalouronic acid (HyA) and have tested them as controlled release agents of human growth hormone (HGH) [142, 143] or as a drug delivery system [149]. Hydrogels used for controlled release of HGH were obtained by mixing a solution of the partners in different ratios: polysaccharide/PVAL (10-40)/(90-60) followed by 8 cycles of freeze-thawing (from –20 °C at 1 hour to room temperature for 30 minutes); the first cycle of freezing at –20 °C lasts 12 hours. The CS/PVAL hydrogel is characterised by an increase in the T_g value up to a 10% content of biopolymer in the blend, then remains constant. A decrease of the storage modulus value, E′ with the increase in chitosan content was noted. Hydrogels have a macroporous structure with a low degree of order. The degree of crystallinity of PVAL from the blends decreases, especially by increasing the chitosan content of the blend.

The dextran/PVAL hydrogel is more compact and homogenous than the chitosan-containing ones. The storage modulus value, E′, and the crystallinity degree of PVAL increase with increasing the dextran content. In the HyA/PVAL hydrogel an increase in the elastic modulus value and thermal stability was seen for a certain HyA/PVAL ratio [142, 143].

The controlled release of HGH depends on the content of the biological component. By increasing the content of the biological component the HGH release is increased. In this way the rate of release could be controlled [142, 143]. The amount and method of HGH release depend, too on the polysaccharide type from the hydrogel. For the CS/PVAL hydrogel with a CS content up to 30%, the total amount of the released HGH decreases, and takes place in two steps [149]. However the total amount of HGH released is in a physiological range (2.125–100 ng/ml), enough to have an osteoblast proliferation [142]. In the dextran/PVAL blend up to 30% content in dextran, the total amount of the released human growth hormone increases.

6.3.4 Other Applications

6.3.4.1 Protein/Polysaccharide Systems

Protein-polysaccharide conjugates prepared by naturally occurring reaction without any chemicals are useful as new functional biopolymers. They have an improved solubility and excellent emulsifying, antioxidant and antimicrobial effects for food applications [150-152]. For instance, insoluble wheat gluten was solubilised and its functional properties enhanced by Pronase treatment followed by dextran conjugation [150]. For casein, the conjugation of protein-polysaccharide was quickly done within 24 hours. The emulsifying properties of this conjugate were found to be much better in acidic and high-salt content [153] than those of commercial emulsifiers. Nakamura and co-workers [151] have reported that the lysozyme-dextran conjugate has bifunctional properties and excellent emulsifying properties with antimicrobial effects against both Gram-positive and Gram-negative bacteria. Hybrid proteins with polysaccharides such as dextran or galactomannan commonly revealed excellent functional properties [154]. Conjugation of a protein with a polysaccharide is desirable for industrial applications, also, in order to improve the surface properties of the proteins.

Blends of lysozyme/galactomannan (or xyloglucan) in a mixing ratio of 1/4 (or 1/8) are dissolved in distilled water, lyophilised and then incubated at 60 °C for 2 weeks at a relative humidity of 78.9%. The protein-polysaccharide conjugates were separated by gel filtration on columns of Sephadex G-50 and CM-Toyopearl 650 M [154]. They have good emulsifying properties and are heat stable forms.

Alginate/protein blends can be used as emulsifying and foaming agents too [68].

6.3.4.2 Starch-Based Blends

Thermoplastic starch (TPS) was blended with biodegradable polymers like polylactic acid [155] or PVAL and plastifying agents (glycerol and its derivatives, sorbitol and its derivatives, etc) in order to obtain *bioplastics*, i.e., biodegradable materials with high water vapour permeability, a good oxygen barrier and not electrostatically chargeable. Bioplastics can be processed by the existing plastic processing industrial production lines. During the production of polymer blends a transesterification reaction takes place *in situ* at different temperature and shearing conditions [156-159]. Bioplastics reduce the waste amounts in an ecological manner by closing the natural material cycles [160]. Their main applications are as compostable packaging (compostable in accordance with DIN V 54900 [161]) or as short-term consumable articles and special products. Biodegradable plastics known on their commercial name as Mater-bi AF10H, prepared from a blend of starch (60%) and natural additives with (40%) modified PVAL and plasticisers are used for packaging and agricultural mulch films [162, 163]. Mater-bi AF10H presents a low biodegradability compared with other biodegradable plastics [164]. By melt-mixing and extrusion of 70% starch/PVAL polymer alloy (Mater-bi) with 30% talc, a paper exhibiting degradation after 8 months was obtained [165].

An eraser for rubbing off letters without damaging paper was obtained by mixing and injection moulding of potato starch (100 parts), PVAL (20 parts), glycerin (50 parts) and water (50 parts) [166].

Starch/PVAL blends are used as adhesive corrugated paperboard [167, 168]. Photo or biodegradable films (with an optimal starch content of 50%) were produced by doctor blade casting process [169].

Arvanitoyannis and co-workers prepared a biodegradable material from starch/MC (78-35)/(17-30) with 5-35 water and/or polyols as plasticisers, by extrusion and subsequent hot pressing or by casting [88]. Gas permeability of CO_2 varies between 7.0×10^{-10} and 6.9×10^{-15} cm²/s Pa [89], depending on the mixing ratio of the polysaccharide and increases proportionally to the plasticiser content. The biodegradability of these systems in the water-soil environment increases with the concentration of starch in the blend.

Starch/alginate blends have applications in the most different fields. For example starch/alginate can be used to produce crunchy products and cellular foods [170] while starch/sodium alginate blend is a thickening agent in printing ink components for local bleaching in grey linen fabric printing [171]. Cellular foods are manufactured by freeze-dehydration of gels [172, 173].

By moulding at 180-190 °C a formulation of waste paper fibre/starch (8-26%)/hydrocolloids (5%-15%) (as modified starch, xanthan or locust bean flour) with water and blowing agent (≤ 3% azodicarbonamide) was obtained giving biodegradable foam materials with dimensional stability for packaging [174]. Thermoplastic foams expanded in a low ratio were produced from protein/starch by extrusion or injection moulding. To improve mechanical properties of the protein/starch based thermoplastic metallic salt hydrate is added. Natural cellulosic fibres such as grass fibres, wood fibre, chopped straw, bagasse, etc., can be used as reinforcing fillers.

6.3.4.3 Systems Used in Purification and Separation Operations

Chitosan is an attractive material for hydrophilic membranes [175, 176] due to the presence of both amino and hydroxy groups in its structure. CS membranes crosslinked with glutaraldehyde [177, 178] and sulfuric acid [179] were applied successfully in the dehydration of alcohol-water mixtures. The membranes of PVAL/CS blends have good performance for separation of ethanol-water mixtures [180]. Addition of hydroxyethylcellulose (HEC) to the composition of the membranes affects their sorption properties; the content of HEC that gives the highest sorption selectivity depends on the concentration (or activity) and the type of mixture. CS/HEC membranes show a higher sorption selectivity than the pure CS membranes, but a membrane with a higher HEC content is not appropriate for separation of the higher water content mixture [181].

CS-based blends were used to prepare membranes with applications in pervaporation of ethanol-water mixtures or isopropanol-water mixtures. Films from CS and HEC were obtained by casting and drying at room temperature from the mixed solutions of 1.2 wt% CS and HEC made in 10 wt% aqueous acetic acid. The crosslinking of the blend was performed with a mixture of 2.5 wt.% urea - 2.2 wt% formaldehyde - 2.5 wt% sulfuric acid containing 50 wt% ethanol at room temperature for 24 hours. Mechanically robust membranes with a thickness of 30-35 μm were obtained. For the pervaporation of a 90 wt% ethanol-water mixture maximum pervaporation characteristics were exhibited by a membrane that contains 25 wt% HEC while for a 90 wt% isopropanol-water mixture a membrane with a 10 wt% content in HEC is suitable. Total sorption and the water sorption increase with increasing of the HEC content. A membrane with a CS/HEC ratio of 3/1 exhibits the highest total and water sorption, and a decrease in the sorption selectivity, as the weight fraction of water in liquid increases. Each membrane shows a different trend of ethanol sorption. All membranes have the highest sorption selectivity at low water content [181].

A membrane with a high permselectivity for separation of aqueous solution of 50-95% ethanol was prepared from CS/cellulose acetate blends [182].

Huang and co-workers prepared membranes from CS and PAA for separation of ethanol-water mixture, too [183]. The membranes were prepared by mixing, at room temperature of PAA (5%) and CS (0.5%) solutions made into 30 wt% acid acetic solution at a molar ratio of mixing 1:1. The mixed solution was cast onto a polyvinyl chloride plate and dried at room temperature for 24 hours. The thickness of this membrane is about 20-40 μm. By washing this membrane with water for 24 hours another type of membrane (membrane thickness is 30 μm) is obtained with a structure of a polyelectrolyte complex. Heating the PEC membrane for 1 hour, at 180 °C in an oven, produced another type membrane, by the reaction between the amino groups of CS and the carboxyl groups of PAA. The PEC membrane exhibits the higher water selectivity, the highest permeability $(1.4 \times 10^{-3} g/m \ h)$ and separation factor at feed solution of high ethanol content (over 2000 at a 95 wt% ethanol aqueous solution of 30 °C) and exhibits a great potential for the dehydration of organic solvents. The pervaporation performance can be improved by using a high molecular weight polymer [184].

A PEC membrane with excellent pervaporation characteristics, influenced by the after treatment methods, was prepared from a CS/PAA blend too by Xi and co-workers for the separation of water/organic liquid mixtures [185].

CS/PEG membranes prepared by dissolving PEG and CS in glutaraldehyde and then evaporating the solution at about 30 °C are proposed, for use with the pervaporation separation technique, i.e., water-alcohol mixture [186]. The mixing ratio of the two polymers could have great effect on the properties of the membrane [187].

Blend membranes (thickness 30 μm) were prepared by casting and drying on a glass plate, a solution mixture in different ratios of polyvinyl pyrrolidone (PVP) (1.5 wt%) and CS (1.5 wt%) (deacetylation degree = 78%) in 1.0 wt% aqueous acetic acid [188]. It was found by infrared spectroscopy that the two polymers are compatible on a molecular level. Membranes were evaluated for pervaporation separation of methanol and methyl *tert*-butyl ether and it was found that methanol preferentially permeates through all the membranes and the partial flux of methanol increases with increasing PVP content.

Researchers tried to obtain films from CS and hydroxypropylcellulose (HPC) blends [189]. CS and HPC in aqueous acetic acid solution were mixed and crosslinked with glyoxal and glutaraldehyde in the presence of hydrogen chloride. Films were obtained by casting from solution. They were amorphous and the Young's modulus and tensile strength of the crosslinked blend films greatly depended on blend composition. The solubility of crosslinked blend film cast from the glyoxal system is greater than that from glutaraldehyde.

Protein separation and purification is generally realised by means of the so called aqueous two-phase systems (ATPS). ATPS are easily applied as downstream procedures because they do not require expensive specialised equipment, they can be scaled up and operated on a continuous basis [190]. Protein separation is influenced by a number of parameters such as the pH of the system, the type and concentration of salts present in the system, the polymer molecular mass and concentration and protein properties (structure, hydrophobicity, molecular mass) [191].

Most of the laboratory work was done with a system composed of fractionated dextran and PEG [192]. Akashi and co-workers [193, 194] focused on dextran/polyvinylalkylamide (thermosensitive polymer) as ATPS and they found that this system has a higher separation ability for both polymers and biomolecules than that of the PEG/dextran system, usually used. Because fractionated dextran has a high cost, in order to obtain ATPS some other polysaccharides were tested: starch derivatives [195, 196], maltodextrin [197], cellulose derivatives [198], PEI [199] agarose [200], guar gum [201], and LBG [202]. By including a starch derivatives (hydroxypropyl starch crude or purified) in polyethylene oxide, Almeida and co-workers [203] obtained an ATPS system used for cutinase separation and purification.

Concluding Remarks

Water soluble polymer blends or formulations have been used in many applications especially in the food, pharmaceutical and cosmetic industries and for medical uses, due to their environmentally safe features and non-polluting technologies of the blends manufacture.

Although the number of water soluble polymer base systems is enormous, until now only a relative low number of such blends (with applications in food industry as thickeners, gels and pharmacy as controlled or sustained drug delivery matrix or coatings) are commercialised. Nevertheless, the interest in such kinds of materials is continuously increasing.

References

1.   B. Philipp, J. Kötz, K-J. Linow and H. Dautzenberg, *Polymer News*, 1991, **16**, 4, 106.

2.   G.V. Samsonov, *Polymer Science USSR*, 1980, **21**, 4, 787.

3.   V.A. Kabanov, *Die Makromolekulare Chemie - Macromolecular Symposia*, 1991, 48/49, 425.

4.   A.V. Kabanov and V.A. Kabanov, *Polymer Science*, 1994, **36**, 2, 157.

5. V.A. Kabanov, *Polymer Science*, 1994, **36**, 2, 143.

6. A.B. Zezin, V.A. Izumrudov and V.A. Kabanov, *Die Makromolekulare Chemie - Macromolecular Symposia*, 1989, **26**, 249.

7. I.Y. Galaev, *Uspekhi Khimii*, 1995, **64**, 5, 505.

8. C.J. Davison, K.E. Smith, L.E.F. Hutchinson, J.E. O'Mullane, L. Brookman and K. Petrak, *Journal of Bioactive and Biocompatible Polymers*, 1990, 5, 3, 267.

9. V.A. Kabanov, *Die Makromolekulare Chemie - Macromolecular Symposia*, 1990, **33**, 279.

10. H. Dautzenberg, F. Loth, K. Fechner, B. Mehlis and K. Pommerening, *Makromolekulare Chemie*, 1985, **9**, Supplement, 203.

11. P.L. Dubin, J. Gao and K. Mattison, *Separation and Purification Methods*, 1994, **23**, 1.

12. J-Y. Shich and C.E. Glatz in *Macromolecular Complexes in Chemistry and Biology*, Eds., P. Dubin, J. Bock, R. Davis, D.N. Schulz, C. Thies and D. Schulz, Springer-Verlang, Berlin, Germany, 1994, 273.

13. Y. Wang, J.Y. Gao and P.L. Dubin, *Biotechnology Progress*, 1996, **12**, 356.

14. A.L. Margolin, S.F. Sherstiuk, V.A. Izumrudov, A.B. Zezin and V.A. Kabanov, *European Journal of Biochemistry*, 1985, **146**, 625.

15. H.F.M. Schoo and G. Challa, *Polymer*, 1990, **31**, 8, 1559.

16. R.M. Ottenbrite and A.M. Kaplan, *Annals of the New York Academy of Science*, 1985, **446**, 160.

17. I.C. Kwon, Y.H. Bae and S.W. Kim, *Journal of Controlled Release*, 1994, **30**, 2, 155.

18. B. Philipp, H. Dautzenberg, K-J. Linow, J. Kötz and W. Dawydoff, *Progress in Polymer Science,* 1989, **14**, 91.

19. M.J. Lysacht in *Ionic Polymers*, Ed., L. Holliday, Applied Science Publishers, London, UK, 1975, 281.

20. J. Stamberg, H. Dautzenberg, F. Loth, M. Benes and A. Kuhn, inventors; ADW Institut fur Polymerenchemie, assignee; DD 218 372, 1985.

21. M. Sittinger, B. Lukanoff, G.R. Burmaster and H. Dautzenberg, *Biomaterials*, 1996, **17**, 10, 1049.

22. J.H. Dong, C.K. Huang and Y.Y. Jiang, *Polymer Bulletin*, 1988, **20**, 6, 521.

23. H. Karakane, M. Tsuyumoto, Y. Maeda and Z. Honda, *Journal of Applied Polymer Science*, 1991, **42**, 12, 3229.

24. T. Takagishi, H. Kozuka and N. Kuroki, *Journal of Polymer Science: Polymer Chemistry Edition*, 1983, **21**, 2, 447.

25. D. Horn, *Progress in Colloid and Polymer Science*, 1978, **65**, 251.

26. E. Kokufuta, H. Shinizu and I. Nakamura, *Polymer Bulletin*, 1980, **2**, 2, 157.

27. N.A. Plate, E.D. Alieva and A.A. Kalachev, *Polymer Science USSR*, 1981, **23**, 3, 720.

28. H. Ito, T. Shibayata, T. Miyamoto, Y. Noishiki and H. Inagaki, *Journal of Applied Polymer Science*, 1986, **31**, 8, 2491.

29. H. Ito, T. Miyamoto, H. Inagaki, H. Iwata and T. Matsuda, *Journal of Applied Polymer Science*, 1986, **32**, 2, 3413.

30. U. Oertel, G. Petzold, H. Buchhammer, S. Geyer, S. Schwarz, U. Mueller and M. Raetzsch, *Colloids and Surfaces*, 1991, **57**, 3-4, 375.

31. J. Kötz and S. Kosmella, *Journal of Colloid and Interface Science*, 1994, **168**, 505.

32. Y-H. Liao, K. Levon, J. Laakso and J.E. Osterholm, *Macromolecular Rapid Communications*, 1995, **16**, 5, 393.

33. H-M. Buchhammer, G. Petzold and K. Lunkwitz, *Colloid & Polymer Science*, 2000, **278**, 841.

34. C.L. McCormick, J. Bock and D.N. Schulz in *Encyclopedia of Polymer Science and Engineering*, Eds., H.F. Mark, N. Bikales, C.G. Overberger, G. Menges and J.I. Kroschwitz, John Wiley and Sons, New York, NY, USA, 1989, Volume 17, p.730.

35. *Industrial Gums: Polysaccharides and their Derivatives*, 3rd Edition, Eds., R.L. Whistler and J.N BeMiller, Academic Press, Inc., San Diego, CA, USA, 1992, 21.

36. M. Puaud, S.E. Hill and J.R. Mitchell in *Hydrocolloids*, Volume 2, Ed., K. Nishinari, Elsevier Science, Oxford, UK, 2000, 147.

37. D.V. Zasypkin, E.E. Braudo and V.B. Tolstoguzov, *Food Hydrocolloids*, 1997, **11**, 159.

38. C. Michon, G. Cuvelier, B. Launay and A. Parker in *Food Colloids: Proteins, Lipids and Polysaccharides*, Eds., E. Dickinson and B. Bergenstahl, The Royal Society of Chemistry, Cambridge, UK, 1997, 316.

39. V.V. Suchkov, V.Y. Grinberg and V.B. Tolstoguzov, *Carbohydrate Polymers*, **1**, 1981, 39.

40. S. Alevisopoulos, S. Kasapis and R.K. Richardson, *Carbohydrate Research*, 1996, **293**, 79.

41. I.C.M. Dea, A.A. McKinnon and D.A. Rees, *Journal of Molecular Biology*, 1972, **68**, 153.

42. V. Carroll, M.J. Miles and V.J. Morris in *Gums and Stabilisers for the Food Industry 2. Applications of Hydrocolloids*, Eds., G.O. Phillips, D.J. Wedlock and P.A. Williams, Pergamon, Oxford, UK, 1983, 501.

43. P.B. Fernandes, M.P. Goncalves and J.L. Doublier, *Carbohydrate Polymers*, 22, 193, 99-106.

44. P.B. Fernandes, M.P. Goncalves and J.L. Doublier, *Food Hydrocolloids,* 1994, **9**, 1.

45. J. Saguy and M. Karel, *Food Technology*, **10**, 1980, 78.

46. E.B. Bagley in *Physical Properties of Foods-2*, Eds., R. Jowitt, F. Escher, M. Kent, B. McKenna and M. Roquers, Elsevier Applied Science, London, UK, 1987, 345.

47. I.C.M. Dea, E.R. Morris, D.A. Rees, E.J. Welsh, H.A. Barnes and J. Price, *Carbohydrate Research*, 1977, **57**, 249.

48. I.C.M. Dea and A. Morrison, *Advances in Carbohydrate Chemistry and Biochemistry*, 1975, **31**, 241.

49. G. Cuvelier and B. Launay, *Carbohydrate Polymers*, 1988, 8, 271.

50. C. Tonon, G. Cuvelier and B. Launay in *Physical Networks, Polymer and Gels*, Eds., W. Burchard and S.B. Ross-Murphy, Elsevier Applied Science, London, 1990, 335.

51. I.H. Smith, K.C. Symes, C.J. Lawson and E.R. Morris, *International Journal of Biological Macromolecules*, 1981, **3**, 129.

52. G. Cuvelier and B. Launay, *Carbohydrate Polymers*, 1986, 6, 321.

53. L. Lundin and A-M. Hermansson, *Carbohydrate Polymers*, 1995, **26**, 129.

54. P.A. Williams and F.B. Ahmad in *Hydrocolloids, Volume 2*, Ed., K. Nishinari, Elsevier Science, Oxford, UK, 2000, 165.

55. C.B. Closs, B. Conde-Petit, I.D. Roberts, V.B. Tolstoguzov and F. Escher, *Carbohydrate Polymers*, 1999, **39**, 67.

56. M. Alloncle, J. Lefebvre, G. Llamas and J.L. Doublier, *Cereal Chemistry*, 1989, **66**, 90.

57. Z.H. Mohammed, M.W.N. Hember, R.K. Richardson and E.R. Morris, *Carbohydrate Polymers*, 1998, **36**, 27.

58. M.L. Fishman, D.R. Coffin, R.P. Konstance and C.I. Onwulata, *Carbohydrate Polymers*, 2000, **41**, 317.

59. P. Annable, M.G. Fitton, B. Harris, G. O. Phillips and P.A. Williams, *Food Hydrocolloids*, 1994, **8**, 351.

60. M. Yoshimura, T. Takaya and K. Nishinari, *Journal of Agricultural Food Chemistry*, 1996, **44**, 2970.

61. C. Ferrero, M.N. Martino and N.E. Zaritzky, *Journal of Thermal Analysis*, 1996, **47**, 1247.

62. K. Kohyama and K. Nishinari, *Journal of Food Science*, 1992, **57**, 128.

63. V. Sudhakar, R.S. Singhal and P.R. Kulkarni, *Food Chemistry*, 1995, **53**, 405.

64. M. Alloncle and J.L. Doublier, *Food Hydrocolloids*, 1991, **5**, 455.

65. M. Yoshimura, T.J. Foster, A. Norton and K. Nishinari in *Hydrocolloids, Volume 2*, Ed., K. Nishinari, Elsevier Science, Oxford, UK, 2000, 197.

66. M. Hendrickx, C. Vanden Abeele, C. Engels and P. Tobback, *Journal of Food Science*, 1986, **51**, 1544.

67. F. Warin, V. Gekas, A. Voirin and P. Dejmek, *Journal of Food Science*, 1997, **62**, 454.

68. K. Clare in *Industrial Gums. Polysaccharides and their Derivatives*, 3rd Edition, Eds., R.L. Whistler and J.N BeMiller, Academic Press, Inc., San Diego, CA, USA, 1993, 97.

69. N. Ptitchkina, N. Panina and L. Khomutov in *Hydrocolloids, Volume 2*, Ed., K. Nishinari, Elsevier Science, Oxford, UK, 2000, 177.

70. D.G. Dalgleish and E.R. Morris, *Food Hydrocolloids*, 1988, **2**, 311.

71. T.H.M. Snoeren, *Kappa – Carrageenan – A Study on its Physico-Chemical Properties, Sol-gel Transition and Interchains with Milk Properties*, University of Wageningen, 1976. [PhD Thesis]

72. B. Launay, G. Cuvelier, C. Michon and V. Langendorff in *Hydrocolloids, Volume 2*, Ed., K. Nishinari, Elsevier Science, Oxford, UK, 2000, 121.

73. S. Guilbert, N. Gontard and G.M. Gorris, *Lebensmittel-Wissenschaft und Technologie*, 1996, **29**, 10.

74. E.A. Baldwin, M.O. Nisperos-Carriedo and R.A. Baker, *Critical Reviews in Food Science and Nutrition*, 1995, **35**, 509.

75. I.S. Arvanitoyannis, E. Psomiadou and A. Nakayama, *Carbohydrate Polymers*, 1996, **31**, 179.

76. M. Krochta and C. De Mulder-Johnston, *Food Technology*, 1997, **51**, 61.

77. *Edible Coatings and Films to Improve Food Quality*, Eds., M. Krochta, E.A. Baldwin and M.O. Nisperos-Carriedo, Technomic, Lancaster, PA, USA, 1994.

78. J.J. Kester and O. Fennema, *Food Technology*, 1986, **40**, 47.

79. I.S. Arvanitoyannis, *Journal of Macromolecular Science - Reviews in Macromolecular Chemistry and Physics*, 1999, **C39**, 2, 205.

80. S. Guilbert, N. Gontard in *Foods and Packaging Materials, Chemical Interactions*, Eds., P. Ackerman, M. Jagerstad and T. Ohlsson, Royal Society of Chemistry, Cambridge, UK, 1995, 159.

81. J.W. Park, R.F. Testin, H.J. Park, P.J. Vergano and C.L. Weller, *Journal of Food Science*, 1994, **59**, 916.

82. H.J. Park, C.L. Weller, P.J. Vergano and R.F. Testin, *Journal of Food Science*, 1993, **58**, 1361.

83. C. Koelsch, *Trends in Food Science and Technology*, 1994, **5**, 76.

84. J.M. Krochta, A.E. Pavlath and N. Goodman in *Engineering and Food, Volume 2, Preservation Process and Related Techniques*, Eds., W.E. Spiess and H. Schubert, Elsevier Science Publishers, Oxford, UK, 1990, 329.

85. J.M. Krochta in *Advances in Food Engineering*, Eds., R.P. Singh and M.A. Wirakartakusumah, CRC Press, Boca Raton, FL, USA, 1992, 517.

86. D.W. van Krevelen, *Properties of Polymers: their Correlation with Chemical Structure; their Numerical Estimation and Prediction from Additive Group Contributors*, 3rd Edition, Elsevier, Amsterdam, The Netherlands, 1990, 189.

87. M. Salame, *Polymer Engineering and Science*, 1986, **26**, 1543.

88. I.S. Arvanitoyannis, E. Psomiadou, A. Nakayama, S Aiba and N. Yamamoto, *Food Chemistry*, 1997, **60**, 593.

89. I.S. Arvanitoyannis and C.G. Biliaderis, *Carbohydrate Polymers*, 1999, **38**, 47.

90. R.F. Westover in *Encyclopedia of Polymer Science and Technology*, Volume 8, Eds., N.M. Bikales, J. Conrad, A. Rucks and J. Perlman, Wiley Interscience, New York, NY, USA, 1967, 533.

91. S. Guilbert in *Food Packaging and Preservation, Theory and Practice*, Ed., M. Mathlouthi, Elsevier Applied Science, London, UK, 1986, 371.

92. M.T. Kalichevsky, J.M.V. Blanchard, P.F. Tokarczyk, *International Journal of Food Science and Technology*, 1993, **28**, 139.

93. C.R. Southward in *Developments in Dairy Chemistry - 4*, Ed., P.F. Fox, Applied Science Publishers, London, UK, 1989, 173.

94. M.N. Moorjani, K.C.M. Raja, P. Puttarajapa, N.S. Khabade, V.S. Mahendrakar and M. Mahadevaswamy, *Indian Journal of Poultry Science*, 1978, **13**, 1, 52.

95. D. Lourdin, G. della Valle and P. Colonna, *Carbohydrate Polymers*, 1995, **27**, 261.

96. G. Cherian, A. Gennadios, C. Woller and P. Chinachoti, *Cereal Chemistry*, 1995, **72**, 1.

97. D.A. Ledward in *Protein Functionality in Food Systems*, Eds., N.S. Hettiarachchy and G.R. Ziegler, IFT Basic Symposium Series, Marcel Dekker, New York, NY, USA, 1994, 225.

98. D.A. Ledward, *Trends in Food Science and Technology*, 1993, **4**, 402.

99. I.S. Arvanitoyannis and C. Biliaderis, *Food Chemistry*, 1997, **62**, 3, 333

100. L. Ming, inventor; no assignee; CN 1, 136,050, 1996.

101. D.R. Coffin and M.L. Fishman, *Journal of Agriculture and Food Chemistry*, 1993, **41**, 1192.

102. M.L. Fishman and D.R. Coffin, inventors; no assignee; US Patent 5,451,673, 1995.

103. M.L. Fishman, D.R. Coffin, J.J. Unruh and T. Ly, *Journal of Macromolecular Science: Pure and Applied Chemistry*, 1996, **A33**, 639.

104. D.R. Coffin and M.L. Fishman, *Journal of Applied Polymer Science*, **54**, 1994, 1311.

105. M. Miya, R. Iwamoto and S. Mima, *Journal of Polymer Science: Polymer Physics Edition*, 1984, **22**, 1149.

106. E. Psomiadou, I. Arvanitoyannis and N. Yamamoto, *Carbohydrate Polymers*, 1996, **31**, 196.

107. R.D. Hagenmaier and P.E. Shaw, *Journal of Agriculture and Food Chemistry*, 1990, **38**, 1799.

108. J.F. Hanlon, *Handbook of Package Engineering*, Technomic Publishers, Lancaster, PA, USA, 1992, 1.

109. P. Sakellariou and R.C. Rowe, *Macromolecular Reports*, 1994, **A31**, 6&7, 1201.

110. P. Sakellariou, A. Hassan and R.C. Rowe, *Colloid and Polymer Science*, 1994, **272**, 1, 48.

111. P. Sakellariou, A. Hassan and R.C. Rowe, *Polymer*, 1993, **34**, 6, 1240.

112. R.C. Rowe in *Advances in Pharmaceutical Sciences, Volume 6*, Eds., D. Ganderton and T. Jones, Academic Press, London, UK, p.65.

113. N.B. Shah and B.B. Sheth, *Journal of Pharmaceutical Science*, **61**, 1972, 412.

114. R.C. Rowe, *Pharmacy International*, 1985, **6**, 14

115. M.C. Bonferoni, C. Caramella, M.E. Sangalli, U. Conte, R.M. Hernandez and J.L. Pedraz, *Journal of Controlled Release*, 1992, **18**, 3, 205.

116. P. Sakellariou and R.C. Rowe, *International Journal of Pharmaceutics*, 1995, **125**, 289.

117. P. Sakellariou, R.C. Rowe and E.F.T. White, *Journal of Controlled Release*, 1988, **7**, 147.

118. K.D.T. Satybaldyeva, M.Y. Mukhamedzhanova, A.A. Sarymsakov and Y.T. Tashpulatov, *Chemistry of Natural Compounds*, 1998, **34**, 3, 322.

119. D. Harris and J.R. Robinson, *Journal of Pharmaceutical Science*, 1992, **81**, 1.

120. M.J. Rathbone and J. Hadgraft, *International Journal of Pharmaceutics*, 1991, **74**, 9.

121. T. Nagai, Y. Machida, Y. Suzuki and H. Ikura, inventors; Trjin Limited, assignee; US 4226848, 1980.

122. T. Nagai, Y. Machida, Y. Suzuki and H. Ikura, inventors; Trjin Limited, assignee; US 4250163, 1981.

123. S. Dumitriu, P. Magny, D. Montané, P.F. Vidal and E. Chornet, *Journal of Bioactive and Compatible Polymers*, 1994, **9**, 184.

124. B.A. Zielinski and P. Aebischer, *Biomaterials*, 1994, **15**, 1049.

125. Polk, B. Amstel, K. De Yao, T. Peng and M.F.A. Goosen, *International Journal of Pharmaceutics*, 1994, **83**, 178.

126. L. Shahabeddin, O. Damour, F. Berthod, P. Rousselle, G. Saintigny and C. Collombel, *Journal of Materials Science: Materials in Medicine*, 1991, **2**, 222.

127. M.N. Taravel and A. Domard, *Biomaterials*, 1993, **14**, 930.

128. G.S. Macleod, J.T. Fell and J.H. Collett, *International Journal of Pharmaceutics*, 1999, **188**, 11.

129. M.J. Fernández-Hervás and J.T. Fell, *International Journal of Pharmaceutics*, 1998, **169**, 115.

130. G.S. Macleod, J.T. Fell, J.H. Collett, H.L. Sharma and A.M. Smith, *International Journal of Pharmaceutics*, 1999, **187**, 251.

131. G.S. Macleod, J.T. Fell and J.H. Collett, *Journal of Controlled Release*, 1999, **58**, 303.

132. G.D. Guerra, P. Cerrai, M. Tricoli, S. Maltinti and R. Sbarbati del Guerra, *Journal of Materials Science: Materials in Medicine*, 1998, **9**, 73.

133. P. Cerrai, G.D. Guerra, M. Tricoli, S. Maldini, N. Barbani and L. Petarca, *Macromolecular Chemistry and Physics*, 1996, **197**, 11, 3567.

134. R. Bodmeier, H. Chen and O. Paeratakul, *Pharmaceutical Research*, 1989, **6**, 413.

135. R. Bodmeier, K.H. Oh and Y. Pramar, *Drug Development and Industrial Pharmacy*, 1989, **15**, 1475.

136. L-S. Liu, S-Q. Liu, S. Y. Ng, M. Froix, T. Ohno and J. Heller, *Journal of Controlled Release*, 1997, **43**, 65.

137. S. Miyazaki, A. Nakayama, M. Oda, M. Takada and D. Attwood, *International Journal of Pharmaceutics*, 1995, **118**, 257.

138. C. Remuñán-López and R. Bodmeier, *International Journal of Pharmaceutics*, 1996, **135**, 63.

139. D. Thacharodi and K. Panduranga Rao, *International Journal of Pharmaceutics*, 1995, **120**, 115.

140. P. Giusti, L. Lazzeri and M.G. Cascone in *The Polymeric Materials Encyclopedia*, Ed., J.C. Salamone, CRC Press, Boca Raton, FL, USA, 1996.

141. M. Seggiani, L. Lazzeri, M.G. Cascone, N. Barbani, S. Vitolo and M. Palla, *Journal of Materials Science: Materials in Medicine*, 1994, **5**, 868.

142. M.G. Cascone, B. Sim and S. Downes, *Biomaterials*, 1995, **16**, 569.

143. M.G. Cascone, L. Di Silvio, B. Sim and S. Downes, *Journal of Materials Science: Materials in Medicine*, 1994, **5**, 770.

144. R.W. Baker, M.E. Tuttle and R. Helwing, *Pharmaceutical Technology*, 1984, **8**, 26.

145. M. Nambu, inventor; no assignee; Japanese Patent, 57/130543, 1982.

146. P. Giusti, L. Lazzeri, N. Barbani, P. Narducci, A. Bonaretti, M. Palla and L. Lelli, *Journal of Materials Science: Materials in Medicine*, 1993, **4**, 538.

147. L. Lazzeri, N. Barbani, M.G. Cascone, D. Lupinacci, P. Giusti and M. Laus, *Journal of Materials Science: Materials in Medicine*, 1994, 4, 862.

148. R. Sbarbati del Guerra, M.G. Cascone, N. Barbani and L. Lazzeri, *Journal of Materials Science: Materials in Medicine*, 1994, 5, 613.

149. M.G. Cascone, *Journal of Materials Science: Materials in Medicine*, 1999, 10, 301.

150. A. Kato, Y. Sasaki, R. Furuta, K. Kobayashi, *Agricultural and Biological Chemistry*, 1991, 54, 1053.

151. S. Nakamura, A. Kato and K. Kobayashi, *Journal of Agriculture and Food Chemistry*, 1991, 31, 647.

152. S. Nakamura, A. Kato and K. Kobayashi, *Journal of Agriculture and Food Chemistry*, 1992, 40, 2033.

153. A. Kato, R. Mifuru, N. Matsudomi and K. Kobayashi, *Bioscience, Biotechnology and Biochemistry*, 1992, 56, 567.

154. Y-W. Shu, S. Sahara, S. Nakamura and A. Kato, *Journal of Agriculture and Food Chemistry*, 1996, 44, 2544.

155. P.J. Wuk, L.D. Jin, Y.E. Sang, I.S. Soon, K.S. Hyun and K.Y. Ha, *Korea Polymer Journal*, 1999, 7, 2, 93.

156. I. Tomka, inventor; Bio-tec Biologische Naturverpackungen GmbH, assignee; EP 0542 155, 1998.

157. I. Tomka, inventor; no assignee; US 5 280 055, 1994.

158. I. Tomka, inventor; Bio-tec Biologische Naturverpackungen GmbH, assignee; EP 596 437, 1997.

159. W. Pommeranz, J. Lorcks and H. Schmidt, inventors; Biotec Biologische Naturerpackingen, GmbH, assignee; WO 9619599A1, 1996.

160. J. Lörcks, *Polymer Degradation and Stability*, 1998, 59, 245.

161. DIN V 54900, *Testing of the Compostability of Polymeric Materials*, 1998.

162. T. Iwanami and T. Uemura, *Japanese Journal of Polymer Science and Technology*, 1993, 50, 767.

163. C. Bastioli, V. Bellotti, L.D. Giudice and G. Gilli, *Journal of Environmental Polymer Degradation*, 1993, **1**, 181.

164. T. Ishigaki, Y. Kawagoshi, M. Ike and M. Fujita, *Journal of Microbiology and Technology*, 1999, **15**, 321.

165. M. Sakamoto, inventor; Japan Kokai Tokkyo Koho, assignee; JP 175288, 1998.

166. K. Suzuki and S. Mitsuhiro, inventors; Japan Kokai Tokkyo Koho, assignee JP 132525, 2000.

167. Z. Youming, R. Xiangyuan, L. Zhimeng, Y. Qingrong, Z.Y.Y. Shiyun and T. Sherong, *Huanxue Yu Nianhe*, 1998, **3**, 154-155, 158.

168. Y.P. Nekrasov, O.N. Belyatskaya, L.I. Bulatnikova and L.V. Bykova, *Plastiche Massey*, 1997, **4**, 29.

169. Y.C.F. Shaoming and G. Mingge, *Henan Huagong*, 1999, **3**, 15.

170. D.K. Rassis, I.S. Saguy and A. Nussinovitch in *Hydrocolloids – Volume 2,* Ed., K. Nishinari, Elsevier Science, Oxford, UK, 2000, 203.

171. E.L. Vladimirtseva, L.V. Sharnina, I.B. Blinicheva and B.N. Mel'nikov, *Izvestiya Vysshikh Uchebangkh Zavedenii Khimiya Ikhimicheskaya Tekhnologiya,* **1998**, **4**, 50-53.

172. D.K. Rassis, I.S. Saguy and A. Nussinovitch, *Journal of Agriculture and Food Chemistry*, 1998, **46**, 2981.

173. A. Nussinovitch, R. Velez-Silvestre and M. Peleg, *Biotechnology Progress*, **9**, 1993, 101.

174. H-J. Steiger, G. Arnold and C. Gass, inventors; IBN GmbH Dresden, assignee; DE 19,751,234, 1999.

175. *Pervaporation Membrane Separation Processes*, Ed., R.Y.M. Huang, Elsevier, Amsterdam, The Netherlands, 1991.

176. R.Y.M. Huang and X. Feng, *Journal of Membrane Science*, 1996, **116**, 67.

177. T. Uragami, T. Matsuda, H. Okuno and T. Miyata, *Journal of Membrane Science*, 1994, **88**, 243.

178. T. Uragami and K. Takigawa, *Polymer*, 1990, **31**, 668.

179. Y.M. Lee, S.Y. Nan and D.J. Woo, *Journal of Membrane Science*, 1997, **133**, 103.

180. L.G. Wu, C.L. Zhu and M. Lu, *Journal of Membrane Science*, 1994, **90**, 199-205.

181. A. Chanachai, R. Jiraratananon, D. Uttapap, G.Y. Moon, W.A. Anderson and R.Y.M. Huang, *Journal of Membrane Science*, 2000, **166**, 271.

182. F. Jun, H. Jicai, C. Liankai and G. Qunhui, *Mo Kezue Yu Jishu*, 1997, **17**, 5, 60.

183. J-J. Shieh and R.Y.M. Huang, *Journal of Membrane Science*, 1997, **127**, 184.

184. H. Karakane, M. Tsuyumoto, Y. Maeda, K. Satoh and Z. Honda in Proceedings of the Third International Conference on Pervaporation Processes in the Chemical Industry, Ed., R. Bakish, Bakish Material Corporation, Englewood, NJ, USA, 1988, 194.

185. Z. Xi, S. Yanqiao and C. Guanwen, *Gongneng Gaofenzi Xuebao*, 1998, **11**, 3, 358.

186. W.H. Jiang and S.J. Han, *Journal of Polymer Science: Polymer Physics Edition*, 1998, 8, 1275.

187. X.P. Wang, F.Y. Zhang and Z.Q. Shen, *Chinese Membrane Science and Technology*, 1996, **16**, 62.

188. S. Cao, Y. Shi and G. Chen, *Journal of Applied Polymer Science*, 1999, 74, 1452.

189. S. Suto and N. Ui, *Journal of Applied Polymer Science*, 1996, **61**, 13, 2273.

190. M-R. Kula, K.H. Kroner and H. Hustedt, *Advances in Biomedical Engineering*, 1982, **24**, 74.

191. P.A. Albertsson, *Partition of Cell Particles and Macromolecules*, 3rd Edition, Wiley, New York, NY, USA, 1986.

192. P.A. Albertsson, G. Johansson and F. Tjerneld in *Separation Processes in Biotechnology*, Ed., J.A. Asenjo, Marcel Dekker, New York, NY, USA, 1990, Chapter 10, 287.

193. A. Kishida, Y. Kikunaga and M. Akashi, *Journal of Applied Polymer Science*, 1999, 73, 13, 2545.

194. A. Kishida, S. Nakano, Y. Kikunaga and M. Akashi, *Journal of Applied Polymer Science*, 1998, **67**, 2, 255.

195. A. Venâncio, J.A. Teixeira and M. Mota, *Biotechnology Progress*, 1993, **9**, 635.

196. F. Tjerneld, S. Berner, A. Cajarville and G. Johansson, *Enzyme and Microbial Technology*, 1986, **8**, 417.

197. D.C. Szlag and K.A. Giuliano, *Biotechnology Techniques*, 1988, **2**, 277.

198. D.R. Skuse, R. Norris-Jones, M. Yalpani and D.E. Brooks, *Enzyme and Microbial Technology*, 1992, **14**, 785.

199. U. Dissing and B. Mattiasson, *Biotechnology and Applied Biochemistry*, 1993, **17**, 15.

200. A.S. Medin and J-C. Janson, *Carbohydrate Polymers*, 1993, **22**, 127.

201. A. Venâncio, M.C. Almeida, L. Domingues and J.A. Teixeira, *Bioseparation*, 1995, **5**, 253.

202. A. Venâncio, M.C. Almeida and J.A. Teixeira, *Journal of Chromatography B*, 1996, **680**, 131.

203. M.C. Almeida, A. Venâncio, J.A. Teixeira and M.R. Aires-Barros, *Journal of Chromatography B*, 1998, **711**, 151.

7 Reactive Blending

Cornelia Vasile

7.1 Introduction

Many reactive processes involving multicomponent systems are known, such as: polymerisation, modification during polymerisation of one or both components, alloying polyolefins (PO) during polymerisation, blending, interpenetrating polymer network (IPN) preparation, filling of polymers with conductive particles, controlled rheology, by which completely new (innovative) products have been obtained and, in some cases starting from existing polymers [1-3].

The performance of such multicomponent systems depends critically, mainly on their phase morphology and polymer-polymer interaction energy at the interface; to a certain extent the two are related via the magnitude of the thermodynamic interaction between segments, and also on the interfacial tension between the two polymers, in the absence of any grafting reactions or even in the presence of such reactions, the nature of the interface between the phases, processing procedures and rheological properties. The interaction energy also determines the interface thickness, which controls the extent of the reactions possible in the interfacial zone. These interrelated aspects play a crucial role in the determination of final morphology of non-reactive and reactive blends and properties of multiphase systems as schematically is shown in **Scheme 7.1** [3].

An unfavourable interaction causes a large interfacial tension in the melt, poor interfacial adhesion in the solid state, makes it difficult to finely disperse the components of a blend and, particles' coalescence leads to phase separation and as a consequence to inferior mechanical properties of the blends relative to its components [4]. For effective compatibilisation, architecture and molecular weight of block or graft copolymers previously synthesised must be carefully controlled. Such tailor-made compatibilisers are rarely available commercially and are quite expensive and therefore have not been extensively used. One difficulty is adequately dispersing a block copolymer near the interface between two phases due to its high viscosity and sometimes the added block copolymer can localise in a homopolymer phase in micelle form rather than at the interface [5, 6].

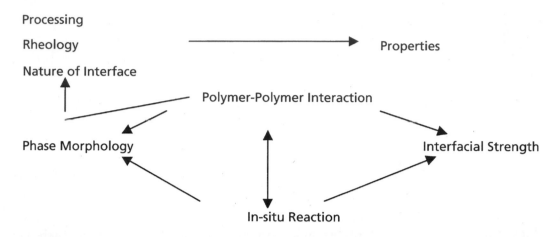

Scheme 7.1 Interrelation between polymer-polymer interaction – morphology - properties

The use of a polymer having hydrogen bonding, ionic bonding and acid-base complex functional groups to improve the miscibility of two immiscible polymer blends has also been studied [7, 8]. These kinds of compatible blends show phase separation when the temperature is higher than the dissociation temperature.

In order to overcome these disadvantages, reactive blending techniques are applied. In particular, the chemical modification of thermoplastics in twin-screw extruders gained a lot of attention in recent years as an advantageous route for the production of new materials. Recycling to solve the growing problem of solid waste because of recycling of post-consumer scrap materials based mixed plastics also needs to solve very difficult situation with compatibilisation; in this context reactive blending is recommended.

Reactive blending can be used to change the structure of polymers and thus increase their attraction for each other. An attractive technique is to synthesise the compatibilisers by a chemical reaction directly in the melt blending process. Chemists refer to this as '*in situ*' synthesis, while engineers generally use terms such as reactive processing or reactive extrusion [9]. The functional units are easily introduced by a copolymerisation reaction step or by grafting reaction during the extrusion process. Compared with conventional monomeric reactions, analogous polymer reactions are hindered by slow diffusion, low concentration of reactive groups, steric hindrance and the short time in the extruder. However the reaction time in the extruder is shorter than in other type of reactors. For example only 4 minutes were required for the reaction between polyethylene terephthalate (PET) and polyamide (PA) in extruder compared with 36 hours in an unstirred reactor. The reaction time can be controlled and minimised by using the catalysts.

Reactive compatibilisation methods are:

- Co-reaction within the blend to generate *in situ*, either copolymers or interacting polymers;

- using IPN technology:

  Crosslinking the blend ingredients,

  Modification of homopolymers e.g. through incorporation of acid/base groups, charge transfer electron donor-electron acceptor complexation, ion-dipole, ionic groups, etc.,

  Irradiative crosslinking,

  Physical crosslinking via crystallisation,

An increasing number of polyblend processors will develop 'in-house compatibilisation' using there own proprietary ingredients and techniques for compatibilisation of the polyblends they sell so, current commercial practice is still new and secretive.

Reactive processing is now the most active aspect of compatibilisation research. Reactive processing is also a method for producing commercially useful blends when components are incompatible by making a controlled stable morphology of a significantly reduced domain size relatively to the incompatible blends primarily by suppressing domain coalescence in the melt and for improving interfacial strength, increasing interphase thickness and strengthen the interface in solid state so preventing mechanical failure due to the weak defects between phases. The process may be defined as a reaction between two or more polymers to form a copolymer so, integrating polymer chemistry with polymer processing. Many commercial blends exist because of reactive processing. The reactive processing during compounding or forming stage assures a controlled chemical reaction between components. The technology opens up new possibilities for old polymers but also allows preparation of blends which could not be economically made before. Super-tough PA/PO, polyphenylene ether (PPE)/PA, polycarbonate (PC)/PA, polybutylene terephthalate (PBT)/PET, polyoxymethylene (POM)/PC are only a few examples. From the economic as well as the performance point of view, the technique is more interesting than the addition of a compatibiliser.

7.2 Requirements and Conditions for Operation

The requirements for an efficient reactive compatibilisation are [10]:

- Intensive and controlled mixing to achieve the desired dispersed morphology of one polymer in another;

- Presence of reactive functionality for covalent or ionic bond formation between components. The reactive groups can be mainly end groups for one component and end- or statistically distributed-groups along the chain of one polymer. If one of the polymers does not bear reactive groups, it should be functionalised;

- High reactivity to react across the phase boundary and within the residence time in the extruder;

- The formed bonds to remain stable in subsequent processing steps;

- Accurate metering, process and quality control;

- No degradation occurs during process and removal of by-products.

Optimal operational conditions must be established for each process and each type of blend. Several ways to perform reactive blending and average values of parameters are described here.

Functionalisation can be performed before the blending process by solution grafting, melt grafting, solid state functionalisation, copolymerisation, end-capping, physical procedures, surface functionalisation, mechanochemical [9-14] and others. A Brabender mixer operating under a nitrogen atmosphere at a rotation speed of ~50 rpm and T = 200 –220 °C may be used. Maleation of PO can be carried out by melt mixing initiated by peroxides (as dicumyl peroxide; DCP) in Brabender mixing chamber at 180 °C. After the polymer is melted, a mixture of monomer and initiator is added [13]. The grafting of PO with maleic anhydride (MA) was done using a co-rotating intermeshing twin-screw extruder (ZSK 30; Werner & Pfleiderer) with screw configuration adapted for grafting. A masterbatch of PO granules and 0.5 wt% di-*tert*-butyl peroxide (Trigonox B; Peroxide Chemie) and 0.2 wt% MA (Merck) was fed to extruder. A melt temperature profile from 180 to 210 °C for PO and 215-260 °C for other polymers, an output of 6 kg h^{-1} and a screw speed of revolutions 150 min $^{-1}$ were used; or feeding the first the polymer then the MA and finally the peroxide; residence time was ~5 minutes. For neatness, these could be carried out at different times but this would add to the complexity of the process and to the final cost. Thus some processors feed the first ingredients to the back of the screw and then inject other ingredients along the screw, to separate and run all reactions in their proper sequence. The simplest is to feed all the ingredients to the back of the extruder.

7.3 Reactive Extrusion or Reactive Compounding

The extruder, considered as another form of a chemical reactor provides extensive and intensive mixing, good temperature control and controlled residence time distribution (residence time ranges from ~100 to 200 seconds) affecting the molecular weight distribution of the final product. The mixing they offer and their superior heat and mass transfer capability [15, 16]. The process can takes place in three, two or one step, e.g., functionalisation of one of the other component or preparing a copolymer, blending and extrusion or feeding together all polymers and functionalising agents. Because of the residence time limitations in the extruder, high conversion rates are required. These can be achieved either by using a relatively high concentration of the reactive groups, e.g., high concentration of chain ends, i.e., low molecular weight of the components, highly reactive functional groups, or efficient catalysts.

Although both single and twin-screw extruder configurations are used in reactive processes, twin screw ones are increasingly favoured over the single screw ones [15].

Moreover, each type of twin-screw extruder (counter- or co-rotating) has a certain uniquess. A co-rotating, intermeshing twin-screw extruder has been found to be the most suitable for many continuous reactive processes. Intermeshing co-rotating twin-screw extruders are available in simple, double and triple start configuration [17] see **Figure 7.1**; the double start variant being most commonly used in reactive extrusion. In

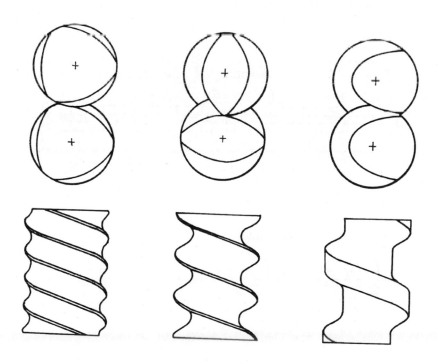

Figure 7.1 Simple, double and triple lobe intermeshing twin screw

fact, this design can allow high speeds and thus strong shearing forces and high outputs. With the co-rotating screw the material is transferred from one screw to the other and undergoes a constant mixing.

The screw configuration has different conveying elements for feeding, melting reaction and discharge, thorough mixing of the reactants was accomplished through a set of kneading blocks just prior to the reaction zone. The reaction zone length estimated from the previous batch studies is more accurately determined experimentally and could be controlled by the barrel temperature and a certain temperature profile. Favourable degassing conditions are easily created. The modular design and assembly arrangement of the screw and barrel sections of the machine, the screw profile, along with the use of special feeding and venting ports provide adequate flexibility for specific reactive processes.

Two screw configurations A and B are depicted in **Figure 7.2** for co-rotating twin-screw extruder (Werner & Pfleiderer ZSK 30) with the following characteristics, screw diameter D = 30 mm, length to diameter ratio (L/D) =42 [18-21].

The operational conditions are: average feed rate 5.5 kg/h; screw speed 75-150 rpm, mean residence time 150 seconds. The residence time has to be correlated with the lifetime of the peroxide used (see **Table 7.1**) [18-21].

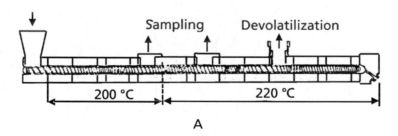

A

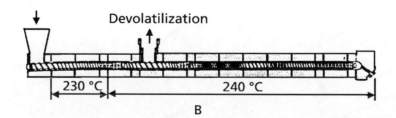

B

Figure 7.2 Screw configurations of a modular co-rotating twin-screw extruder recommended for grafting (A) and reactive blending (B). Redrawn from [18-21]

Table 7.1 The grafting temperature and lifetime of several peroxides		
Peroxide	Grafting temperature (°C)	Half-life time (seconds)
1,3-Bis (*tert*-butyl-1-peroxyisopropyl) benzene	218-220	5-10
2,5-Bis (*tert*-butylperoxy)-2,5-dimethylhexane (DHBP Trigonox 101)	200	6.1
Benzoyl peroxide T_m = 105 °C	180 111	3 3600
2,5-Dimethyl-2, 5-di (*tert*-butylperoxyl) hexane T_b = 249	180 111	75 3600
T_m – *melting temperature* T_b – *boiling temperature*		

Before feeding the materials into the extruder, the monomer(s) and peroxide were absorbed into the small porous polymer pellets for 30 minutes at room temperature. Reactive blending may carried out in the same twin-screw extruder as is used for grafting (A) but screw configuration are different (B). The line may be equipped with a vacuum pump to remove products as residual monomer (devolatilisation zone) that could alter the final performance of the product. Before blending all materials must carefully dried.

7.4 Compatibiliser Efficiency and Problems Associated with the Concentration of the Reactive Groups

The ordering processes are the basis of morphological architecture for multiphase polymeric materials based on reacting polymer blends. Reactive polymers each bearing a large number of reactive groups are important for new compatibilised blends with the graft copolymers generated *in situ*. The availability of more and more functional polymers, whether they are made by copolymerisation or chemical modification provides new opportunities to make new alloys by coupling two reactive polymers. The complexity of the graft copolymers grows as the grafting reaction continues and gelation occurs when the extent of grafting reaction goes beyond a certain limit. The amount that can be formed is limited by the gelation, but it is a need for enough graft copolymers to be produced for a homogeneous microphase size. It is therefore necessary to stop the grafting reaction at some point if gelation is not desirable. The material properties such as mechanical properties, transport properties, optical properties, etc., depend very much on phase size and phase homogeneity. The amount and the structure of the copolymers play a major role in that regard. Certain degrees of demixing between the chain segment of graft copolymers and the solubilised homopolymers within the microdomains can appear, even with homogeneous solubilisation. At low grafting conversion, i.e., < 10 wt% graft copolymers in the products, the graft copolymers act as a compatibiliser. The reactive polymers are often highly polydispersed so it is also important to know the effect of polydispersity [22-24].

The molecular architecture of compatibiliser may significantly affect the final morphology by changing the particle size of dispersed phase (see **Figure 7.3**) as even compatibiliser exhibits various morphological aspects [25].

It is well known that block and graft copolymers are very efficient as compatibilisers for a high number of polymer blends of decreasing particle size. However, in the case of the polyolefin modification with acrylic acid (AA) is has been established [25] that in random

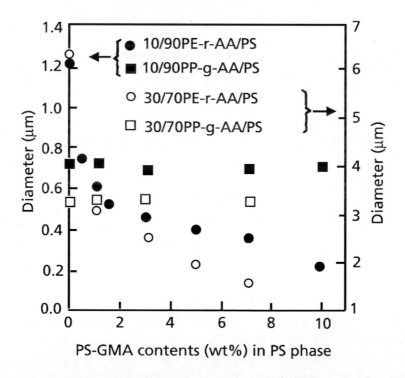

Figure 7.3 The number-average diameter of the dispersed phase *versus* the amount of PS-GMA for four different blend systems indicated in the legend

Redrawn with permission from S. Kim, J.K. Kim and C.E. Park, Polymer, 1997, 38, 8, 1809, Figures 1 and 3. Copyright 1997, Elsevier

copolymers the AA units are well distributed across the entire chain and thus the grafting reaction between polyethylene (PE)-r-AA and PS-GMA occurs uniformly giving rise to a significant decrease in domain size. Contrary, because the AA units in PP-g-AA exist at very limited positions, they are not well distributed, the reaction between PP-g-AA and PS-GMA is highly constrained at certain places and thus the graft copolymer has a lower ability to considerably reduce the domain size.

The generation of morphology in multiphase polymeric systems through chemical reactions is a dynamic process [26, 27]. The process has both chemical and physical limitations. From a kinetic point of view, the limitation is due to the residence time in compounding extruders that is about one minute and high density can physically hinder some groups. Moreover, the interface could become too crowded with copolymer before significant reduction in interfacial tension to occur, therefore the predominant mechanism of reducing particle size is the suppression of droplet-droplet coalescence through steric hindrance. The size of the dispersed particles represents a critical factor in controlling the fracture toughness, or more correctly the interparticle distance, which is lower as the particle size decreases, or as the volume fraction of the dispersed phase increases. It has been shown that a critical value of this parameter exists above which a transition from a ductile to brittle behaviour occurs in impact tests [28, 29].

Critical surface concentration (CSC) is the amount of compatibiliser needed for absorption of one monolayer of compatibiliser molecules on the dispersed polymer surface. The calculations are based on the effective area of adsorption for compatibiliser molecule calculated from the hydrodynamic radius and the total area to be stabilised in θ conditions: CSC >1 means multiplayer adsorption and CSC <1 means surface is not completely covered or stabilised. CSC can be used to establish the compatibiliser efficiency [30].

The interfacial area (A) stabilised per molecule for a graft copolymer located at the interface can be estimated by [31]:

$$A = 3\phi \bar{M}n /(RWcN) \tag{7.1}$$

Where ϕ is volume fraction of the dispersed phase, $\bar{M}n$ is the average molecular weight of the compatibiliser, R is the particle radius, Wc is the mass of compatibiliser per unit volume of blend and N is Avogardo's number.

Only low concentrations of block or graft copolymers are required to saturate the interface and produce optimum compatibilisation. As little as 0.5-2 wt% of such copolymer is sufficient for optimisation of blends performance but more frequently 10-20 wt% are used.

The effect of the compatibiliser can also be established by a rheological Cole-Cole plot [30] By this method, the loss modulus is plotted as a function of the elastic modulus or more commonly, the imaginary viscosity is plotted as a function of the real part. A continuous or homogeneous system is represented as one uniform half-circle while for immiscible blends the continuity is broken.

The impact strength is the most sensitive to blend compatibilisation. The addition of the compatibiliser changes the morphology of the blends considerably from a coarse morphology with average domain size as large as tens of microns (incompatibility) to much finer, with the average particle size of about 1 micron [32]. When finely dispersed, the rubbery particles act as effective stress concentrators initiating both crazing and shear yielding from many sites of the matrix, favouring the dissipation of large amounts of impact energy and slowing down the growth of cracks, hence increasing significantly the toughness and impact strength. In the case of the polypropylene (PP)/PBT blend reactive compatibilised with PP-*g*-oxazoline both impact strength and elongation at break significantly increased with graft copolymer content, see **Figure 7.4**.

Sometimes phase separation is associated with chemical reactions [9, 33-35]. Miscible polymer blends are destabilised by crosslinking reactions and undergo phase separation. These phenomena are often observed during the preparation of IPN where the morphology resulting from phase separation plays a crucial role in modifying the physical properties of these polymer materials.

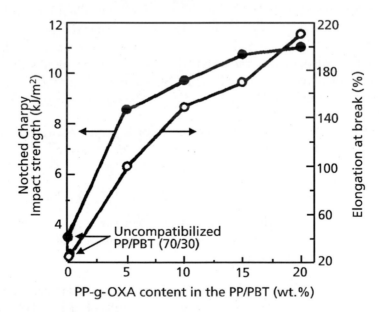

Figure 7.4 Notched Charpy impact strength and elongation at break *versus* graft copolymer content [18-21]

Reproduced with permission from T. Vainio, G-H. Hua, M. Lambla and J. Seppala, Journal of Applied Polymer Science, 1997, 63, 7, 883, Figure 5. Copyright 1997, Wiley

7.5 Chemical Reactions Occurring During Reactive Blending

7.5.1 Polymer Functionalisation

When a polymer does not have enough natural reactivity for reactive processing, the polymer may be modified ahead of time to add the functional groups that will be needed (functionalisation). This may be done during polymerisation by adding a comonomer to supply the functional groups or it may be done by specific post-polymerisation reactions for example olefin-acrylic copolymer, MA or acrylic onto a PO backbone [35-38].

The incorporation of carboxyl, epoxy, oxazoline, isocyanate, carbodiimide and bismaleimide groups are studied mostly.

7.5.1.1 Carboxyl Groups

Carboxyl groups are introduced by grafting or copolymerisation with MA, fumaric acid and related compounds, acrylic acid (AA) and methacrylic acid, acrylic esters, etc.

Maleation is industrially highly important modification of polymers. Maleic anhydride (T_m = 55 °C, T_{nb} (normal boiling temperature) = 200 °C) attachment is usually performed by grafting most polymers as PO and their copolymers, PPE, acrylonitrile-butadiene-styrene (ABS), etc. The MA is attached to PO by reactive extrusion or other mixing device at elevated temperatures in presence of suitable initiator (or peroxides) [39-49]. Grafting yield depends on peroxide and monomer concentration and also as a function of screw position. It is an evolution of grafting of MA or oxazoline (OXA) onto PE, EPDM and PP along the screw axis (see **Figure 7.5**) [50].

Locations A to F refer to successive collection ports along the screw. Melting takes place between locations A and B. The grafting profiles are similar, showing high conversion rates from melting up to L/D=15, where the peroxide decomposition is completed. The highest value of grafting was obtained for PE (1.7 wt%) and the lowest for PP (0.64 wt%). For PE and EPR a crosslinking was observed after melting and for PP degradation.

Owing to reversibility of the reaction with the equilibrium shifted toward MA, the extent of grafting is small. MA unit can be attached either in form of single succinic anhydride ring (in PP and ethylene-propylene (EP) copolymers) as well as short oligomers grafts, at the chain end (as in PPE).

Modifications with maleic acid anhydride and maleic esters substantially change the polymer's reactivity and chemical and mechanical properties.

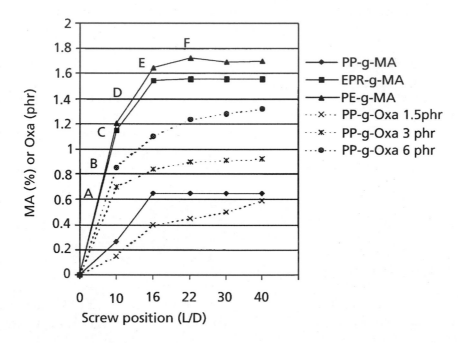

Figure 7.5 MA content along the extruder for modification of PE, EPR and PP [50] and oxazoline (Oxa) content along the screw for grafting of PP, [ROOR] 0.3 phr [18-21]

Redrawn with permission from T. Vainio, G-H. Hua, M. Lambla and J. Seppala, Journal of Applied Polymer Science, 1997, 63, 7, 883, Figure 4. Copyright 1997, Wiley

Due to their hydrophobic nature, poor dispersability with inorganic fillers and poor miscibility in blends even with themselves, PO have restricted use in several new engineering technologies. One way to overcome these drawbacks is the introduction of a small amount of polar groups onto polymer backbone [51]. By controlling the extent of reaction, properties can be changed in a targeted way to improve biocompatibility, fire retardancy, adhesion or their ability to blend with other polymers. Grafted copolymers of PO are widely used as compatibilisers in blends of PO with PA and polyesters (PES) [52].

Initiator concentration was found to be the predominant parameter controlling the flow properties of the polymer. A commercial product from BP Chemicals, Polyblend is maleated selectively at the chain ends. Ionic crosslinked rubber-like polymers were obtained from maleic anhydride grafted onto a polyolefin (MA-*g*-PO) with some alkali metal compounds. Gaylord [40] obtained alternating styrene-maleic anhydride copolymer (S/MA) chain onto PO backbone. The succinic anhydride formed is highly reactive being a starting point for synthesis of new polymers [53-55].

An alternative route to solution and bulk (melt grafting) methods has been proposed by Lee [56] and Borsig [11, 12, 57-59] who used a solid phase polymerisation (modification) technique. This technique offers the advantage of low temperatures and low operating pressures.

7.5.1.2 Photofunctionalisation

Photofunctionalisation is considered an environmentally friendly technology (it minimises the use of the solvents and hazardous reagents) and is a cost effective method. PE was photo-oxidised and then melt-blended with polyamide [60]. The introduced carboxyl groups react with amine groups of polyamide. Photo-oxidised PE from waste can similarly react.

A PP-*g*-AA copolymer is a commercial one with trade name Polybond 1003 used as compatibilising agent. The acrylate group can be either in the main backbone or can be introduced by grafting. Many random copolymers and terpolymers of MA are tested as compatibilising agents as: S/MA, S/MMA/MA, S/AN/MA, MA/VA [61] for polyvinyl chloride (PVC)/collagen blends [62-63] and polysaccharide-based blends [64, 65] and also graft copolymers such as SEBS-*g*-MA, styrene-butadiene-styrene three block copolymer (SBS) MA, ABS *g* glycidyl methacrylate (GMA), styrene-butadiene copolymer (SB)-MMA-*g*-GMA used for PA/PES toughening. They react rapidly with amine groups of polyamides, with hydroxyl groups of polyesters, polysaccharides, epoxy or oxazoline groups, etc., of other polymers, [61-65].

Unfortunately MA is very corrosive in extruders and in aqueous suspension, less reactive maleic acid is formed. Sometimes other monomers are preferred such as monoethylmaleate that can be converted in anhydride form by simple heating (see **Scheme 7.2**).

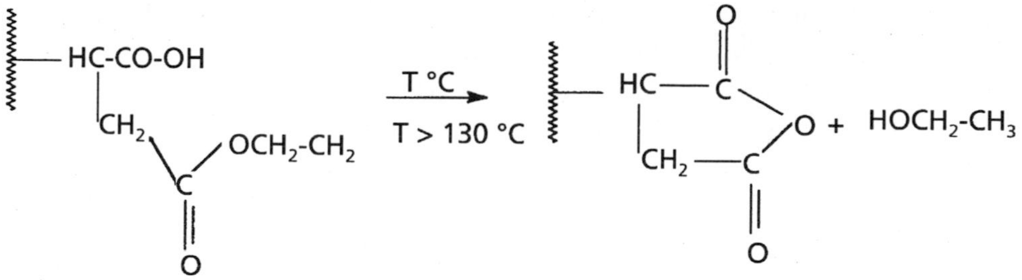

Scheme 7.2

Functionalisation of styrene-ethylene/butene-styrene three block copolymer (SEBS) and SBR random copolymer with diethylmaleate (DEM) by free radical process depends on feed composition and in particular on the DEM/DCP ratio. The high number of double bonds in styrene-butadiene copolymers does not allow the direct use of DEM in combination with free radical initiators, because it rapidly crosslinks [66]. Such functionalisation takes place preferentially at the aliphatic carbons of the PO blocks. To favour functionalisation and crosslinking in certain position versus degradation, some different approaches have been used. Among these, the reaction of SBR with thioglycolic acid and its aliphatic ester occurs mainly by addition at the vinyl double bonds of the 1-2 butadiene units with high selectivity. Thiols are very suitable because they combine a very high transfer capacity of hydrogen to formed radicals with the capacity of the free radicals formed on the sulfur atom to add a double bond [66].

7.5.1.3 Epoxy Groups

Epoxy groups are mainly introduced by grafting of GMA (T_{nb} = 189 °C) onto PO. The grafting of GMA onto ABS or PPE leads to good compatibilising agents [67, 68]. The GMA functionalised polymers exhibit two different, independent kinds of functionality: that of free radical of the methacrylate group and of the epoxy group of glycidyl. Epoxy group reacts very easily with carboxyl, amine or other hydrogen active groups, end-groups of PA, PES, etc., giving good compatibilising agents for PS/PO, PBT/PS, PA blends, etc.

Epoxy groups can be also introduced by reaction with epichlorohydrin onto natural polymer backbone (lignin or starch) to improve their compatibility with synthetic polymers (PO) [68, 69].

7.5.1.4 Oxazoline Groups

Oxazoline groups are introduced by copolymerisation or grafting of vinyl oxazoline compounds. Oxazoline is highly reactive with strong and weak nucleophiles by a rapid kinetics (few minutes) above 200 °C. It is easy detectable by Fourier-Transform Infra Red (FTIR) and Nuclear Magnetic Resonance (NMR) spectroscopy. Vinyl or isopropenyl oxazoline has been grafted on PO or copolymers with styrene and then reacted with carboxylic acid or anhydride to produce compatibilisation. Long chain oxazolines have been reported to be less toxic than MA and GMA. Oxazolines react fast with carboxyl, amino, phenol, mercaptan groups [18-21, 70-73] - **Scheme 7.3**.

PP, PE, EPR or SEBS react with ricinol oxazoline maleinate on a co-rotating twin screw extruder at 170-260 °C at a screw speed of 50-65 rpm in the presence of peroxides. Both batch and continuous mixing can be used, grafting time 5 minutes, under a nitrogen atmosphere. The monomer and peroxide were absorbed into the powdered matrix material at room temperature for at least 15 minutes. Degradation of PP limits the degree of its grafting with oxazoline [18-21].

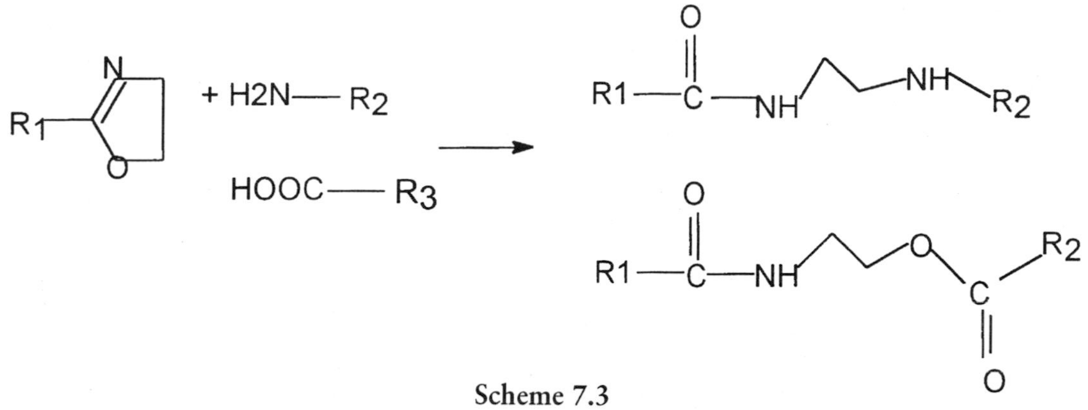

Scheme 7.3

7.5.1.5 Bismaleimide Groups

By virtue of its tetra-functionality, *bismaleimide* (BMI) was [74] used as a reactive modifier for polymer blends and as a crosslinking additive for elastomers [75]. The effect of BMI on the morphology and properties of PP/EPDM blends is particularly noteworthy [76]. The BMI is activated to form a charge transfer complex with an accelerator (poly (2,2,4-trimethyl-1,2-dihydroquinoline)). This system was found to increase the interfacial adhesion between the dispersed EPDM phase and the PP continuous phase enhanced nucleation of PP matrix. Functionalised PO with maleimides (1%) showed significantly improved hydrophilicity [47, 77].

Melt state chemical reactions carried out in the extruder have been studied and commercialised at Exxon Chem. Co., for over 30 years, under the name EXXELOR.

7.5.2 Degree of Functionalisation (FD)

Degree of Functionalisation (FD) (or grafting yield) can be determined by FTIR on compression-moulded films of purified samples, from the ratio of the intensities of the characteristic absorbance bands using pre-established calibration curves.

Purification of functionalised polyolefins is achieved by dissolving them in boiling xylene or toluene and then by precipitation in acetone at room temperature followed by drying at T > 100 °C under vacuum.

The characteristic bands for the monomer are: 1671 cm⁻¹ for –C=N stretching and oxazoline ring bending; 1737 cm⁻¹ for ester group, 1200-1100 cm⁻¹ epoxy, 1090 cm⁻¹ -C-O-C asymmetric stretching is related to either chemical reactions or very strong physical

229

interactions such as hydrogen bonding. The band, 720 cm^{-1}–CH$_2$-; 700 cm^{-1} is characteristic for grafted PS. Reference 2732 cm^{-1} for E/P copolymer, 1601/1463 block of styrene. The grafted MA content was determined by quantitative IR spectroscopy using the intensity of the absorption band at 1792 cm^{-1}. It commonly varies between 0.5 wt% and 4.7 wt%.

Carboxyl groups can be also determined by a simple titration by KOH in *n*-butanol [13, 14, 56-59].

7.5.3 Compatibilisation Reactions

Compatibilisation reactions may conveniently be classified by reactive group type such as: free radical, acid + amine, acid + hydroxyl and a broad variety of other miscellaneous reactions [78, 79].

The chemical processes by which inter-chain copolymers can be formed [80-84] are:

- chain cleavage and recombination resulting either block or random copolymers (PES/PES, PA/PA, PA/PES),

- reaction between end groups resulting in block copolymers,

- reaction between end group of a polymer with pendant functionality of the other copolymer resulting in a graft copolymer (PA/PO and their copolymers),

- PA/polybutadiene and styrene copolymers, PPE/PA, PPE/PES, PES/PO, others),

- reaction between pendant groups or main chain of the two polymers, as a result either graft copolymer or crosslinked network are formed (E-co-S/PP, natural rubber/ PE, PVOH/PE, PPE/PP or E or S copolymers), and

- ionic bonds formation leading to graft copolymer or a crosslinked system.

The radicals of the two polymers recombine with each other producing block or graft copolymers. Free radical reactions can be produced by thermal decomposition of initiators (peroxides, azo compounds), redox sources, photochemical decomposition, high-energy radiation (γ-rays), thermal and/or mechanical shear. For saturated PO, reactivity is limited primarily to the tertiary hydrogen atoms at branch point. These can be abstracted by peroxides or radiation, producing free radicals, which can then graft to each other or to unsaturated monomers or polymers in the vicinity (LLDPE-*g*-PS, HDPE/LLDPE/PS block copolymers, LDPE-crosslinks or graft-PP, LDPE or PP/EPDM, ethylene-methyl acrylate copolymer/EPDM or graft LDPE/PS, LDPE/HDPE/PP, HDPE/LDPE/EPDM, PO-*g*-MA or other polar monomers) [43-49, 79].

Chain cleavage and recombination is usually accompanied by secondary reactions leading to random copolymers thus is difficult to be controlled and is not always useful.

The reaction of the two chain ends leads to diblock copolymers and seems to be the most desirable one but reaction is not always limited to just this. The block and graft copolymers formed during processing have molecular weights close to the sum of the two reacting chains.

More typically a polymer has multiple reactive groups along its chains so graft copolymers result when it reacts with another polymer having a single reactive chain end. The number of grafts can be controlled by the number of functional groups. If both chains have multiple reaction sites then crosslinking occurs by extensive mixing.

Most combinations of polymers derived from a condensation reaction (PA and PES) which exhibit nucleophilic, (i.e., electron donor), end groups such as NH_2 (or NH), COOH (or derivatives) and OH. They can react with electrophilic groups such as cyclic anhydride, epoxy, isocyanate, carbodiimide, etc., placed along the chain by copolymerisation, end capping or grafting. The possible reactions leading to the block or graft copolymers during processing or reactive blending classified by the reacting groups are [26-30, 36, 84-85]:

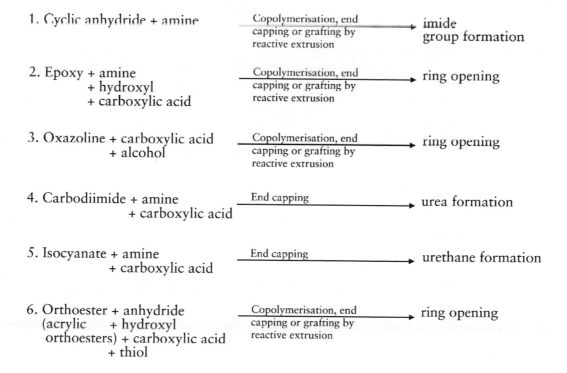

1. Cyclic anhydride + amine
   → (Copolymerisation, end capping or grafting by reactive extrusion) → imide group formation

2. Epoxy + amine
   + hydroxyl
   + carboxylic acid
   → (Copolymerisation, end capping or grafting by reactive extrusion) → ring opening

3. Oxazoline + carboxylic acid
   + alcohol
   → (Copolymerisation, end capping or grafting by reactive extrusion) → ring opening

4. Carbodiimide + amine
   + carboxylic acid
   → (End capping) → urea formation

5. Isocyanate + amine
   + carboxylic acid
   → (End capping) → urethane formation

6. Orthoester + anhydride
   (acrylic + hydroxyl
   orthoesters) + carboxylic acid
   + thiol
   → (Copolymerisation, end capping or grafting by reactive extrusion) → ring opening

231

Three pairs of reactive groups are presently used in industry: anhydride/amine (primary and secondary); carboxylic acid/epoxy; oxazoline/anhydride or carboxylic acid [86, 87]. The reaction of cyclic anhydride with polyamides has been the basis for numerous polyamide-based blends (PA 6/PA 12, PP/PA, PE/PA, PA 6/PC, PA 6/SEBS, PA 5/styrene-acrylonitrile (SAN), PA/ liquid crystalline polymer (LCP)) [3, 9, 15, 26, 27, 83].

The following schemes illustrate some representative routes of specific reactions of functional groups taking place with interchain copolymer formation.

a) *Grafting* [26, 27] of MA-functionalised polymers on polyamides or on liquid crystalline polymers – **Scheme 7.4**

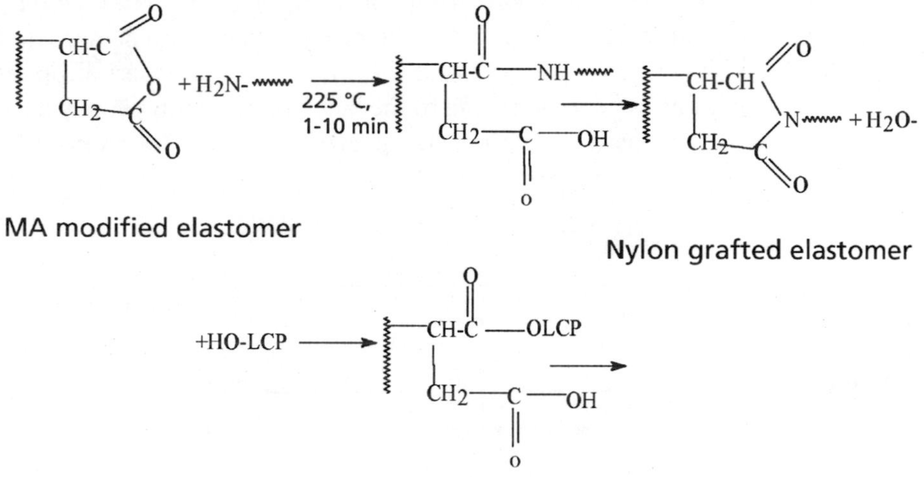

CH-C
O
$+H_2N-$
225 °C,
1-10 min
CH-C
NH
OH
CH-CH
N
$+H_2O-$
MA modified elastomer
Nylon grafted elastomer
+HO-LCP
OLCP

Some workers consider that the MA content in the system is much more important than that of the compatibilising agent [88, 89]

b) *Hydrolysis* - **Scheme 7.5**

$$-\overset{O}{\overset{|}{C}}-NH- + H_2O \longrightarrow -\overset{O}{\overset{|}{C}}OH + H_2N-$$ grafting then via way of Scheme 7.4

Acid + Amine Reactions. PA/PO are immiscible. PO offer melt strength, flexibility, lubricity, impact strength, electrical resistance, low dielectric constant and loss, water resistance and low price. PA offer melt fluidity, rigidity, strength, high temperature performance and solvent resistance. MA-functionalised PO have been widely used as compatibilising precursors during the reactive blending of various PO and PA. Carboxylic groups are grafted onto PO and then reacted with PA end groups leading to PA-*g*-PO copolymer. At high PO/PA ratio the PA generally contributes to impermeability to gasoline while at high PA/PO ratios, PO contributes impact strength and water resistance. Examples: PP, HDPE, LDPE, LLDPE, EPR, EPDM, metallocene PO, ethylene-vinyl alcohol (EVOH); carboxylating agents PO-*g*-MA or with acrylic acid, methacrylic acid, succinic acid, PO-ionomers.

Acid + Hydroxyl reactions–esterification. PP-*g*-MA or acrylic acid, succinic acid, SEBS-*g*-MA, styrene-maleic anhydride (S-MA) are reacting with hydroxyl containing polymers as EVOH, hydroxylated EPR, hydroxyl terminated polymers such as PPE, PET, PBT, LCP. A large number of studies aim: a) to facilitate recycling of mixed PO + PES packaging polymers by compatibilising them together; b) PP + engineering plastics; c) increase impact of PP and PBT.

c) *Amide-anhydride reaction* – **Scheme 7.6**

d) *Epoxide reactions* – **Scheme 7.7**

233

GMA is one of most popular process of reactive compatibilisation with polyamides or polyesters, etc.

e) *Oxazoline* reacts with carboxylic groups, amine, phenol, and mercaptan to form an esteramide so it is an effective compatibiliser. The oxazoline-modified PO have been used for compatibilisation of PO with PA or PET in a one or two-step process. The blends exhibited high impact strength and elongation at break.

The reactions occurring are shown in **Scheme 7.8**:

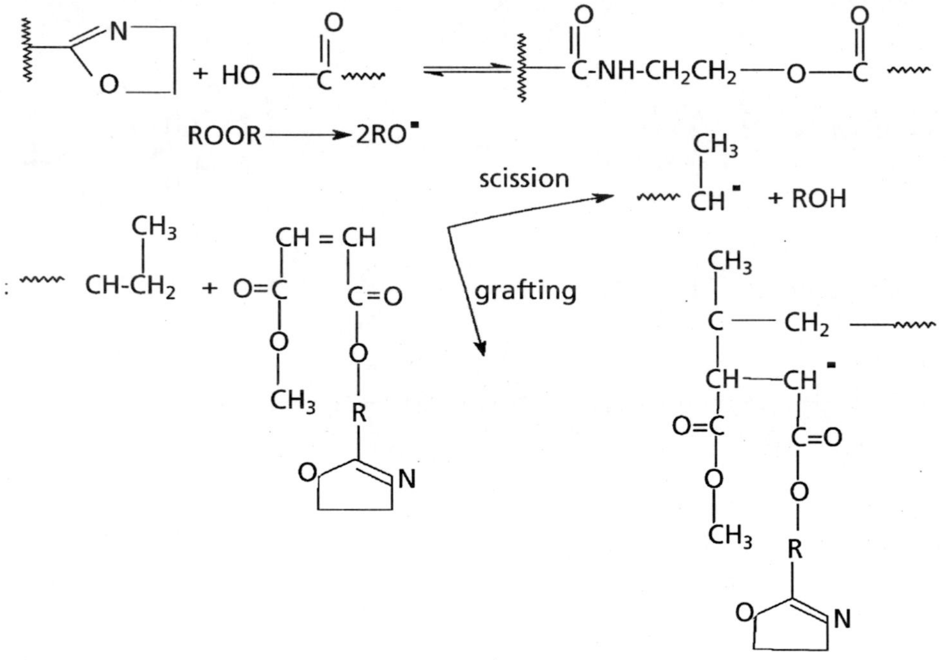

Scheme 7.8

f) *Carbodiimide* - **Scheme 7.9**

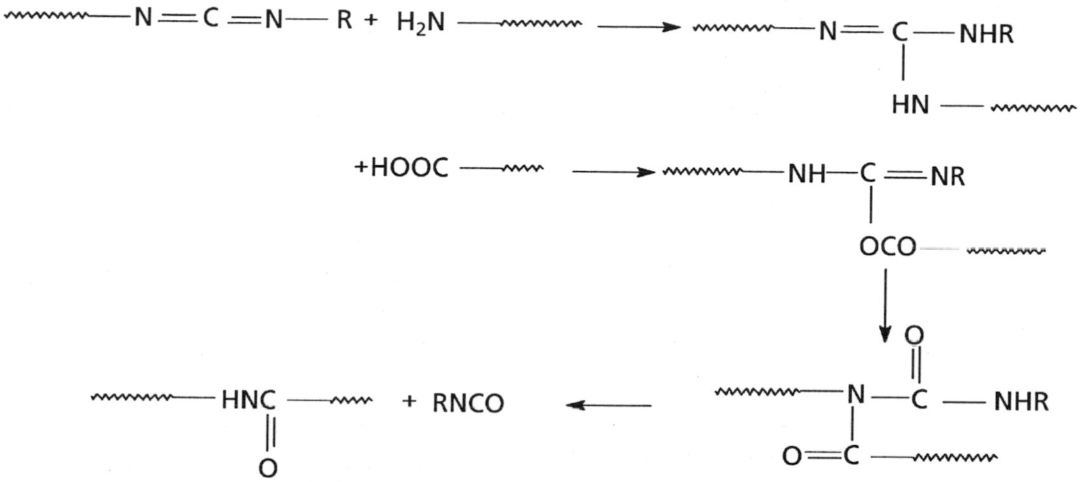

g) *Urethane* - **Scheme 7.10**

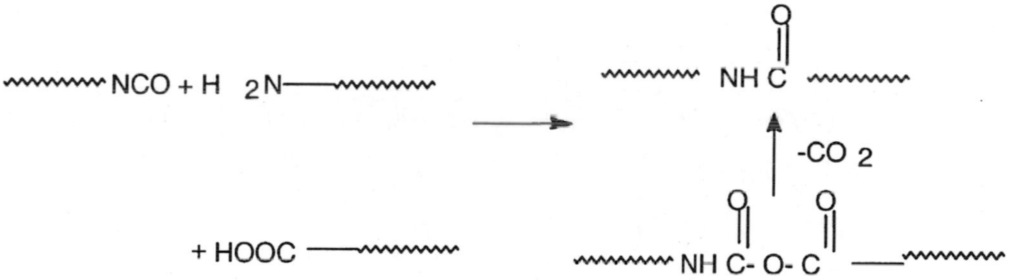

Trans-reactions or exchange reactions such as amide-amide, amide-ester interchange and transesterification are also frequently used to compatibilise of PES and PA. The process is known as ***redistribution.*** Redistribution is a slow statistical process sometimes occurring as a reversible reaction in the melt. It could leads to destruction of crystallinity and/or lowering of T_g, as was observed for PET/PC or PET/PA 6 [15].

Transesterification has been used for years to manufacture polyesters or to modify the properties of miscible polyester blends comprise of at least two components as: PES, PC, polyethylene glycol (PEG), aromatic polyamides (PAr), LCP and polymers with side-chain ester groups (PMMA, EVA, acrylic impact modifiers, etc). Transesterfication may involve chain cleavage followed by recombination of the end groups to change molecular weight distribution (MWD) and randomise the monomeric sequences in the polymers. It is catalysed by metal compounds, e.g., organo-tin, SnO_2, $Ti(OBu)_4$ and hindered triphenyl phosphite, *p*-toluene sulfonic acid [80]. Compatibilisation of PET with PBT depends on the extent of transesterification. In a twin-screw extruder the compatibilisation of PC with PAr occurs to a greater extend than in a single extruder or solution casting. In another procedure PET has been mixed in Brabender plasticorder with sodium salt of *n*-butyl-2-sulfobenzoate resulting new types of chain ends [90].

Redistribution between **two PA** is a trans reaction between amine and amide groups. Trans-amidation products of PA 6 and polyamide 66 (PA 66) are commercially available. They show improved processability, enhanced mechanical and barrier properties. Blends of aromatic PA are usually immiscible. Compatibilisation by trans-reaction may reduce crystallinity. The reaction is controlled by temperature and intensity of mixing. Commercial examples are PA 6 and PA 46 and PA 6 with poly (*m*-xylene adipamide).

Redistribution between *PA and PES* must be strictly controlled by process variables such as screw speed, T, P, residence time, and catalysts. The reaction can be catalysed by *p*-toluene sulfonic acid, dimethylol propanoic acid, etc. The reaction product should be impact modified.

The reactions during reactive blending of PBT and PA 66 [91] – **Scheme 7.11** - are:

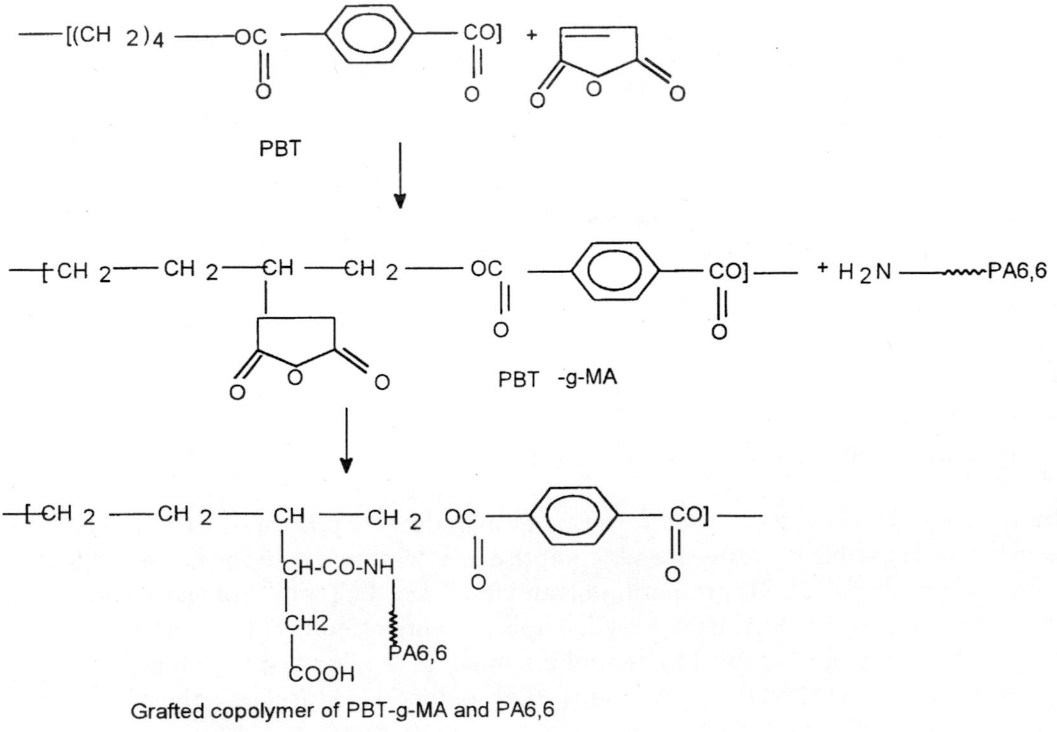

Grafted copolymer of PBT-g-MA and PA6,6

Commercial examples are: a) PET/PA 66 processed as a monofilament exhibits good processability, zero rejects, zero shrinkage, enhanced mechanical properties; b) PET-*g-p*-toluene sulfonic acid groups/PA 46 blend has good impact resistance and heat deflection temperature (HDT); c) Polyether-b-amide copolymer/LCP (Vectra) polyester compatibilised by trans-reaction exhibit high strength as fibres [15] The role of the random S-MA copolymer in compatibilisation of PA with LCP can be explain both by hydrogen bonding with PA 6 and LCP and/or an esterification reaction with LCP [92, 93].

Miscellaneous compatibilisation reactions are: epoxy-GMA with OH-containing polymers leading to ether linked graft copolymers, hydroxyl + amine end groups of Polyamide lead to ester linked graft copolymer; phenolic resins as compatibilisers – cure reaction with diene rubber lead to IPN.

7.6 Morphology

The morphology evolution during compounding or processing steps needed to be controlled. With this aim in mind a mathematical model was devised by combining the chemical kinetics with flow and heat transfer and fluid mechanics to give predictions and control of copolymer formation and morphology evolution during processing. The theory of the process has been developed on the basis of reaction kinetics between the reactive end group polymer 1/polymer 2 system by O'Shaughnessy and Sawhney [94], or controlled by diffusion [95]. To generate the blend formulations, an experimental response surface modelling can be used as. The model was designed using MODDE 3.0 software from Umetrics AB in Sweden,

The final morphology of a blend obtained by reactive processing results from two competitive phenomena: reduction in particle size and coalescence.

Wu [96] proposed the following equation:

$$C_a = (G\eta_m d)/(\gamma) = 4\,(\eta_r)^{\pm 0.84} \tag{7.2}$$

where C_a is capillary number or Weber number, γ is interfacial tension, G is the shear rate, η_m is the viscosity of the matrix phase and $\eta_r = \eta_d/\eta_m$ with η_d the viscosity of the dispersed phase. The plus sign is applied for $\eta_r > 1$ and the minus sign for $\eta_r < 1$. The equation could be applicable to both nonreactive and reactive systems. The Wu model gives a master curve with a sharp minimum at a viscosity ratio equal to unity which means that the smallest particles are formed when the viscosities of the matrix and the dispersed phase are equal. Commonly, in reactive systems (mainly grafting), the viscosity of the mixture increases. The coalescence phenomenon is possible when the particles' number and their density increase. In fact, some of factors that favour drop breakup such as high shear rates and reduced dispersed phase viscosity, lead also to increased coalescence of polymer blends. Reactive compatibilisation lowers interfacial tension, producing a steric stabilisation by reducing the size of dispersed phase particles. Due to these competitive phenomena the relationship (Equation 7.1) is not always valid.

For example, the dispersed-phase particle size dramatically decreases as the amount of MA grafted onto PP increases in spite of the fact that the viscosity ratio is 0.1 [97-99]. The decrease is very important at a MA content below 1-2 wt% or compatibilising agent below 2-5 wt% in various polymer blends as it appears from examination of **Figure 7.6** (as Nylon 6/PP-*g*-MA, PPE/PA-*g*-MA, SEBS/SEBS-*g*-MA (2%)/PA 6). Further increase of MA concentration in copolymer or of compatibiliser in blends resulted in only minimal additional decrease in particle size [26, 27].

For ABS-containing blends (ABS/PA, ABS/PES, ABS/PC, ABS/PBT), that are multiphase blends, a functionalised polymer that is miscible with one of the phases is recommended [26, 27, 100-106].

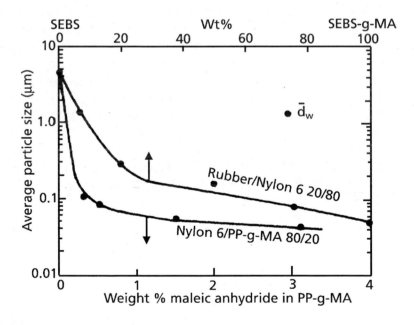

Figure 7.6 Effect of the amount of MA content in PP-g-MA on the particle size of the dispersed phase in PA matrix and of SEBS-*g*-MA content on particle size of the dispersed hase in Nylon 6 matrix and dispersed phase weight average particle size for mixture of PA 6 with SEBS and SEBS-*g*-MA (2% MA). Redrawn from [26, 27]

The block and graft copolymers act by both lowering interfacial tension (or reducing dispersed-phase particles) and making coalescence of particles that makes steric stabilisation more difficult. In the latter, it is considered that coalescence preventing lowering of interfacial tension is more important than the former. In the presence of a reactive compatibiliser, the particle does not grow with concentration or in time the morphology being stabilised by chemical bonds formed [26, 27, 104, 105] the coalescence process being retarded.

The functionality or the number of reactive groups per chain is an important factor in generation the morphology in reactive systems. Sometimes, reactive compatibilisation is based on the reactive chain ends resulting from condensation. For example, Nylon 66 is difunctional having two amine groups as end chains, while Nylon 6 is monofunctional. Some kinds of Nylon-6-based materials could have a greater amine functionality [107]. Comparing the blends of Nylon 6 and Nylon 66 with SEBS-g-MA or ABS [29], Paul and co-workers [22, 26, 108, 111] found that the rubber particles were extremely small and spherical in Nylon 6 containing blends, while in the Nylon 6,6 containing blends, the rubber domain were much larger and more complex in shape. The explanation could be that the monofunctional chains simply graft onto the maleated rubber particles with one attachment per chain, while the difunctional chains can form bridges or loops by crosslinking so the polyamide can be entrapped within rubber particles resulting in complex morphologies. To produce the fine dispersion of maleated rubber in difunctional PA needed for efficient toughening it is necessary to use intense mixing capabilities of a fully intermeshing, co-rotating twin-screw extruder, while very fine dispersions with monofunctional PA can be achieved in single-screw extruder [111]. On the contrary, it has been shown [112] that reactive compatibilisers with epoxide groups are not very effective in Nylon 6/ABS blends, because they react with both the amine and acid end group. In this case, the Nylon 6 is difunctional and it is difficult to obtain a good dispersion with ABS. Therefore, there are severe limits in respect with functionality for PA and PES, too.

In the PA 6/SEBS blends, the particle size decreases with increase of molecular weight of PA 6, increase of MA content of SEBS-g-MA and increase of intensity of mixing, the dependence is continuous both for maleated and unmaleated SEBS [113-115] without a minimum as predicted by Wu's relationship [96].

A widely used technique in commercial systems is to incorporate a compatibiliser precursor into one of the phases that can react with the functional groups of the second phase when the two meet at the interface [113-115]. The compatibiliser should be selected by the type of functional groups, the number of functional groups per chain, the manner in which these groups are incorporated (grafted, comonomer or terminal groups), molecular weight and the level of miscibility with the phase into which it is to be incorporated.

If the compatibiliser (C) is fully miscible with the rubber phase (B), one may expect it to reside primarily within the rubber phase, with its functional groups forming chemical bonds to thermoplastic chains (A) at the interface. This situation is expected to provide the most efficient reduction in rubbery domain size through a decreased interfacial tension and increased steric stabilisation.

If the functionalised compatibiliser resides at the rubber interface because it has some thermodyamic affinity for this phase, but not enough to be miscible with it, than the reduction of domain size can be expected, but it is not very effective. The third case is when compatibiliser is even less attached to the rubber phase and it remains in the thermoplastic matrix away from the interface, but it can increase the matrix viscosity which, in turn, will lead to change in the morphology of the blend. That is why, the reactive compatibilisation depends both on the content of reactive groups and also on the order of the mixing steps.

The reactive polymer is a precursor for reactive compatibiliser formation and this at its turn determines (by its structure and distribution between phases) the morphology and the end-use of the compatibilised blend. In the case of the ABS/PA 66 system, in presence of styrene-acrylonitrile (SAN)-*co*-GMA, the epoxy groups of GMA react with amine end-groups of PA 66, so a grafted SAN-*co*-GMA-*g*-PA 66 copolymer is formed. Two ways of mixing have been tested [116]: (1) Premixing of the SAN-*co*-GMA with PA 66 to form grafted copolymer followed by mixing with ABS. If the SAN-*co*-GMA contains only up to 2% GMA, the grafted copolymer will be easily formed and it will contain many free segments of SAN and SAN-*co*-GMA, which will migrate in ABS phase in the mixing step. The grafted copolymer chains across traversing the interface efficiently improve the interface adhesion and interface will have a large width. Alternatively, if the GMA content is ~ 10%, a high content of grafted copolymer will be formed with high molecular weight and low mobility, so it will reach with difficulty the interphase to penetrate the ABS phase. The length of the free fragments is smaller and the interface weaker than in the first case. If this mixing way is preferred, a copolymer with low degree of grafting should be chosen.

(2) The second way consists in premixing of SAN-*co*-GMA with ABS and then mixing with PA 66. The ABS is miscible with SAN-*co*-GMA at low GMA content and the distribution will be uniform. The copolymer molecules in the vicinity of the interface are able to directly contact and react with amino-end groups of PA 66. The grafting reaction takes place at interface so the number of the grafting points is low and interface is weak. The copolymer with high GMA content will be much efficient in this case.

Morphology of the blends evolves as a function of time in batch mixers and along the length of extruders [117, 118]. It was found that the morphology evolved in a complex manner, accompanied by phase inversion or even increase of particle size at intensive mixing beyond the melting reaction of the extruder. In the case of Nylon 6/SAN blends a steady decrease in the dispersed phase particle size was observed along the screw for nonreactive blends and as the screw speed was increased. However for ternary blends with a reactive compatibiliser (imidised acrylic polymer), a dramatic drop in the dispersed phase particle size was observed in the initial region of the screw immediately after melting, followed by some growth caused by coalescence in the latter stages, therefore the intensive mixing beyond of certain point might not be beneficial [119].

The interphase thickness is much increased by reactive blending compared to other compatibilisation procedures, see **Table 7.2**.

Studies showed that interphase thickness increased with time to a plateau, whose level depends on temperature and concentration of reactive sites.

Table 7.2 Interphase thickness [15, 120-123]	
Blend type	Interphase thickness (nm)
Immiscible	2
Block copolymers	4 to 6
Polymer/copolymer	30
Reactive Compatibilisation	30 to 60
Electron donor-electron acceptor complexation	80-90

7.7 Reactive Compatibilised Polymer Blends

Some reactive compatibilised blends are listed in **Table 7.3**.

7.7.1 Rubber-Toughened PA

In 1979, Epstein from DuPont [200] and Houghton [201] patented the blends of polyamides with functionalised elastomers. This opened a way for commercially successful 'supertough' Nylon products that means products with Izod impact strength higher than 800 J/m, value that is obtained for a minimum particle size of about 1 μm [26, 27, 113-115] and for a certain interparticle distance [108-110, 202].

Table 7.3 Reactive Compatibilised Polymer Blends [9, 10, 15, 78, 79]			
Blend	Reactive processing procedure conditions and compatibiliser used or synthesised *in situ*	Some characteristics and enhanced properties*	Ref.
PP/EVA	*PP-g-MA*	*Reduced permeability to gases and liquids*	
PO/Elastomers		*Improved impact resistance*	
PA 6 or PA 66/PP or EPR or EPDM, or nitrile rubber	*Poly(E-co-acrylic ester-MA) and SEBS-MA. PP-g-MA-PA 6, oxazoline-functionalised NBR*	*Recyclable, high impact strength, particle size ~ 1 μm*	[124]
PO/PA or PES	PO-g-GMA or MA	Rigid	
PO/EPDM	Melt grafting 200-210 °C, DCP or BPO, roll-mill/press method. PO functionalised oligomers with MA diallyl maleate, AA, ethylenglycol dimethacrylate (EGDM), bismaleimide (BMI)	Three times increased impact strength	[43-45]
PA 6/PO	Melt grafting 200-210 °C, DCP or BPO, roll-mill/press method. Functionalised PO and oligomers with MA diallyl maleate, AA, EGDM, BMI	Increased thermal properties	[46]
iPP/LLDPE or HDPE, Mixed plastics	Maleated pyrolysis products of PE/PP blends	Increased degree of crystallinity	[47-49]
80 PA 6 (62 000; NH_2 content 32 meq/kg/20 VLDPE (885.6 kg/m³, 1.77 g/10 min; C_3/C_4 = 9:4, PA 6/ maleated PE octene elastomer	Extrusion functionalisation: 35 rpm; temperature profile 110/130/145/150 °C Blending: 110 rpm; twin screw, temperature profile: 170/210/235/230/220 °C 0.8 pph MA, 8 pph DEM VLDPE-g-MA or DEM 0.43 %mol DEM;	Improved toughness, impact strength; fracture behaviour; particle size ~ 5 μm; ε = 120-140% in respect with 5% for unfunctionalised	[125, 126]

Table 7.3 Continued			
Blend	Reactive processing procedure conditions and compatibiliser used or synthesised *in situ*	Some characteristics and enhanced properties*	Ref.
25 PA/75 – 40 PP	Intermeshing co-rotating twin-screw extruder, 150 rpm, 250-265 °C, 2.5 min 0-35 PP-g-PFA; viscosity ratio $\approx$ 1; shear rate 200-500 s^{-1}, inert atmosphere.	Dispersed PA particles – from 5.3 μm to 0.3-0.6 μm. The most efficient LMW PP backbone and high content of HMWPF polymer.	[30]
PA 6/Maleated PE 9.5% octane elastomer or EPDM	*Multifunctional epoxy resin 20 wt% EPDM-MA-g-PA 66*	*Supertough; impact strength 25 times higher, low water absorption; reduced cost, producer DuPont Co. (Zytel)*	[83]
PO/PA 6	PO-g-DES; FD= 1.7 mol %		[127, 128]
75 wt%HIPS/25 wt% PA 10,10	2.5, 5, 10 wt % HIPS-g-MA (1.6 or 4.7 wt% MA)	Impact strength, tensile strength; refrigerator inner box, TV housing, etc.	[31]
PE/MPA MPE/PA	Corotating twin-screw extruder; 195/215/230/250 °C; 200, 400, 600, 800 rpm 30% Zn neutralised E-acrylic acid copolymer	Durable containers for packaging agriculture chemicals, paint thinner, gasoline	[129]
PE/PS	Co-rotating twin screw; ditertbutylperoxide, styrene dissolved into PE		[130]
PP(Borealis)/PBT (Hüls); PP/10 – 30 wt%PPE	25 mm Clextral BC 21, 150 rpm; 250-270 °C for PBT blends; 255-285 °C for PPE blends; 2.5 min		[30, 32]
PP(amorphous)/PA (isophthalic acid,4,4′ diamino3,3′-dimethyldicyclohexylmethane and laurolactone)	PP-g-MA $\overline{M}n = 6000$	Finer particle size, better morphology stability, thicker interface ~40 nm, lamellar crystals	[131]

Table 7.3 Continued			
Blend	Reactive processing procedure conditions and compatibiliser used or synthesised *in situ*	Some characteristics and enhanced properties*	Ref.
75 Nylon 10,10/25PP (1 g/10 min)	PP-*g*-AA (1.02%) 95 wt%/ 5 wt% AA, 0.01 wt% DCP 175 °C, vacuum dried before blending, 80 °C, 12 h, single screw L/D = 20, 210-215 °C	Increased adhesive energy 100-250 J/m², viscosity, strong interaction, decrease particle size, increased Izod impact strength and tensile strength	[132]
PBT/PA 66 PBT/PA 6	2.5 wt% PBT-*g*-MA, multifunctional epoxy resin	Enhanced mechanical properties	[91]
PA 6/PC/PPO	Extrusion: 240 °C, 250 rpm PA 6-b-PC	Increased impact strength	[133]
PA/PPO	PP-*g*-SMA		[134]
PA 6/PC	PP-*g*-SMA		[135, 136]
PC/cyclic anhydride	1,2-Cyclohexanedicarboxylic, succinic, pyromellitic anhydrides	Succinic anhydride is more reactive compared with phthalic and pyromellitic anhydride	[137]
PP/Wood flour (40-45%)	Twin screw ZSK 25; L/D = 42, 200 °C, 300 rpm; masterbatches; PP/5-10% compatibiliser; PP-*g*-MA (5-6%); 3 wt%-10 wt% SEBS-*g*-MA; EPDM-*g*-MA	Improved surface hardness; 20% increase of tensile strength and stiffness; increased notched strength.	[138-140]
PA/PP relative brittle	SEBS-*g*-MA EPR-*g*-MA, MA~ 1% PP-*g*-MA SEBS-*g*-DEM; SEBS (reinforcing filler) are more efficient than EPR-*g*-MA.	Low water absorption, better dimensional stability, low cost, lower ductile-brittle transition, toughening.	[22-24]
PA 6/PP	*In situ* polymerisation of ε-caprolactam in PP matrix; PP-*g*-PA	Nano-PP/PA blends, 142% elongation at break	[23]

Table 7.3 Continued			
Blend	Reactive processing procedure conditions and compatibiliser used or synthesised *in situ*	Some characteristics and enhanced properties*	Ref.
PA/ABS	S/AN/MA imidised acrylic polymers, poly(S-co-AA) or poly (S-co-AA-co-MA)	Compatibilised blend is supertough and low temperature toughness	[100-103, 141-142]
PA/PO	EPR-g-MA	Imide linkages	[143-145]
PA 6/EEA by trans reaction; PA 6/LDPE	*LDPE-g-BA PA with Zn neutralised ionomer, EPDM-g-MA, EPR-g-MA, EVA-g-MA, NBR-g-MA, PE-g-GMA, SEBS-g-MA*	*Toughened PA*	
PA/PVDF	*PVDF-g-acid group;*	*Enhanced processability, increased modulus*	
PA/SEBS	SEBS-g-MA; MA + amine groups of PA 6	Remarkable increase of impact strength	[26, 27]
a) Ductile PA 6 (M_n= 33,000)/brittle S-MA 95/5 – 60/40 (M_w=90,000 – 180,000); b) Epoxy resin (Bisphenol A type) modified PA 6/SMAc with liquid rubber amino end-groups; c) PA 6 /PS/SEBS; d) PA 6/SMAc/-EPR/SEBS; e) PA 6/PC/PPO	Brabender plasticorder at 250 °C, 50 rpm, 10 min; injection moulding a) SMA or 12.5-100% fatty amine (C18); modified SMA b) Crosslinking reaction, significant torque increase c) SEBS d) EPR – MA (0.6% MA) and SEBS (29% S) with 2% MA	a) Dispersed particle size ~ 0.1 μm – 100 nm; higher strength and stiffness; unchanged toughness; cold drawing determines plastic deformation of SMAc particles, The highest toughness exhibits PA 6/50% amine modified SMAc; b) Simultaneously could drawing; increased toughness, decreased crystallinity; c) reactivity control d) rigid and elastomer toughening	[146-150]

	Table 7.3 Continued		
Blend	Reactive processing procedure conditions and compatibiliser used or synthesised *in situ*	Some characteristics and enhanced properties*	Ref.
PA 66 (M=25 000) /PMMA	PBAc-PMMA core-shell impact modifier; epoxy resin compatibiliser	Epoxy has the effect of increasing the values of the properties; ϕ = 0.2-0.3 μm, for 3 wt% DGEBA; increased impact strength 15 times, viscosity increases	[151]
PA 6/PBT PA 6/EEA, PA/LDPE	*S-g-MA or SMAc; LDPE-g-BA*	*Increased impact properties and tensile elongation.*	
PET/PA PET/HDPE/ PET/LCP/PP/ PET/PC/	*EPR-GMA SEBS-g-MA EEA-g-GMA SEBS-g-MA*	*Increased HDT, fracture strain by a factor of 10; Impact improving; Toughening; Dynamic vulcanisation*	
PES/Elastomers	*Reactive compatibiliser*	*Toughening*	
PBT/LDPE	*EVA-g-MA*		
PC/LLDPE	*LLDPE-MA, EEA+SEBS or EGMA*	*Impact and solvent resistance*	
PA/PC or PAr	*EGMA, ABS-GMA or ABS/MA or SEBS-g-MA*	*Improved processability, impact strength, HDT > 200 °C, low mould shrinkage and solvent sensitivity, adhesion and mechanical properties*	
PC/ABS	*SAN-g-PC amine functionalised SAN. 1-(2-aminoethyl)piperazine+ S/AN/MA*	*Improved processability and toughness of thick parts*	[106, 152]
	1 phr Poly(alkyl-g-caprolactone) 0.5-2 phr solid epoxy+MA-g-PP low cost compatibiliser	*Applications: automotive industry improved impact strength, elongation at break, high ductility, smoother PC/ABS surface.*	[152-154]

Table 7.3 Continued			
Blend	Reactive processing procedure conditions and compatibiliser used or synthesised *in situ*	Some characteristics and enhanced properties*	Ref.
ABS/PA or PES	*Reactive compatibiliser*		
PBT/ABS	*MGE terpolymer; MMA/GMA/EAc. MMA miscible with SAN, GMA reactive functionality; acrylate – stabilisation of MMA backbone against unzipping*	*Good tensile properties, abrasion and chemical resistance; brittle during impact tests. Improved low temperature toughness, broaden compatiblity window*	[26, 27, 100-103, 155]
PBT/SBS, SBS, HIPS or HIPS/SBS	EBA-GMA; MMA-g-GMA, S-co-GMA	*Good impact at low temperatures, toughened PBT*	
PPE/PA	Grafting of MA, fumaric, citric, itaconic acid onto PE; *In situ* generation of functional groups (aldehydes or acids) *via* thermal oxidation	Particle size required for toughening: < 0.6 μm	[26, 27, 129, 141, 142, 152, 156-159]
PPE-g-MA/SEBS-g-MA/PA 66	*PPE-g-MA/SEBS-g-MA/*		
PA/PPE	*PPE-g-epichlorohydrin or glycerol*	*Good impact, tensile and flexural strength, solvent and chemical resistance, low moisture absorption, dimensional stability, etc.*	
PO/Silicone copolymers	Polycaprolactone-polydimethylsiloxane-polycaprolactone block copolymer	Low T_g, low surface energy, good thermal and UV resistance, biocompatibility, high gas permeability	[160]
PC/PBT (50/50); PC/PET	MBS containing GMA, AAM, MAA; MBS, EPDM-g-MA-impact modifier;	Easy processability, good size stability, heat resistance, solvent resistance, improved impact strength	[161]

Table 7.3 Continued			
Blend	Reactive processing procedure conditions and compatibiliser used or synthesised *in situ*	Some characteristics and enhanced properties*	Ref.
PBT/EVOH, functionalised EPR or end-capped polymers as polyethers	Blending concurrent with polymerisation of PBT. EPR-*g*-Ma (1.5wt%) EPR-*g*-DBS	Improved impact strength	[51]
PPE/PES	PPE-*g*-GMA, PO-*g*-GMA Styrene polymers-*g*-GMA, PPO-*g*-carbodiimide [163, 164], PPE end-capped with epoxytriazines [168] and phenoxyresins [167], PPE-*g*-acrylate orthoester [165]	Lower water absorption/low heat resistance	[162-168]
PPE/PPS	*Functionalisation of one or both of the phases with orthoesters [172], epoxy [170, 173], maleic anhydride [174] in PPE phase, amide or acids [171]; addition of a functionalised or unfunctionalised impact modifier E-GMA copolymers*	*Lower moisture absorption, improved ductility, chemical and thermal performance; electric and automotive application*	[169-173]
PPE/PS/Nylon	*SES triblock, sulfonated PS ionomer*		
PCL/Liquid polybutadiene rubber	Hydroxy-terminated PCL + aminosilane coupling agent		[174]
PC/PA 6 PA 6/PBT	Epoxy resin (bisphenol A type) MA-*g*-ABS MA-*g*-PP	Epoxy resin promotes interfacial adhesion, finer and more uniform distribution of the dispersed phase.	[9, 26, 27]
PP/PBT–PP/AN-B--acrylic acid	PP-*g*-recinoloxazoline maleinate, PP-*g*-isopropenyloxazoline	Improved impact strength and elongation at break	[18-21]

Table 7.3 Continued			
Blend	Reactive processing procedure conditions and compatibiliser used or synthesised *in situ*	Some characteristics and enhanced properties*	Ref.
PA6/20wt%LCP, Vectra A 950	Brabender plasticorder/extrusion/ injection moulding, 285-290 °C, 50 rpm. Up to 10 wt% random SMA (5 33% mol MA)	Toughening, improved fibrillation and dispersion of LCP in PA 6 matrix, improved stiffness, tensile strength, prolonged crack initiation time.	[1, 92, 93, 175-178]
LCP/PP/PBT	MA-*g*-PP 5%acrylic acid functionalised PP epoxy functionalised PP	Improved impact strength, reduced size of the dispersed domains.	[179, 180]
PS/PE and NBR PP/NBR (0.2-3% amine terminated liquid NBR), PP/PC	PS functionalised with vinyl oxazoline, PP-2-isopropenyl-2-oxazoline, 10% PP-*g*-MA), phenolic curing agent + SnCl$_2$	Geolast, producer Monsanto Co. Oil resistant thermoplastic elastomer	[181-186]
LLDPE or PB/PS	SEBS (19 mol% S) and SBR (27 mol % S) g DEM		[26, 27]
PP/20-45%WF	PP-*g*-MA SEBS-*g*-MA EPDM-*g*-MA	100% increase of notched impact strength	[138-140, 187, 188]
PVC/LDPE	PO-*g*-DEM	High thermal stability	[189]
PVC/S-MA	terpolymer	High thermal resistance	
PO/PS	PE-*g*-AA PP-*g*-AA, M$_n$ 73000 P E-r-AA: M$_n$ 20 000; PD 8.7, 8.3 AA units/chain, PE (or PP) *g*-PS, PS-GMA (2 wt% GMA, 6-4 epoxy groups/chain, M$_n$ = 46 000, PD = 2.5)	PE-r-AA the most effective in reducing disperse phase domain size	[25]
LDPE/PS HDPE/PS PP/PS	SEBS-*g*-DEM SEBS followed by γ-irradiation	Asymmetric membranes with satisfactory mechanical properties	[190-192]
PPO/PS	Carboxylated PS, ~1% carboxylation degree	Satisfactory mechanical properties	[193]

Table 7.3 Continued			
Blend	Reactive processing procedure conditions and compatibiliser used or synthesised *in situ*	Some characteristics and enhanced properties*	Ref.
PCL/Biopol PHB/PHV (0-100%); PBAc/Biopol	Organic peroxide, 100 °C, 32 rpm. *In situ* polymerisation of polybutyl acrylate PBAc	Toughening of Biopol, improved mechanical performance, biodegradability, biocompatible, increased impact resistance	[194, 195]
Cellulose acetate/SMA copolymer		Phase homogeneity appears when grafting conversion of ~ 50% SMA	[196]
PPE/Poly(2,6 dimethyl-1,4-pheny-lene ether)/epoxy resin as reactive solvent		PPE continuous matrix decrease of processing temperature avoiding degradation, chemical induced phase separation, epoxy dispersed spheres in PPE matrix	[197]
80 wt% Thermoplastic polyurethane (TPU)/20 wt% PP	210 °C, 150 rpm, 50 s	Improvement of mechanical properties	[198]
EPR/PA	EPR-*g*-MA could be Ma, add also during hydrolytic polymerisation of caprolactam		[50]
PP/Elastomers		Better toughening, uniform distribution of the particle size	[199]
PVC/Collagen	EPR-*g*-MA, MA-*co*-VA, MA-*co*-MMA, MA-*co*-S	Improved compatibility, biocompatibility, superficial properties	[62, 63]
Polysaccharides/synthetic polymers	Maleic copolymers	Controlled drug delivery systems	[65]

* Commercial blends are in italics. Other examples have been given in the other chapters of this book. EVA: ethylene vinyl acetate BPO: benzoyl peroxide DES: diethylsuccinate iPP: isotactic PP MPA: modified polyamide MPE: modified polyethylene Co: copolymer PVDF: polyvinylidene fluoride NBR: acrylonitrile-butadiene rubber C_3/C_4: ratio of branches containing 3 or 4 carbon atoms LMW: low molecular weight M_n: number average molecular weight M_w: weight average molecular weight EEA: ethylene-ethyl acetate copolymer PD: degree of polydispersity PBAc: polybutyl acrylate Ma: maleic anhydride SMAc: styrene-maleic anhydride copolymer

Polyamides like Nylon 66 or Nylon 6 are frequently toughened by blending with hydrocarbon elastomers, which have been grafted with MA. Since there are both upper and lower limits on particle diameter for toughening, control of these reactions is crucial in designing optimally toughened PA. The change in rubber particle size with polyamide type seems to be greater than can be accounted for by assuming that the grafting reaction between PP-*g*-MA or SEBS-*g*-MA and PA changes the interfacial tension and the extent of steric stabilisation. Maleation reduced the ethylene-propylene rubber (EPR) drop diameter by a factor of ~ 5. The extent of the reaction increases as the hydrocarbon character of the PA (CH₂/NHCO ratio) increases. The estimated interfacial thickness increases in the same sense (see **Figure 7.7**).

The size range for supertough PA widens as the molecular weight of the matrix increases [113-115]. The worsening of properties observed in the region of lower particle size were explained either by the changes in the orientation of polyamide crystallites [203] or by their incapacity to cavitate [113-115, 203].

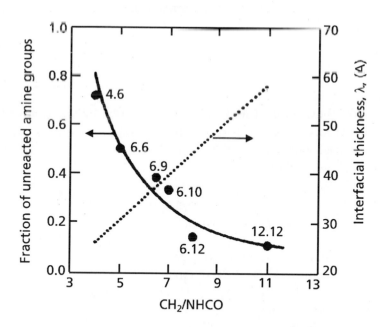

Figure 7.7 Extent of reaction in blends of Nylon $_{x,y}$ with SEBS-*g*-MA and estimated interfacial thickness [3]

Reproduced with permission from D.R. Paul, Macromolecular Symposia, 1994, 78, 83, Figure 5. Copyright 1994, Wiley-VCH

To obtain commercially useful blends, all of the characteristics have to be carefully designed such as rubber particle size, graft ratio, rubber content, viscosity of components, and especially to optimise the composition and functionality of the compatibilisers, end group type (monofunctional versus difunctional).

Toughened PA with maleated EPDM or EPR elastomers and PP-g-MA were developed first, than the second generation of PA blends such as PA/PPE, PA/ABS, and PA/PC were introduced. As a common feature, these blends synergistically combine the mechanical, thermal and chemical resistance typical of PA with superior toughness, higher dimensional stability at high temperature and lower moisture absorption provided by the second polymer [204]. When grafting degree increases crosslinking of elastomers appears (EPR-g-Ma (1.3%) crosslinked with hydroxyl-terminated PB). Elastomeric network could be formed with thermoreversible linkages. These networks are receiving more interest, as it is possible to process them several times by simple reheating.

The internal carbonate groups of PC may react with the amine terminal groups of PA 6 during the melt mixing process at 240 °C. The formation of block PA 6-b-PC copolymers suggested that these copolymers can act as compatibilisers, inducing compatibility of the heterogeneous blends of PA 6/PC [83].

7.7.2 PC/ABS

PC offers, transparency, dimensional stability, flame retardance, high HDT and high impact strength. ABS is a successful rubber-toughened thermoplastic, low cost, notch insensitive/low thermal stability, poor thermal and chemical resistance. Their combination gives materials with good properties.

7.7.3 Reactive Core – Shell Impact Modifiers

Many blends are prepared by simultaneous compatibilisation and impact modification by addition of multicomponent modifiers. Some examples of compatibilised and impact modified blends are: PE/PA 6/SEBS-MA (<10 wt%), PS/PVC/CPE with SEB-GMA, or random SEB-GMA (compatibilisation depends on GMA content and chlorine content of CPE), high impact polystyrene (HIPS)/PBT with PS-GMA [15, 26, 27] that exhibit improved impact strength.

Reactive core-shell impact modifiers are based for example, on crosslinked butadiene or *n*-butyl acrylate rubber as core, while the shell consists of grafted chains that chemically interact with the matrix. They are developed mainly for toughening polyamide-based systems, which require strong interactions. Usually, the reactive grafts are chemically incorporated during the final stage of emulsion polymerisation. The reacting groups can be chemically bonded either in the shell or are part of an added polymer such as S/MA copolymer, which is miscible with the shell chains. Due to the independent control of particle size, supertough polyamide compositions can be obtained [151, 202-209].

Another example is: PBAc-rubbery core, PMMA-glassy shell compatibilised with epoxy resin DGEBA [208]. The compatibilisation of digycidyl ether of bisphenol A (DGEBA) is achieved by the reaction of glycidyl groups with amine group of Nylon 6 and hydrogen bonds may be also generated between the hydroxyl groups and the carbonyl groups on PMMA. That leads to a dramatic increase of the impact strength and a finer dispersing of the core-shell particles in the Nylon 6 matrix. ABS is a core-shell particle impact modifier [100-103].

When a core-shell impact modifier is used, the problem of compatibility between the matrix and the polymer shell arises. In several cases, the shell is compatible with the matrix and the core-shell particle well dispersed. Otherwise, the particles exist as lumps instead of individual particles, which resulted in poor mechanical properties.

Compatibilisation of condensation type polymers, i.e., PA, PC or PES, usually involves a chemical reaction either with chain ends, modified chain ends or a trans-reaction. Initially this was accomplished by adding a compatibiliser-cum-impact modifier, ABS-MA (core BR with d = 0.3 to 0.8 µm, shell having thickness ≥ 0.25 µm grafted BR with S and other monomers. Modified butadiene styrene rubber (MBS) demonstrated a significant-modifying effect at low temperatures, but when they are used in very large amount (~ 20 wt%). This large amount is required probably because it has no suitable interaction between the interface of the MBS and the PC/PBT alloy. Furthermore, such a large amount of MBS affects some physical properties, such as HDT. A suitable interaction is provided if functional-containing MBS is used. Such impact modifier is composed of three layers. The inner layer is SBR, which mainly absorbs the outer impact energy. The middle layer is PS, which mainly connects the inner and outer layers. The outer layer includes a copolymer MMA/GMA or acrylamide or MMA [161].

The Monsanto Company patented a low cost method as reactive extrusion process for complete recycling of post consumer carpets, including Nylon 66 face fibres, PP backing and other components in presence of 3-10% PP-*g*-MA and SEBS as impact modifiers (reactive fillers), to produce pellets for use in injection moulding counter-rotating non-intermeshing extruder [210]. No fibre separation or latex removal steps are required.

Post-consumer carpets (PCC) account for ~ 1 wt% of municiple solid waste in the US. They contain Nylon 6, PES or PP. PP-*g*-MA (1-1.5%), eliminated the PP/Nylon 66 delamination and improved physical properties. Glass-filled compositions of PPC compete with several commercial resins. With SEBS-MA (20%), a significant increase in Izod resulted, so the recyclate can replace the parts of PP or replace 30% of virgin polyamide in many automotive products applications without sacrificing product quality.

7.7.4 PA/PO Blends

One major problem that limits the direct blending of PO with engineering polymers is the nonpolar structure of PO. However, reactor-made olefin based co- and terpolymers with functional groups are commercially available and many functionalities can be introduced into the structure of POs through post-modification [9, 51]. The presence of branched olefin blocks favours a high degree of functionalisation (FD) values due to the easier formation of macromolecular free radicals by hydrogen abstraction from tertiary C-atoms (LLDPE, very low density PE (VLDPE), EPR, PP). In presence of double bonds (as in EPDM), FD increases due to a combination of 'ene' reaction between the double bonds present on the polymer and the monomer; in the same time the crosslinking also takes place. The PO which degrade by chain scission, like as PP or SEBS showed the lowest FD values due to low permanence time of free radicals on the backbone [211].

Popular PO/PA blends are usually reactively compatibilised in three steps: (1) modification of PO by grafting acid or anhydride moieties; (2) mixing the PO-adduct with PA; (3) addition of main PO and PA components. Two-step blending by addition of the first-stage acidic copolymer, e.g., PO-MA, or SEBS-MA, to a PA/PP mixture has been carried out as well. The method can be extended to simpler system, e.g. impact modification of PA, PC or PES, or to more complex systems e.g. PO/PA/PES.

To improve the impact resistance, lower water absorption and reduce material costs of PA, various elastomers containing acrylate (methyl, butyl), acrylic acid or MA, diethylmaleate, have been blended in the PA melts [9, 212-214].

The compatibilising effect of PP-*g*-phenolformaldehyde (PFA) [30] is due to the strong interactions as hydrogen bonding or covalent binding between PA end-group carboxylic acids and the PFA end group hydroxyls.

The graft copolymer yield in reaction of functionalised (2-diethylsuccinate) PO with PA 6 is determined by the limited diffusion not by kinetic factors, because some catalysts did not change the compatibility and the degree of functionalisation [215].

Mixtures of various PO have been compatibilised in presence of maleated oligomers, which act as lubricants and compatibilisers improving processability, aiding to the recovery of polymer waste [48, 49].

7.7.5 Rubber Toughened PP

Reactive blending concurrently with the polymerisation reaction is very convenient method in obtain rubber-modified thermoplastic materials, because it is a single step process saving time tooling of materials, dispersion of rubber component in matrix is finer than that obtained with melt-mixing and unusual morphologies (core-shell, 'salami', co-continuous, etc., structures) are frequently obtained. Application of PP as an engineering thermoplastic is limited because of its relatively poor impact resistance, especially at room and low temperature. Rubber toughening in PP can solve these problems. PP/PP-g-MA/PEI with PEG of PPG backbone reactively extruded have improved printability for exterior automotive parts such as automotive fascia and body panel at low cost [216].

7.7.6 PPE/PA

PPE is an amorphous resin with $T_g = 210$ °C with good rigidity, high melting point. PPE offers high rigidity, high T_g, good heat resistance, dimensional stability, outstanding electrical properties required for automotive and electronics but it exhibits poor processability, susceptibility to oxidative, UV and visible light degradation, low impact strength, etc, so it can not be used as a neat resin. These last properties are offered by PA. PA absorb water and have poor dimensional stability; reactive compatibilisation of PPE has been at the forefront of its technology in the blends such as: PPE/PA/PPE-MA, PPE/PBT/PC in 1970 and with PA in 1979. In 1985, Torray introduced high performance blends for automotive applications PPE/PBT/PC/E-co-GMA. Hydroxyl-terminated PPE was end capped with trimellitic anhydride acid chloride. The resulting copolymer was used as compatibiliser for PPE/PA/SEBS blend. Mouldings showed high impact and tensile strength [26, 27].

7.7.7 PP/Wood Flour-Reactive Extrusion

During compounding, the succinic anhydride-functionalised polymers react with hydroxyl groups of the wood flour to form graft copolymers. PP-g-MA is the most common coupling agent for PP/wood flour (WF) [138-140]. Dalvag and co-workers [187] dispersed 45 wt% WF in PP and observed a 100% increase in modulus. With 6 wt% PP-g-MA on 30 wt% WF, they reached a 20% increase in tensile strength.

Oksman [188] used SEBS-g-MA, PP-g-MA and EPDM-g-MA as compatibilisers in compounds with 40% WF; with 10 wt% SEBS-g-MA, they obtained 100% notched impact strength.

7.7.8 Polyalkanoates-based Blends

Poly (hydroxybutyrate) (PHB) and its copolymers are environmentally degradable and their mechanical properties are comparable with iPP and PET. Their toughening must be achieved keeping their biodegradability (Biopol material) [194, 195]. The suitable components for such purpose are poly (butyl acrylate) and polycaprolactone, both being biodegradable polymers. First was polymerised in presence of PHB and the second was melt blending in presence of peroxide. In both cases grafted species were formed, the end products being completely biodegradable materials.

7.7.9 Complex Blends

Reactive blends of **PPE/SEBS/PET/PC/E-*g*-GMA** were developed for automotive applications. Low molecular weight maleated PO have been successfully used as melt flow improvers and reactive compatibilising agents for recovery of light autofluff [217].

PA/ABS/SAN-MA, PPE/PA/SBR-MA, S-GMA, SMAc or EMAA, PP/PA/maleic, fumaric, acrylic or methacrylic acid grafted PP or SEBS, PES/ABS/ABS-MA or ABS-GMA, PET/PO/PO-GMA or ethylene ethacrylate (EEA)-GMA with enhanced chemical and solvent resistance have been obtained [9, 79].

7.8 Compatibilisation by Using Ionomers

Compatibilisation by using ionomers has been used for impact modification of PA with multifunctional ionomeric acrylic elastomers. The incorporation of ionic groups in a polymer backbone induces drastic modifications in physico-chemical properties of the original materials; molecular flexibility and network structure may be strongly modified, providing a wide range of properties.

The ionic crosslinks are temporary and exist only at low temperatures. Poly (*p*-phenylene terephthalamide) K-amide salt with P-4 vinylpyridine (VP) or P-2-VP by formation of clusters and PA 6 and PA 66 with sulfonated styrenics containing 10 mol% of $-SO_3H$ are examples. The ethylene copolymers with acrylic and methacrylic acid partially neutralised with Zn^{+2} or Na^+ can be used to control the compatibility degree of non-miscible polymers such as PO with PA, PET and PC. Incorporation of ionic groups (PS with 5 mol% - SO_3H groups; ethyl acrylate with 5 mol% VP; sulfonated polyisoprene + PS 5 mol% VP); ion-dipole (P(S-*co*-lithium methacrylate) with 9.5 mol% of ionic groups + polyalkylene oxides). Feng and co-workers [218-220] found for the miscible lightly (4.4 – 9.5 mol % sulfonate groups) sulfonated PS ionomers and *N*-methylated polyamide (poly-*N,N*'-dimethylethylene sebacamide) system that the strength of the ion-amide (ion-dipol) complex increased with increasing electron-widthdrawing power of the cation in the order Zn, Cu, Mn, Cd, Li. The blends exhibit lower critical solution temperature (LCST ~ 150 °C) phase behaviour.

7.9 Charge Transfer Electron Donor-Electron Acceptor Complexation

Charge transfer complexation by formation of an intermolecular thermoreversible crosslinking can be achieved by incorporation in some incompatible blends such as PS/PMMA, PMA/PMMA, PBAc/PMMA, PS/polyisoprene of a certain number (30% mol) of electron donor (*N*-ethylcarbazol-3-yl)methylmethacrylate and the electron acceptor {2-[(3,5-dinitrobenzoyl)oxy]ethyl methacrylate} [121-123, 221-226]. The complexation determines the topology of the system, orientation, increased glass transition temperature, increase in polymer stiffness, reduced mobility. Decomplexation and phase separation occurs around LCST ~ 200 °C.

7.10 Thermoplastics/Thermoset Systems and IPN

Some areas of the phase separation in such systems are outside the scope of this chapter, so only several aspects are presented.

Thermosets and thermoplastics can be successfully combined to bring out the best properties of each system [227]. At high enough loading levels of reactive thermoplastics in a thermoset matrix, the systems exhibit toughness and solvent insensivity properties with retaining the processability of the neat thermoset.

Thermosetting polymers like phenolics, epoxies, unsaturated polyesters, etc., are frequently used in formulations containing a low or high-molecular mass-rubber, a thermoplastic polymer, etc., in an amount of 2-50 wt% compared to the thermoset. This extra component, called the modifier, may initially be miscible or may phase separate during cure (reaction induced phase separation). Toughnening of the polymer network means increased values of fracture energy but decrease in the elastic modulus and yield stress. This situation is typical of rubber-modified epoxies used as coatings, structural adhesives or matrices of composites materials. When engineering thermoplastics like polyetherimide or polyether sulfone are used the decrease of mechanical properties is avoided [228]. The brittleness of thermosets (epoxies and polyesters) at room temperature can be overcome by adding reactive low molecular mass rubbers, by reactive blending procedures, to the brittle matrix [51, 229] especially to enhance the toughness of tri- and tetra-functional epoxies and of unsaturated polyester resins.

To obtain the highest modifying effect of an additive strong adhesive, interaction between included particles and matrix is often a necessity. Obviously, the interfacial interaction is the strongest when chemical bonds form between these phases. Therefore the modifiers must contain functional groups that can react with one of the components of the oligomer system being cured. Modifying additives are usually miscible with the initial oligomer, however, such a single-phase system undergoes phase separation during the cure. As a result, two phases are formed: a matrix phase, which is enriched with the cross-linked polymer and a disperse phase rich in modifying additive. It is quite obvious that chemical bonds between the phases of a heterophase polymer must appear before the phase decomposition takes place in the course of curing. The rate of the reaction involving functional groups on the surface of particles that have separated from solution during the cure decreases sharply, as compared to the reaction rate of these groups in solution.

Controlling kinetics of phase separation during the formation of the IPN (or semi-IPN) provide the method of generation of desired morphology and properties. One of the most important characteristics of the morphological structure of a heterophase system is the interfacial interaction. The miscibility with monomers may be varied significantly by changing the nature of end-groups. Increasing polarity of rubber through incorporation of polar comonomers in butadiene/acrylonitrile copolymers increased miscibility with DGEBA-based epoxy resin [228] in the order amino, epoxy, carboxyl and non-functional end-capped copolymers. Carboxylic acid terminated PS or PB increased energy per unit area required to fracture at epoxy-polymer interface. Core-shell particles bearing functional groups such as GMA can help to disperse thermoplastic phase in the thermosetting matrix.

The reaction becomes diffusion-controlled and the concentration of the reacting groups becomes lower. Therefore, the rate of reaction tends to zero. The second condition that must be satisfied is that the reactivity of functional groups of modifying additive with one of the components of the oligomer system must significantly exceed that in reactions leading to the formation of a network.

Low molecular weight polybutadiene and butadiene-acrylonitrile (B-AN) copolymers terminated with carboxyl, vinyl, amine, epoxy, phenol and hydroxyl groups have been widely used as toughening agents both for epoxy and polyester resins. Thermally reactive isoprene-acrylonitrile and ethylacrylate-butyl acrylate copolymers have been also used. The toughening effect of several liquid rubbers has been tested such as: carboxyl terminated B-AN, vinyl terminated B-AN, hydroxyl terminated polyether, polyepichlorohydrin, on polyester resins etc. [230, 231].

Toughening agents based on polyetherimide (M_W = 17000), amine-terminated PEI (M_W = 9000) have been used into epoxy resin/diaminodiphenylsulfone systems, by a solventless process. A mixture of PEI and amine-terminated PEI gave the best result in the epoxy resin. High performance material must meet the following criteria: high T_g, high modulus, good adhesion to a variety of substances, processing easy, resistance to solvents.

7.11 Reactive Extrusion of Water Soluble Polymers

Diallyldimethylammonium chloride (PDADMAC)/polyacrylamide was reactively compatibilised on a non-intermeshing counter-rotating tangential twin-screw extruder using glycerol as plasticiser and 2,5 dimethyl-di-(t-butylperoxy) hexane-3 (Lupersal 130) as initiator [232]. Grafted copolymers showed remarkable improvement in flocculation and sludge dewatering performance over homopolymers, random copolymers and their blends.

7.12 Other Applications of the Reactive Blending

Grafting of light stabilisers as hindered amine light stabilisers and antioxidants [233, 234], good adhesion filler/matrix (example EPR-g-DEM with silica are some examples of various applications of reactive blending) [234].

7.13 Future Trends

The problems related to morphology stabilisation against damage during processing and adhesion between the phases in the solid state still need basic research. In order to a better understanding, control and modelling both of functionalisation and reactive blending as an interdisciplinary research (chemistry, physics and engineering) should be developed. To improve theoretical insights kinetics of process, mass and heat transfer for many types of systems has to be known. Search for new functional, very reactive groups is in progress. New solutions for scrap reclamation could be found.

References

1. *Handbook of Polyolefins,* 2nd Edition, Ed., C. Vasile, Marcel Dekker Inc., New York, NY, USA, 2000, Chapter 1, p.17; Chapter 23, p.633; Chapter 26, p.723.

2. K.J. Ganzeveld and L.P.B.M. Jenssen, *Industrial Engineering Chemistry and Research,* 1994, 33, 2398.

3. R.D. Paul, *Die Makromolekulare Chemie - Macromolecular Symposia,* 1994, 78, 83.

4. L. Leiber, *Macromolecular Chemistry, Macromolecular Symposia*, 1988, **16**, 1.

5. K. Shull, *Macromolecules*, 1993, **26**, 9, 2346.

6. R.J. Roe, *Macromolecules*, 1986, **19**, 731.

7. C. Painter, Y. Park and M.M. Coleman, *Macromolecules*, 1989, **22**, 2, 570.

8. E.J. Moskala, S.E. Howe, P.C. Painter and M.M. Coleman, *Macromolecules*, 1984, **17**, 9, 1671.

9. *Reactive Extrusion: Principles and Practice*, Ed., M. Xanthos, Hanser Publishers, Munich, Germany, 1992.

10. *Polymer Blends and Alloys*, Eds., M.J. Folkes and P.S. Hope, Blackie Academic, London, UK, 1993, 491.

11. M. Ratzsch, H. Bucka, A. Hesse, N. Reichelt and E. Borsig, *Macromolecular Symposia*, 1998, **129**, 53.

12. M. Ratzsch, H. Bucka, A. Hesse, N. Reichelt and E. Borsig, Proceedings of ANTEC '96, Indianapolis, IN, USA, 1996, Volume 2, 1616.

13. E. Passaglia, J.B. de Veiga, M. Aglietto and F. Ciardelli, 1st International Conference on Polymer Modification, Degradation and Stabilisation, Palermo, Italy, 2000, (MoDeSt), Symposium 7, Paper No. P7Th1400.

14. V.P. Volkov, A.N. Zelenetskii, M.D. Sizova, N.Y. Artemeva, N.A. Egorova and V.P. Nikolovskaya, 1st International Conference on Polymer Modification, Degradation and Stabilisation, Palermo, Italy, 2000, (MoDeSt), Symposium 7, Paper No. P7Th2100.

15. L.A. Utracki, Miscibility and Compatibilisation, 1st International Conference on Polymer Modification, Degradation and Stabilisation, Palermo, Italy, 2000, (MoDeSt), Symposium 7, Paper No. SL17M1715.

16. A. Addeo, C. Morandi and A. Vezzoli, *Die Makromolekulare Chemie - Macromolecular Symposia*, 1994, **78**, 313.

17. D.B. Todd in *Reactive Extrusion: Principles and Practice*, Ed., M. Xanthos, Carl Hanser Publishers, Munich, Germany, 1992, Chapter 5, 203.

18. T. Vainio, G-H. Hu, M. Lambla and J.V. Seppala, *Journal of Applied Polymer Science*, 1996, **61**, 5, 843.

19. T. Vainio, G-H. Hu, M. Lambla and J.V. Seppala, *Journal of Applied Polymer Science,* 1997, **63**, 7, 883.

20. P. Hietaoja, M. Heino, T. Vainio and J.V. Seppälä, *Polymer Bulletin,* 1996, **37**, 3, 353.

21. H. Cartier and G-H. Hu, *Journal of Materials Science,* 2000, **35**, 8, 1985.

22. A. J. Oshinski, H. Keskkula and D. R. Paul, *Polymer,* 1992, **33**, 2, 268 and 284.

23. G.H. Hu, H. Cartier and C. Plummer, *Macromolecules,* 1999, **32**, 14, 1713;

24. J. Pieglowski, M. Trelinska-Wlazlak and I. Gancarz, *Angewandte Makromolecular Chemistry,* 1999, **269**, 61

25. S. Kim, J.K. Kim and C.E. Park, *Polymer,* 1997, 38, 8, 1809.

26. R. Majumdar and D.R. Paul in *Polymer Blends,* Volume I, Formulation, Eds., D.R. Paul and C.B. Bucknall, John Wiley & Sons, New York, NY, USA, 2000, Chapter 17, 540-673 and references herein.

27. E. Passaglia, S. Ghetti, F. Picchioni and G. Ruggeri, *Polymer,* 2000, **41**, 12, 4389.

28. S. Wu, *Polymer,* 1985, **26**, 12, 1855.

29. K. Dijkstra, J. ter Laok and R.J. Gaymans, *Polymer,* 1994, 35, 1399.

30. K.L. Borve, H.K. Kotlar and C.G. Gustafson, *Journal of Applied Polymer Science,* 2000, 75, 3, 355.

31. G. Chen and J. Liu, *Journal of Applied Polymer Science,* 2000, **76**, 6, 799.

32. K.L. Borve, H.K. Kotlar and G.C. Gustafson, *Journal of Applied Polymer Science,* 2000, 75, 3, 361.

33. A. Harada and Q. Tran-Cong, *Macromolecules,* 1997, 30, 6, 1643.

34. S.B. Brown and C.M. Orlando in *Encyclopedia of Polymer Science and Engineering,* Volume 14, John Wiley & Sons, New York, NY, USA, 1988, 169.

35. J.L. Lee in *Comprehensive Polymer Science,* Eds., G.G. Allen and J.C. Bevington, Pergamon Press, Oxford, UK, 1989, Volume 7, 379.

36. N.C. Liu and W.E. Baker, *Advances in Polymer Technology,* 1992, **11**, 4, 249.

37. M. Xanthos and S.S. Dagli, *Polymer Engineering and Science,* 1991, **31**, 3, 929.

38. L. Mascia and M. Xanthos, *Advances in Polymer Technology,* 1992, **11**, 4, 237.

39. N.G. Gaylord and M.K. Mishra, *Journal of Polymer Science, Polymer Letters Edition,* 1986, **21**, 23.

40. N.G. Gaylord and R. Mehta, *Journal of Polymer Science: Polymer Chemistry,* 1989, **26**, 4, 1189.

41. N.G. Gaylord, M. Mehta and R. Mehta, *Journal of Applied Polymer Science,* 1987, **53**, 7, 2549.

42. W. Heinen, C.H. Rosenmoller, C.B. Wenzel, H.J.M. Degroot, J. Lugtenburg and M. van Duin, *Macromolecules,* 1996, **29**, 4, 1151.

43. M. Mihaes and C. Vasile, Proceedings of BRAMAT '99 International Conference, Volume IV, Materials and Environmental Protection, Brasov, Romania, 1999, 246.

44. L. Neamtu, M. Dascalu, H. Darie and C. Vasile, Proceedings of BRAMAT '99 International Conference, Volume IV, Materials and Environmental Protection, Brasov, Romania, 1999, 240.

45. L. Neamtu, M. Esanu, M. Dascalu and C. Vasile, Proceedings of BRAMAT '99 International Conference, Volume IV, Materials and Environmental Protection, Brasov, Romania, 1999, 251.

46. M. Dascalu, H. Darie, C,M. Tabarna, R. Darie and C. Vasile, *Bulletin of the Polytechnical Institute of Iasi,* 1999, **XLV(IL)**, 1-2, 281.

47. C. Vasile, A. Kozlowska, R. Darie, M. Kozlowski, M. Sava and R. Darie, Proceedings of the Europolymer Congress, Eindhoven, The Netherlands, 2001, Paper No. PL3-13.

48. C. Vasile, R.D. Deanin, M. Mihaies, C. Roy, A. Chaala and W. Ma. *International Journal of Polymeric Materials,* 1996, **37**, 3-4, 173.

49. C.Vasile, R. D. Deanin, M. Mihaies, M. Leanca and T. Lee, *International Journal of Polymeric Materials,* 1998, **41**, 335.

50. A.V. Machado, J.A. Covas and M. van Duin, 1st International Conference on Polymer Modification, Degradation and Stabilisation, Palermo, Italy, 2000, (MoDeSt), Symposium 7, Paper No. 07M1645.

51. M. Avella, P. Laurienzo, M. Malinconinco, E. Martuscelli and M.G. Volpe in *Handbook of Polyolefins*, Ed., C. Vasile, Marcel Dekker Inc., New York, NY, USA, 2000, Chapter 26, 723.

52. S. Datta and D.J. Lohse, *Polymeric Compatibilisers: Uses and Benefits in Polymer Blends*, Carl Hanser, Munich, Germany, 1996.

53. N.G. Gaylord, *Journal of Polymer Science, Part C* 1970, **31**, 247;

54. N.G. Gaylord, H. Andripova and B.K. Patnak, *Journal of Polymer Science, Polymer Letters*, 1971, **9** 387.

55. N.G. Gaylord, H. Andripova and B.K. Patnak, *Journal of Polymer Science, Polymer Letters*, 1972, **10**, 2, 95.

56. R. Rengarajan, V.R. Parameswaran, S. Lee, M. Vicic and L. Rinaldi, *Polymer*, 1990, **31**, 9, 1703.

57. M. Lazar, L. Hrckova, A. Fiedlerova, E. Borsig, M. Ratzsch and A. Hesse, *Angewandte Makromolekulare Chemie*, 1996, **243**, 57.

58. E. Borsig and L. Hrckova, *Journal of Macromolecular Science*, 1994, **A31**, 10, 1447.

59. E. Borsig and L. Hrckova, *Journal of Macromolecular Chemistry and Applied Chemistry*, 1995, **A32**, 2017.

60. D. Curto, A. Valenza and F.P. La Mantia, *Journal of Applied Polymer Science*, 1990, **39**, 4, 865.

61. B.C. Trivedi and B.M. Culbertson, *Maleic Anhydride*, Plenum Press, New York, NY, USA, 1982, 296.

62. M. Lungu, G.G. Bumbu, G.C. Chitanu and C. Vasile inventors, Romanian Academy, assignee; Romanian Patent A/00557/2000.

63. M. Lungu, G.G. Bumbu, M. Precup, I. Moldovanu, G.C. Chitanu, H. Darie and C. Vasile, 1st International Conference on Polymer Modification, Degradation and Stabilisation, Palermo, Italy, 2000, (MoDeSt), Symposium 7, Paper No. 02Tu1145.

64. G.G. Bumbu, C. Vasile, G.C. Chitanu and A. Carpov, *Polymer Degradation and Stability*, 2001, **72**, 99.

65. A. Carpov, G.G. Bumbu, H. Darie and C. Vasile, 5th Mediterranean School on Science and Technology of Advanced Polymer-Based Materials, Eds., E. Martuscelli, P. Musto and G. Ragosta, 1998, Naples, Italy, 513.

66. F. Ciardelli, M. Aglietto, E. Passaglia and F. Picchioni, *Polymers for Advanced Technologies*, 2000, **11**, 8-12, 371.

67. P. P. Gan and D. R. Paul, *Polymer*, 1994, **35**, 16, 3513.

68. C. Vasile, Milton Downey, Betty Wong, M. M. Macoveanu, M. Pascu, J-H. Choi, C. Sung and W. Baker, *Cell Chemistry and Technology*, 1998, **32**, 61.

69. Cr.I. Simionescu, M.M. Macoveanu, C. Vasile, F. Ciobanu, M. Esanu, A. Ioanid, P. Vidrascu and N. Georgescu Buruntea, *Cell Chemistry and Technology*, 1996, 30, 411.

70. J.A. Framp, *Chemical Review*, 1997, **71**, 483

71. U. Anttila, C. Vocke and J. Seppala, *Journal of Applied Polymer Science*, 1999, **72**, 7, 877.

72. U. Anttila, C. Vocke and J. Seppala, *Journal of Applied Polymer Science*, 1999, **72**, 43.

73. F.P. La Mantia, R. Scaffaro, C. Colletti, T. Dimitrova, P. Magagnini, M. Paci and S. Filippi, 1st International Conference on Polymer Modification, Degradation and Stabilisation, Palermo, Italy, 2000, (MoDeSt), Symposium 7, Paper No. 07Tu71300.

74. T. Inoue and T. Suzuki, *Journal of Applied Polymer*, 1999, **56**, 9, 1113.

75. C.M. Liauw, V. Khunova, G.C. Lee and R.N. Rothon, *Macromolecular Materials and Engineering*, 2000, **279**, 6, 34.

76. T. Inoue, *Journal of Applied Polymer Science*, 1994, **54**, 6, 709.

77. P. Sulek, S. Knaus, R. Liska, P. Sattler and J. Wendrinsky, 1st International Conference on Polymer Modification, Degradation and Stabilisation, Palermo, Italy, 2000, (MoDeSt), Symposium 7, Paper No. P7Th1900.

78. R.D. Deanin, M.A. Manion, C.H. Chuang and K.N. Tejeswi in *Handbook of Polyolefins*, Second Edition, Ed., C. Vasile, Marcel Dekker Inc., New York, NY, USA, 2000, Chapter 22, 613.

79. R.D. Deanin and M.A. Manion in *Handbook of Polyolefins*, Second Edition, Ed., C. Vasile, Marcel Dekker Inc., New York, NY, USA, 2000, Chapter 23, 633.

80. A. Ajji and L.A. Utracki in *Progress in Rubber and Plastics Technology*, 1997, **13**, 3, 153.

81. A. Ajji and L.A. Utracki, *Polymer Engineering and Science*, 1996, 36, 12, 1574.

82. S.B. Brown in *Reactive Extrusion: Principles and Practice*, Ed., M. Xanthos, Hanser Publishers, Munich, Germany, Chapter 4, 1992, 75.

83. L.A. Utracki, *Polymer Alloys and Blends. Thermodynamics and Rheology*, Hanser Publishers, Munich, Germany, 1989.

84. D.R. Paul in *Thermoplastic Elastomers*, Eds., G. Holden, N.R. Legge, R.P. Quirk and H.E. Schroeder, Hanser Publishers, Munich, Germany, 1996, Chapter 15.

85. J.H. Anastasiadis, I. Gancarz and J.T. Koberstein, *Macromolecules*, 1989, **22**, 3, 1449.

86. P.Guegan, C.W. Macosko, T. Ishizone, A. Hirao and S. Nakahama, *Macromolecules*, 1994, 27, 18, 4993.

87. J.K. Kim and H. Lee, *Polymer*, 1996, 37, 306.

88. J. Duvall, C. Sellitti, C. Myers, A. Hiltner and E. Baer, *Journal of Applied Polymer Science*, 1994, **52**, 2, 195.

89. J. Duvall, C. Sellitti, C. Myers, A. Hiltner and E. Baer, *Journal of Applied Polymer Science*, 1994, **52**, 2, 207.

90. C. Berti, A. Celli, M. Colonna, P. Fabbri, M. Fiorini and E. Marianucci, 1st International Conference on Polymer Modification, Degradation and Stabilisation, Palermo, Italy, 2000, (MoDeSt), Symposium 7, Paper No. P7Th0500.

91. J. John and M. Bhattacharya, *Polymer International*, 2000, **49**, 8, 860.

92. S.C. Tjong, R.K.Y. Li and X.L. Xie, *Journal of Applied Polymer Science*, 2000, 77, 9, 1964.

93. K-C. Chiou and F-C. Chang, *Journal of Polymer Science: Polymer Physics*, 2000, 38, 1, 23.

94. B. O'Shaughnessy and U. Sawhney, *Macromolecules*, 1996, **29**, 22, 7230.

95. N.A. Plate, A.D. Litmanovich, V.V. Yashin, I.V. Ermakov, Y.V. Kudryavtsev and E.N. Govorun, *Polymer Science Series A,* 1997, **39**, 1, 3.

96. S. Wu, *Polymer Engineering and Science,* 1987, **27**, 5, 335.

97. A. Gonzales-Montiel, H. Keskkula and D.R. Paul, *Polymer,* 1995, **36**, 24, 4587.

98. A. Gonzales-Montiel, H. Keskkula and D.R. Paul, *Polymer,* 1995, **36**, 24, 4605.

99. A. Gonzales-Montiel, H. Keskkula and D.R. Paul, *Polymer,* 1995, **36**, 24, 4621.

100. B. Majumdar, H. Keskkula, and D.R. Paul, *Polymer,* 1994, **35**, 15, 3164.

101. B. Majumdar, H. Keskkula, D.R. Paul and N.G. Harvey, *Polymer*, 1994, **35**, 20, 4263.

102. B. Majumdar, H. Keskkula, and D.R. Paul, *Polymer,* 1994, **35**, 20, 5453.

103. B. Majumdar, H. Keskkula, and D.R. Paul, *Polymer,* 1994, **35**, 25, 5468.

104. W. Hale, H. Keskkula, and D.R. Paul *Polymer,* 1999, **40**, 21, 365, 3665.

105. W. Hale, L.A. Pessan, H. Keskkula, and D.R. Paul *Polymer,* 1999, **40**, 15, 4237.

106. G. Wildes, H. Keskkula and R.D. Paul, *Polymer,* 1999, **40**, 20, 5609.

107. A.J. Oshinski, H. Keskkula, and D.R. Paul, *Journal of Applied Polymer Science,* 1996, **61**, 4, 623.

108. H. Keskkula and R.D. Paul in *Nylon Plastics Handbook*, Ed., M.I. Kohan, Hanser Publishers, Munich, Germany, 1995.

109. F.C. Chang in Toughened Plastics II – Novel Approaches in Science and Engineering, Eds., C.K. Riew and A.J. Kinlock, ACS, Washington, DC, USA, 1998.

110. C-C. Huang and F-C. Chang, *Polymer,* 1997, **38**, 17, 4287.

111. B. Majumder, B.H. Keskkula and D.R. Paul, *Journal of Applied Polymer Science,* 1994, **54**, 3, 339.

112. R.A. Kudva, H. Keskkula and D.R. Paul, *Polymer,* 1998, **39**, 12, 2447.

113. O.J. Oshinski, H. Keskkula and D.R. Paul, *Polymer*, 1996, **37**, 22, 4891.

114. O.J. Oshinski, H. Keskkula and D.R. Paul, *Polymer*, 1996, **37**, 22, 4909.

115. O.J. Oshinski, H. Keskkula and D.R. Paul, *Polymer*, 1996, **37**, 22, 4919.

116. F.C. Chang in *Polymer Blends and Alloys*, Eds., M.J. Folkes and P.S. Hope, Blackie, London, UK, 1993, 491-521.

117. C. Maier, M. Lambla and K. Ilham, Proceedings of Antec 95, Boston, MA, USA, Volume 2, 2015.

118. A. De Loor, P. Cassagnau, A. Michel and B. Vergnes, *International Polymer Processing*, 1994, **9**, 3, 211.

119. B. Majumdar, D.R. Paul and A.J. Oshinski, *Polymer*, 1997, **38**, 8, 1787.

120. M.C. Piton and A. Natansohn, *Macromolecules*, 1995, **28**, 1197.

121. M.C. Piton and A. Natansohn, *Macromolecules*, 1995, **28**, 1598.

122. M.C. Piton and A. Natansohn, *Macromolecules*, 1995, 28, 1605.

123. G. Cho and A. Natansohn, *Chemistry of Materials*, 1997, **9**, 148.

124. J. Piglowski, I. Gancarz and M. Wlazlak, *Polymer*, 2000, **41**, 3671.

125. A. Lazzeri, M. Malanima and M. Pracella, *Journal of Applied Polymer Science*, 1999, **74**, 3456.

126. Z.Z. Yu, Y.C. Ke, Y.C. Ou and G.H. Hu, *Journal of Applied Polymer Science*, 2000, **76**, 1285.

127. E. Passaglia, M. Aglietto, G. Ruggeri and F. Picchioni, *Polymer Advanced Technology*, 1998, **9**, 273.

128. A. Lazzeri, M. Malanima and M. Pracella, *Journal of Applied Polymer Science*, 1999, **74**, 3455.

129. J.T. Yeh, C.C. Chao and C.H. Chen, *Journal of Applied Polymer Science*, 2000, 76, 1997.

130. D.J. van der Wal and L.P.B.M. Janssen, Proceedings of ANTEC '94, San Francisco, CA, USA, 1994, 46.

131. H. Li, T. Chiba, N. Higashida, Y. Yang and T. Inoue, *Polymer*, 1997, 38, 15, 3921.

132. X.M. Zhang and J.H. Yin, *Polymer Engineering and Science*, 1997, 31, 1, 197.

133. P. Hing and T.M. Ko, *Polymer Engineering and Science*, 1997, 37, 1326.

134. K. Hodd, *Trends in Polymer Science*, 1993, 1, 129.

135. G. Montaudo, C. Puglisi, F. Samperi and F.P. La Mantia, *Journal of Applied Polymer Science*, 1996, 34, 1283.

136. G. Montaudo, C. Puglisi, F. Samperi and F.P. La Mantia, *Journal of Applied Polymer Science*, 1994, 32, 15.

137. C. Berti, M. Colonna, M. Fiorini and E. Marianucci, 1st International Conference on Polymer Modification, Degradation and Stabilisation, Palermo, Italy, 2000, (MoDeSt), Symposium 7, Paper No. P7Th0200.

138. H. Nitz, P. Reichert, H. Römling and R. Mülhaupt, *Macromolecular Materials Engineering*, 2000, 276/277, 51.

139. C. Joly, M. Kofman and R. Gauthier, *Journal of Macromolecular Science, Pure and Applied Chemistry*, 1996, A33, 1981.

140. J.M. Felix and P. Gatenholm, *Journal of Applied Polymer Science*, 1991, 42, 609.

141. B. Majumar, H. Keskkula and D.R. Paul, *Journal of Polymer Science, Part B: Polymer Physics*, 1994, 32, 2127.

142. B. Majumar, H. Keskkula and D.R. Paul, *Journal of Polymer Science, Part B: Polymer Physics*, 1994, 857.

143. S. Cimmino, L. D'Orazio, R. Greco, G. Maglio, M. Malinconico, C. Mancarella, E. Martuscelli, P. Musto, R. Palumbo and G. Ragosta, *Polymer Engineering and Science*, 1985, 25, 193.

144. R. Greco, M. Malinconico, E. Martuscelli, G. Ragosta and G. Scarinzi, *Polymer*, 1988, 29, 1418.

145. S. Cimmino, F. Coppola, L. D'Orazio, R. Greco, G. Maglio, M. Malinconico, C. Mancarella, E. Martuscelli, and G. Ragosta, *Polymer*, 1986, 27, 1874.

146. I. Kelnar, M. Stephan, E, Jakisch and I. Fortelny, *Journal of Applied Polymer Science*, 1997, 66, 555.

147. I. Kelnar, M. Stephan, E, Jakisch and I. Fortelny, *Journal of Applied Polymer Science*, 1999, **74**, 1404.

148. I. Kelnar, M. Stephan, E, Jakisch and I. Fortelny, *Journal of Applied Polymer Science*, 2000, **78**, 1597.

149. I. Kelnar, M. Stephan, E, Jakisch and I. Fortelny, *Polymer Engineering and Science*, 1999, **39**, 6, 985.

150. D.A. Costa and C.M.F. Oliveira, *Journal of Applied Polymer Science*, 1998, **69**, 857.

151. X. Wang and H. Li, *Journal of Applied Polymer Science*, 2000, **77**, 24.

152. G. S. Wilder, T. Haroda, H. Keskkula, D.R. Paul, V. Jamarthanan and A.R. Padwa, *Polymer*, 1999, **40**, 3069.

153. C.G. Cho, T.H. Park and Y.S. Kim, *Polymer*, 1997, **38**, 4687.

154. S.C. Tjong and Y.Z. Meng, *European Polymer Journal*, 2000, **36**, 123.

155. G.L. Mantovani, L.C. Canto, E.H. Junior and L.A. Pessan, *Macromolecular Symposia*, submitted.

156. M.K. Akkapeddi, B.V. Buskirk and G.J. Dege, Proceedings of ANTEC '94, San Francisco, CA, USA, 1994, Volume 2, 1509.

157. C.R. Chiang and F.C. Chiang, *Journal of Applied Polymer Science*, 1996, **61**, 2411.

158. C.R. Chiang and F.C. Chiang, *Polymer*, 1997, **38**, 3807.

159. T. Nishio, T. Sanada and T. Okada, inventors; Sumitomo Chemical, assignee; US Patent, 5,159,018, 1991.

160. E. Yilor, I. Yilgor and T. Sumazcelik, *Polymer Preprints*, 1998, **39**, 1, 552.

161. W.T.W. Tseng and J.S. Lee, *Journal of Applied Polymer Science*, 2000, **76**, 1280.

162. C.Y. Han and W.I. Gately, inventors; General Electric, assignee; US Patent 4,689,372, 1987.

163. F.F. Khouri, R.J. Halley and J.R. Yates, inventors; General Electric, assignee; US Patent, 5,212,255, 1993.

164. F.F. Khouri, R.J. Halley and J.R. Yates, inventors; General Electric, assignee; US Patent 5,247,006, 1993.

165. M.K. Akkapeddi and B. Van Buskirk, *ACS Polymer Preprints,* 1992, **67**, 317.

166. A. Sugio, M. Okabe and A. Amagai, inventors; Mitsubishi Gas Kagaku KK, assignee; EP 148,774, 1985,

167. S.B. Brown and R.C. Lowry, inventors; General Electric, assignee; US Patent 5,153,267, 1992.

168. S.B. Brown, C.F.R. Hwang, F.F. Khouri, S.T. Rice, J.J. Scobbo and J.B. Yates, inventors; General Electric, assignee; US Patent 5,223,351, 1996.

169. C.Y. Han, S.B. Brown, E.W. Walles, T. Takekoshi and A.J. Caruso, inventors; General Electric, assignee; US Patent, 5,122,578, 1992.

170. H. Izutsu, K. Yamamura and E. Tanigawa, inventors; Dainippon Ink & Chemical Inc., assignee; JP 6166783, 1994.

171. S.B. Brown, K.H. Dai, C.F.R. Hwang, S.T. Rice, J.J. Scobbo and J.B. Yates, inventors; General Electric, assignee; US Patent, 5,504,165, 1996.

172. S.B. Brown, K.H. Dai, C.F.R. Hwang, S.T. Rice, J.J. Scobbo and J.B. Yates, inventors; General Electric, assignee; US Patent, 5,612,401, 1997.

173. No inventors; Sumitomo Chem, assignee; JP 532879, 1993.

174. M. Okamoto and T. Inoue, *Polymer Engineering and Science,* 1993, **3**, 176.

175. S.C. Tjong and Y.Z. Meng, *Polymer,* 1997, **38**, 4609.

176. S.C. Tjong and Y.Z. Meng, *Polymer International*, 1997, **42**, 209.

177. S.C. Tjong and Y.Z. Meng, *Polymer,* 1998, **39**, 99.

178. S.C. Tjong and Y.Z. Meng, *Polymer,* **39**, 1845.

179. A. Datta and D.G. Baird, *Polymer,* 1995, **36**, 505.

180. H.J. O'Donnell and D.G. Baird, *Polymer,* 1995, **36**, 3113.

181. W.E. Baker and M. Salem, *Polymer*, 1987, **27**, 1634.

182. W.E. Baker and M. Salem, *Polymer*, 1987, **27**, 2037.

183. W.E. Baker and M. Salem, *Polymer Engineering and Science,* 1988, **28**, 1427.

184. W.E. Baker and M. Salem, *Journal of Applied Polymer Science,* 1990, **39**, 655.

185. N.C. Liu, H.Q. Xie and W.E. Baker, *Polymer,* 1993, **34**, 4680.

186. N.C. Liu, H.Q. Xie and W.E. Baker, *Polymer,* 1994, **35**, 988.

187. H. Dalvag, C. Klason and H.E. Stromvall, *International Journal of Polymer Materials,* 1985, **11**, 9.

188. K. Oksman and C. Clemans, *Journal of Applied Polymer Science,* 1998, **67**, 1503.

189. B. Rojas, M. Vivas, C. Rosales, J. Gonzales and R. Perera, *Revista de Plasticos Modernos,* 1995, **69**, 468, 555.

190. E. Passaglia, S. Ghetti, F. Piechioni and G. Ruggeri, *Polymer,* 2000, **41**, 4389.

191. M. Chiriac, G.G. Bumbu, V. Chiriac, C. Vasile and M. Burlacel, *International Journal of Polymeric Materials,* 2001, **49**, 419.

192. M. Chiriac, B.S. Munteanu, G.G. Bumbu, M. Burlacel, A. Ioanid and C. Vasile, *Macromolecular Materials Engineering,* 2000, **283**, 26.

193. I. Kelnar, M. Spephan, L. Jakisch, M. Janata and I. Fortelny, 1st International Conference on Polymer Modification, Degradation and Stabilisation, Palermo, Italy, 2000, (MoDeSt), Symposium 7, Paper No. P7Th1100.

194. B. Immirisi, M. Malinconinco, E. Martuscelli and M.G. Volpe, *Macromolecular Symposia,* 1994, **78**, 243.

195. M. Avella, B. Immirzi, M. Malinconico, E. Martuscelli and M.S. Volpe, *Polymer International,* 1996, **39**, 3, 191.

196. L. Niq and R. Nareayan, *Polymer,* 1999, **35**, 20, 4334.

197. R.W. Venderbosch, H.E.H. Meijer and P.J. Lemstra, *Polymer,* 1994, **35**, 4349.

198. K-J. Eichhorn, I. Hopfe, P. Pötschke and P. Schmidt, *Journal of Applied Polymer Science,* 2000, **75**, 1194.

199. J.Z. Liang and R.K.Y. Li, *Journal of Applied Polymer Science,* 2000, **97**, 409.

200. B. N. Epstein, inventor; EI DuPont de Nemours & Company, assignee; US 4,174,358, 1979.

201. J. Houghton, inventor; no assignee; US 174,859, 1979.

202. A. Margolina and S. Wu, *Polymer,* 1988, **29**, 12, 2170.

203. O.K. Morantoglu, A.S. Argon, R.E. Cohen and M. Weinberg, *Polymer,* 1995, **36**, 25, 4771.

204. F. Speroni, *Die Makromolekulare Chemie - Macromolecular Symposia,* 1994, 78, 299.

205. R.J.M. Barggreve and R.J. Saymans, *Polymer,* 1989, **30**, 63.

206. R.J.M. Barggreve and R.J. Saymans, *Polymer,* 1989, **30**, 71.

207. R.J.M. Barggreve and R.J. Saymans, *Polymer,* 1989, **30**, 78.

208. M. Lu, H. Keskkula and D.R. Paul, *Polymer,* 1993, **34**, 9, 1874.

209. M. Lu, H. Keskkula and D.R. Paul, *Polymer Engineering and Science,* 1994, **34**, 1, 33.

210. C.G. Hagberg and J.L. Dickerson, Proceedings of ANTEC '96, Indianapolis, IN, USA, 1996, Volume 1, 288.

211. F. Ciardelli, M. Aglietto, E. Passaglia and G. Ruggeri, 1st International Conference on Polymer Modification, Degradation and Stabilisation, Palermo, Italy, 2000, (MoDeSt), Symposium 7, Paper No. 07Tu71045.

212. M. van Duin and R.J.M. Borggreve in *Reactive Modifiers for Polymers*, Ed., S. Al-Malaika, Kluwer Academic Publishers, Dordrecht, The Netherlands, 1997, Chapter 3, 133.

213. M. Aglietto, E. Benedetti, G. Ruggeri, M. Pracello, A. d'Alessio and F. Ciardelli, *Macromolecular Symposia,* 1995, **98**, 1101.

214. J.Z. Liang and R.K.Y. Li, *Journal of Applied Polymer Science,* 2000, 77, 408.

215. W. Passaglia, F. Piccioni, M. Aglietto, G. Ruggieri and F. Ciardelli, *Turkish Journal of Chemistry,* 1997, **21**, 262.

216. R.J. Clark, Proceedings of ANTEC '94, San Francisco, CA, USA, 1994, Volume 3, 2572.

217. C. Vasile, S. Woramongconchai, R.D. Deanin, M. Mihaes, A. Chaala and C. Roy, *International Journal of Polymeric Materials*, 1997, **38**, 263.

218. Y. Feng, A. Schmidt and R.A. Weiss, *Macromolecules*, 1996, **29**, 3909.

219. Y. Feng, R.A. Weiss and C.C. Han, *Macromolecules*, 1996, **29**, 3925.

220. Y. Feng, R.A. Weiss, A Karim, C.C. Han, J.F. Ankner, H. Kaiser and D.G. Peiffer, *Macromolecules*, 1996, **29**, 3918.

221. M.C. Pascu, G. Popa, M. Grigoras and C. Vasile, *Synthetic Polymer Journal*, 1998, **2**, 158.

222. U. Epple and H.A. Schneider, *Thermochimica Acta*, 1987, **112**, 123.

223. U. Epple and H.A. Schneider, *Thermochimica Acta*, 1990, **160**, 103.

224. M. Bolsinger and H.A. Schneider, *Macromolecular Chemistry and Physics*, 1994, **195**, 2683.

225. M. Bolsinger and H.A. Schneider, *Journal of Thermal Analysis*, 1998, **51**, 643.

226. M. Bolsinger and H.A. Schneider, *Journal of Thermal Analysis*, 1998, **52**, 115.

227. C.D. Athanasiou and R.J. Farris, Proceedings of ANTEC '95, Boston, MA, USA, 1995, Volume II, 2766.

228. R.J.J. Williams, B.A. Rozenberg and J.P. Pascault, *Advances in Polymer Science*, 1997, **128**, 97.

229. E. Martuscelli, P. Musto, G. Ragosta and G. Scarinzi, *Angewandte Makromolecular Chemistry*, 1994, **217**, 159.

230. B.A. Komarov, E.A. Dzhavadyn, V.I. Irzhak and B.A. Rozenberg, *Polymer Science Series A*, 1997, **39**, 2, 153.

231. B.A. Komarov, E.A. Dzhavadyn, V.I. Irzhak and B.A. Rozenberg, *Polymer Science Series A*, **39**, 2, 1997, 237.

232. R. Subramanian, Y.H. Huang, S. Zhu, A.N. Hrymak and R.H. Pelton, *Journal of Applied Polymer Science*, 2000, **77**, 1154.

233. S. Pantschov, S. Knaus and H. Gruber, 1st International Conference on Polymer Modification, Degradation and Stabilisation, Palermo, Italy, 2000, (MoDeSt), Symposium 7, Paper No. P7Th1300.

234. S. Al-Malaika in *Reactive Modifiers for Polymers*, Ed., Al-Malaika, Kluwer Academic Publishers, Dordrecht, The Netherlands, 1997, Chapter 6, 266.

235. E. Passaglia, J.B. Beirao de Veiga, M. Aglietto and F. Ciardelli, 1st International Conference on Polymer Modification, Degradation and Stabilisation, Palermo, Italy, 2000, (MoDeSt), Symposium 7, Paper No. P7 Th1140.

8 Advanced Polymers: Interpenetrating Networks

Leonard Ignat and Aurelian Stanciu

8.1 Short History

The term interpenetrating network (IPN) was introduced for the first time by Millar in 1960s in a scientific study about polystyrene networks [1]. Since that time, the field of IPN has expanded dramatically. Reviews of the field include numerous articles and books by Sperling [2, 3] and Kim [3], Klempner and co-workers [4, 5], Lipatov and Sergeeva [6] and Bischoff and Cray [7]. However, it is difficult to identify quickly, many IPN-related works because many articles, patents and commercial products on IPN do not explicitly contain the terms 'interpenetrating polymer networks' or 'IPN'.

One can ask: why IPN? Is it not enough to know how many classes of polymers, blends, alloys and composites exist? One can successfully search for simple polymeric materials with all the properties that one needs for one's applications. Furthermore, IPN complexity does not mean that the polymer will be more expensive or sensitive to deterioration.

Examples of IPN are seen everyday without realising it. Like Jonas Aylsworth, Thomas Edison's chief chemist, who in 1914 mixed rubber and sulfur with a crosslinked phenol-formaldehyde resin (which was brittle and easily broken), in the hope of obtaining a material with more toughness for phonograph records. It was a great success, his lucky star rising in that day from the constellation of interpenetrating polymer networks, but no one named this constellation until the 1960s and even today many look at it, without knowing his name.

However, the extraordinary properties of what Millar first called IPN could be of great interest to someone who was experiencing difficulty in using traditional polymeric materials for certain applications.

The extreme broadening of the range of polymeric material applications needs new, simple, flexible and efficient solutions. Surprisingly or not, these solutions are quite frequently found by using complex and rigid IPN structures.

Some hybridation features could be extrapolated from the live organisms to these polymeric systems. So, an IPN resulting from the combination of two different polymers could exhibit not only a mixture of the individual, but also a synergism of their properties. Furthermore, an IPN may look like one of the constitutive polymers but act as the other, or it may be quite different from both. It is also important to know that controlling the IPN synthesis one can easily control the properties of the end product. Moreover, IPN are intercompatible, self-priming, self-repairing and very stable over time. As consequence, one IPN can replace with high efficiency, many competitive and expensive materials for a particular job.

8.2 Introduction

Interpenetrating polymer networks are macromolecular assemblies comprising two or more networks at least partially interlaced on a molecular scale but not covalently bonded to each other, which cannot be separated unless chemical bonds are broken.

Reciprocally permanent mutual entanglements that occur when at least one network is synthesised and/or crosslinked in the intimate presence of the other will prevent the total phase separation and stabilise the morphology by keeping the networks together. A schematised spatially distribution of networks chains in an IPN is depicted in **Figure 8.1**.

Because there is no chemical bonding between the networks, each network may retain its individual properties independently of its individual proportion in the blend. As result an improvement can be attained in properties such as mechanical strength, impact resistance, toughness etc, with respect to the corresponding homopolymer networks by use of IPN.

Interpenetrating polymer networks can be distinguished from the other multiphase systems through their bicontinuous structure ideally formed by two crosslinked polymers that are in intimate but not chemical contact, yielding materials with properties ranging from elastomeric to high impact plastics depending on the composition and the degree of crosslinking. So, IPN are not simple, macroscopically homogeneous, physical mixtures of different polymer species, like polymer blends. They are also different from graft copolymers and polymer complexes that involve either chemical bonds and/or low degree of crosslinking. From this point of view only (the preparative steps and most of the structure and properties presented being quite different), IPN can be generically named 'polymer alloys' through which polymer blends can be made chemically compatible to achieve the desired phase morphology. Furthermore, IPN swell and creep and flow are suppressed, features which are not presented by other macromolecular assemblies. Despite the general occurrence of some degree of phase separation because of the combined

Figure 8.1 Schematic representation of the entanglements between IPN networks

action of kinetic and thermodynamic factors, interpenetrating polymer networks represent a successful approach to solve the problem of mutual incompatibility of polymers.

8.3 IPN Characterisation

Morphology and structure of IPN depends on the method of synthesis, composition, compatibility of the monomers and polymer systems used, crosslinking density of each polymer network and the degree of crosslinking reached in IPN, extent of inter-network grafting, reaction conditions and polymerisation rates. IPN have a complex internal friction behaviour and character, as well as some degree of thermodynamical incompatibility. The properties are dependent on the two-phase morphology that develops during the polymerisation process. Therefore the control of the physical properties can be accomplished by establishing the relationships between the synthetic variables and the resultant morphology. By controlling the factors such as relative polymerisation rates and grafting reactions between the components many distinct morphological types can be obtained from the same set of starting ingredients.

The IPN network structure resulting through either chemical or physical crosslinks among the polymer molecules consists of a complex architecture of crystalline, glassy or ionic phase domains that contain multiple segments of polymer chains. While most IPN do not interpenetrate at the molecular level, when properly engineered they may form finely divided co-continuous phase domains in the nanometer size range, behaviour often responsible for their unique properties. Independent of the chosen method of synthesis, the miscibility decreases, and a phase separated IPN will be obtained during the polymerisation of the monomers. Since crosslinking reduces phase separation and, in particular, reduces domain size, it often results in finely dispersed phase domains ranging from about 10 to 100 nm. The kinetics and extent of crosslinking and phase separation have great influence over the phase morphology [2, 3].

Various phase morphologies, i.e., droplets or fibres dispersed in a continuous matrix, could be attributed to the different volume fractions of the immiscible mixed polymers, the shear viscosity of the phases, the degree of interfacial compatibility and the relative rates of crosslinking and phase separation.

The co-continuous nature of IPN can lead to materials having a greater toughness than each of the constituent polymers. When microheterogeneous phase domains are in the range from ~10 to ~20 nm, the whole material is essentially a interphasic one, so the glass transition temperatures (T_g) tend to be very broad, intermediate between those of the two component polymers, giving materials that can absorb energy or are useful as damping materials over a broad temperature and frequency range.

A variety of methods and techniques are used for studying the main IPN features such as miscibility, morphology, synthesis and curing processes, physical, thermo-mechanical, swelling and electrical properties.

The most common procedures for studying miscibility in IPN consist of macroscopic determination of the glass transition temperature and compatibility behaviour by differential scanning calorimetry (DSC) and dynamic mechanical analysis (DMA) techniques.

When IPN constitutive polymers are completely miscible, a single glass transition intermediate between the T_gs of the homopolymers appears, which indicates an infinite number of phases of a different composition. Partially miscible polymers may show one broad glass transition or two sharp transitions, shifted inward compared with those of the homopolymers, each phase containing a certain amount of the other component. For immiscible polymers, two glass transition temperatures can be observed, separate and similar with those of the homopolymers. Thus, the presence of one or two glass transition temperatures, their sharpness and positions gives an indication of characteristic macroscopic properties of IPN materials and provides a qualitative manner of gaining information on miscibility [8]. Also, significant changes in glass transition values of the IPN components compared to the pure polymers indicates the extent of interpenetration in system. Using DSC has shown that the compatibility of polymers in IPN formation depends on the molecular weight of the prepolymers. Compatible IPN systems show reduced swelling behaviour and high-density values.

To observe more directly the compatibility and interpenetration between networks it is of interest to see the tensile fracture surface of IPN using scanning electron microscopy (SEM). However, is difficult to establish the appropriate domain size and interface between phases of different compositions and to associate them with characteristic macroscopic properties.

DSC techniques can also be used, together with Fourier transform infra red spectroscopy (FT-IR), for controlling the curing processes. So, the heats of reaction, together with the variations in T_g value and shape during IPN formation gives information about curing evolution and thermal characteristics of both the final material, and the starting and intermediate components. The DSC data can be correlated with the IR evolution of the functional groups characteristic bands, allowing the determination of the reaction rates and transformation efficiency of the starting materials.

DSC has been also used in studying the thermal stability of the systems, while the heat of photocure can be measured by the differential photocalorimetry (DPC). DMA has been applied to determine the viscoelastic properties of cured IPN. Thermal-modulated differential scanning calorimetry (M-TDSC) could be used for the analysis of IPN interphases and quantitative calculation of interphase mass fraction, interfacial thickness composition and distribution, for the estimation of vitrification during isothermal curing. M-TDSC results could be correlated with transmission electron microscopy (TEM) and dynamic mechanical thermal analysis (DMTA) data.

DMTA analysis of thermal relaxation, associated with DSC and DMA data, gives information on elasticity modulus changes that, compared with those obtained from modulus/composition theories [9, 10], allow relative heterogeneity and mechanical damping determination.

Damping is the most sensitive indicator of all kinds of molecular motions displayed by a polymer in a solid state and can be correlated to many transitions, relaxation processes, structural heterogeneities, and to the morphology of multiphase and crosslinked systems [11].

The integration of the area under the loss modulus curve has been suggested as a quantitative measurement of the damping capacity (tan δ) of the material by many authors [12-14], supported by the fact that the loss modulus is directly proportional to the heat dissipated per cycle of dynamic deformation. The experimental T_g data could be correlated using the equation developed by Brekner [15]:

$$\frac{\left(T_g - T_{g1}\right)}{\left(T_{g2} - T_{g1}\right)} = \left(1 + K_1\right)w_{2c} - \left(K_1 + K_2\right)w_{2c}^2 + K_2 w_{2c}^3 \tag{8.1}$$

where: T_g = glass transition temperature of IPN;

T_{g1}, T_{g2} = glass transition temperatures of pure components;

K_1 = equation parameter reflecting the difference between the interaction energies of the binary hetero-contacts and the homo-contacts;

K_2 = equation parameter reflecting the energetic contributions of the conformational entropy changes during binary hetero-contacts formation;

w_{2c} = the weight fraction of the component with the higher glass transition temperature (T_{g2}), that can be expressed as:

$$w_{2c} = \frac{K_{GT}w_2}{w_1 + K_{GT}w_2} \tag{8.2}$$

where K_{GT} is the Gordon-Taylor parameter.

Prediction of the physico-mechanical properties of phase-separated systems like IPN, i.e., tensile modulus and elongation, could be done through the so-called 'self-consistent model'. This model comprises theories which fit more or less with a particular system, such as those developed by Kerner, Hoshin, Shtrikman, Budiansky and Nielsen, more or less appropriate to a given system. Despite the limitations related to evaluation of synergistic and antagonistic effects, the Budiansky theory could be considered more reasonable and gives the dependence of the shear modulus on the Poisson ratio of the system [16, 17]:

$$\frac{1}{G} = \frac{1}{G_m} + \left(1 - \frac{G_f}{G_m}\right)\frac{\varphi}{G + \beta\left(G_f - G\right)} \tag{8.3}$$

where $\beta = \dfrac{2\left(4 - 5v\right)}{15\left(1 - v\right)}$, v is the Poisson ratio of the system;
G – shear modulus;

G_f, G_m – shear modulus for the minor component polymer chains viewed as filler and matrix network, respectively;

φ– volume fraction of the minor component (network).

The Budiansky model can also be expressed for a rigid/soft type composition of the IPN through the following equation:

$$\frac{v_1}{\left[1+\beta\left(\frac{G_1}{G-1}\right)\right]} + \frac{v_2}{\left[1+\beta\left(\frac{G_2}{G-1}\right)\right]} = 1 \tag{8.4}$$

where: G – shear modulus of the system,

G_1 – shear modulus of the rigid polymer

G_2 – shear modulus of the soft polymer

v_1 – volume fraction of the rigid polymer

v_2 – volume fraction of the soft polymer

The change in chemical potential of individual IPN, $\Delta\mu_2$, may be estimated with the equation of Gibbs-Duhem:

$$w_1 \frac{\delta\Delta\mu_1}{\delta w_1} + w_2 \frac{\delta\Delta\mu_2}{\delta w_2} = 0 \tag{8.5}$$

where: $\Delta\mu_1$ – the change in the chemical potential of the solvent, $\Delta\mu_1 = \frac{RT}{M}\ln\frac{P}{P_0}$,

M is the molecular mass of the solvent and $\frac{P}{P_0}$ is the relative vapour pressure of the solvent in the system;

w_1, w_2 – the mass fraction of the solvent and polymer in a swollen system, respectively.

The values of the polymer-filler interactions in filled IPN, Δg_{P-F}, can be calculated based on the analysis of the solvent sorption by an unfilled polymer, the filler and filled polymer according to the equation [18-20]:

$$\Delta g_{P-F} = n\Delta g_1 + m\Delta g_{11} + \Delta g_{111} \tag{8.6}$$

where: Δg_1 – free energies of the filled polymer interaction with a large amount of solvent;

Δg_{11} – free energy of the filler with a large amount of solvent;

Δg_{111} – free energies of the unfilled polymer interaction with a large amount of solvent;

n – mass fraction of the polymer

m – mass fraction of filler in filled polymers

Temperature dependence of the peak frequency of the dielectric loss for the α- and β-relaxation mechanisms of the IPN can be determined by using the Vogel-Tammann-Fulcher equation [21, 22]:

$$\tau = \tau_0^{[B/(T-T_0)]} \tag{8.7}$$

where τ is the characteristic relaxation time, whereas τ_0, B and T_0 are constants.

From this equation can be calculated the apparent activation energy, E [22, 23]:

$$\frac{E}{k} = \frac{\partial \ln \tau}{\partial \left(\frac{1}{T}\right)} = \frac{BT^2}{(T - T_0)^2} \tag{8.8}$$

The characteristic relaxation parameters for β-relaxation in IPN can be extracted from a single Havriliak-Negami function [22, 24]:

$$\varepsilon^*(\omega) - \varepsilon_\infty = \frac{\Delta\varepsilon}{\left[1 + (i\omega\tau)^{1-\alpha}\right]^\beta} \tag{8.9}$$

where: ε^* is the complex dielectric function, $\varepsilon^* = \varepsilon' - i\varepsilon'$;

$\varepsilon'(\omega)$ are the experimental data, ω is the angular frequency;

ε_∞ is the permitivity real component at high frequencies, $\varepsilon_\infty = \varepsilon'(\omega)$ for $\omega \gg 1/t$;

$\Delta\varepsilon$ is the relaxation strength, $\Delta\varepsilon = \varepsilon'(0) - \varepsilon'(\infty)$;

τ is the relaxation time;

α and β are shape parameters describing the slope of the $\varepsilon'(\omega)$ curve below and above the frequency of the peak ($0 \leq \alpha < 1$, $0 < \beta \leq 1$).

The contribution of each relaxation can be separated by fitting the experimental curves to a model consisting of the sum of two Havriliak-Negami equations, one for each overlapped relaxation:

$$\varepsilon^*(\omega) - \varepsilon_\infty = \frac{\Delta\varepsilon}{\left[1+\left(i\dfrac{\omega}{\omega_0}\right)^b\right]^a} \qquad (8.10)$$

where: ω_0 is the characteristic Havriliak-Negami frequency;

a and b are the shape parameters.

The degree of segregation in IPN can be calculated from the maximum relaxation parameters [25]:

$$SG = \left(h_1 + h_2 - \frac{h_1 l_1 + h_2 l_2 + h_m l_m}{L}\right)\Big/(h_1^0 + h_2^0) \qquad (8.11)$$

where: h_1, h_2 – mechanical loss for each component at different degree of segregation;

h_1^0, h_2^0 – maximum values of mechanical loss for the pure components;

h_m – maximum values of mechanical loss for the relaxation transition;

L – the interval between the glass transition temperatures of the pure components;

l_1, l_2, l_m – the shifts in temperature scale.

Theoretical conversion to gelling for each IPN phase could be determined using the Flory statistical theory [8, 26]:

$$\alpha_{gel} = \left[\frac{1}{(f_{m1}-1)(f_{m2}-1)}\right]^{\frac{1}{2}} \qquad (8.12)$$

where: α_{gel} is the conversion to gelling;

f_{m1} and f_{m2} are the functionality of the given phase constitutive monomers.

The segmental mobility in IPN can be estimated by the calculation of θ parameter with the Fox and Frisch equations [27, 28]:

$$\frac{T_g - T_{g(est.)}}{T_{g(est.)}} = \frac{\theta}{\theta + 1} \qquad (8.13)$$

$$T_{g(est.)} = W_1 T_{g1} + W_2 T_{g2} \qquad (8.14)$$

283

$$\theta = \left(\frac{F_x}{F_m} - 1 \right) X_c'$$

(8.15)

where: T_g = measured values for IPN glass transition;

$T_{g(est)}$ = predicted values for IPN glass transition;

T_{g1}, T_{g2} = glass transition temperatures for pure components;

W_1, W_2 = weight fractions of components;

F_x/F_m = ratio between segmental mobility of components in IPN material and those of components mixed in a non-IPN form, without entanglements;

x_c' ($0 < x_c' < 1$)= increasing of crosslink density by interpenetration.

Small angle neutron scattering (SANS), small angle x-ray scattering (SAXS), atomic force microscopy (AFM), TEM, and nuclear magnetic resonance (NMR), are also used to characterise the morphology, particularly the extent of mixing related to the domain size and to the extent of interfacial mixing, of IPN made by a wide variety of synthesis procedures.

NMR can be successfully used to probe the dimensions of phase-separated regions and also the intimately mixed fraction of the IPN materials. Meanwhile, the electron microscopy techniques provide enough details about the structural features and distances of the order of ~100 nm to allow a comparison with the NMR results given by NMR spin diffusion techniques. Mulder and co-workers [29] have used rapid ^{1}H NMR spin diffusion to bridge distances up to 100 nm coupled with high resolution ^{13}C NMR that affords the selection and detection of ^{13}C label polarisation. ^{13}C NMR spectroscopy is furthermore required to discriminate between rigid IPN material components, ^{1}H NMR resolution being generally not sufficient due to the strong dipolar couplings and limited chemical shift dispersion.

8.4 Types of IPN

A wide variety of IPN architectures have been formulated in an effort to improve material properties. These macromolecular systems differ mainly by synthesis pathway, phase segregation extent, type of polymer network, number and types of crosslinks that occur. However, there are two major types of IPN, named full-, and semi-IPN, and two basic methods of making them, sequential or simultaneous.

When crosslinking affects only one polymer component, the resulting material is a semi-IPN, while a full-IPN is made when all species are in network form. If all constitutive polymers are treated at the same time, one in the presence of the other(s), a simultaneous IPN is obtained; otherwise the IPN is called sequential. These procedures of synthesis are comparatively presented in **Figure 8.2.**

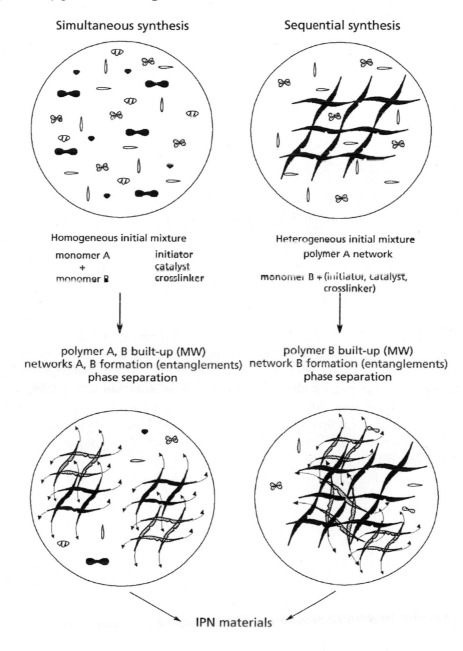

Figure 8.2 The two major pathways for synthesis of an IPN system

The IPN constitutive polymer species can be crosslinked during or after the mixing, in a post-curing process.

These major categories can be furthermore categorised into specific groups like thermoplastic, hydrogel, latex, grafted, hybrid, filled, conductive, elastomeric, gradient and homo IPN, based on their characteristic features. In many cases the borders between different types of IPN are also 'interpenetrated', making an exhaustive classification difficult. The most commonly polymer networks used in IPN formation are those based on polyurethanes, polysiloxanes, polyesters and epoxy resins, polyacrylates, carbohydrate polymers and rubbers.

8.4.1 Simultaneous IPN

When polymer networks are simultaneously formed and subsequently crosslinked one in the presence of the other, in different and non-interfering reactions, a simultaneous IPN (sim-IPN) is obtained. Compatibility is greater in a mixture of monomers and prepolymers than in a mixture of polymers, so sim-IPN generally exhibit a high degree of intermixing and require crosslinking of two mostly compatible polymers because the degree of thermodynamic incompatibility of simultaneously growing chains greater affects the extent of phase separation.

In the preparation of sim-IPN the degree of phase segregation and thus the final morphology attained (ranging from very fine phase dispersion to more or less individualised domains) are principally influenced by the relative rate of the two competitive kinetic processes of phase separation and network formation. Both processes are controlled by the reaction conditions such as composition and miscibility of components, amount of catalyst, temperature, viscosity and thermodynamic compatibility of contained polymers, all of which determine boundary layer composition and the properties of the resulting materials. A selective adsorption of some components by the others takes place simultaneously, a phenomenon which is responsible for incomplete phase separation at boundary layers and no uniform polymer distribution in the thickness.

While equilibrium thermodynamics describes the overall tendency for a system to remain miscible or to realise a phase separation, it is dynamics that determines the speed of a possible phase separation. Spinodal decomposition together with nucleation and growth are the two types of phase separation dynamics generally involved. The average size of the molecules increases with increasing time and therefore the diffusion coefficients decrease rapidly, so the dynamics of the phase separation become very slow and may facilitate an irreversible freezing of the spinodal decomposition in an stationary state. As a consequence, the formation of homogeneous sim-IPN can be interpreted as a freezing of spinodal decomposition by irreversible chemical reactions [30-33].

Three important events take place in the process of sim-IPN synthesis, gelation of the first polymer, gelation of the second and phase separation of former from the latter polymer. By controlling their order of occurrence one can manage the resulting sim-IPN morphology and properties. Based on flow-gelation characteristics, turbidity and FT-IR analysis of the extent of polymerisation of each component, metastable phase diagrams can be constructed which describe the condition under which gelation of one polymer may be caused to precede or follow the other, and under which conditions phase separation precedes or follows one or both gelations [34, 35].

The time of network formation, associated with the physical interlock of components, is the time when both polymers reach the gel point and thus the phase domain size and surface segregation cannot increase much further beyond this point [36, 37]. By reaching simultaneously the gel point of the two essentially independent polymerisation reactions, simultaneous network formation provides the best interchain contact for a given composition. Some examples of sim-IPN are presented in **Table 8.1**.

8.4.2 Sequential IPN

Sequential IPN (seq-IPN) are obtained by firstly polymerising a mixture of monomer, crosslinking agent, initiator or catalyst to form a network, and secondly by swelling the preformed network in another combination of monomer, crosslinking agent, initiator or catalyst that is subsequently polymerised *in situ*. The network swollen in the other component of the mixture may restrict the second network polymerisation and limit the

Table 8.1 Examples of IPN made by the simultaneous method				
Simultaneous IPN	Example	Type	Subtype	Refs
Acrylate/acrylate	PAAm/PNIPAM	Semi	Hydrogel	[38, 39]
Urethane/polyester	PU(MDI/PPG)/ PES(Pa/Ma/PPG)	Full	Thermoplastic	[8]
Acrylate/carbohydrate	PNIPAM/alginate	Full	Hydrogel	[40]
Hydroxy acrylate/carbohydrate	PEGM/chitosan	Full	Hydrogel	[41]

PAAm = polyacrylamide; PNIPAM = poly(N-isopropylacrylamide); PU = polyurethanes; MDI = 4,4′-methylene bis(phenyl isocyanate); PPG = polypropylene glycol; PES = polyester; Pa = phthalic anhydride; Ma = maleic anhydride; PEGM = poly(ethylene glycol) macromer

compositions and properties of resulting seq-IPN. The first formed network is stretched due to the swelling step and tends to form the matrix, whereas in sim-IPN both networks are more or less relaxed and the polymer that gels first forms the more continuous phase.

Sequential crosslinking is usually applied to polymers incompatible to some degree and gives a smaller degree of intermixing than simultaneous method. Specific interactions between non-miscible components increase the cohesiveness and the connectivity between phase-separated polymers, while the intramolecular crosslinking density increases the cohesiveness of each phase. When the crosslinking density of the first polymerised network is small, the second network growing chains will tend to push apart the already existing chains, giving a seq-IPN with significant phase separation. As crosslinking density of the first polymerised network increases, a forced compatibilisation phenomenon of the two polymers takes place, allowing formation of a fine, controllable and stable dispersed phase, hence a more homogenous IPN. The degree of phase separation, together with composition and miscibility of polymer components, polymerisation sequence and kinetics of polymerisation affect the seq-IPN phase morphologies [42]. Some examples of seq-IPN are illustrated in **Table 8.2**.

Table 8.2 Examples of IPN made by the sequential method					
Sequential IPN	**Example**	**Network I**	**Type**	**Subtype**	**Refs**
Urethane/aniline	FL-PU/Pan	FL-PU	Semi	Conductive	[43]
Acrylate/acrylate	PMAA/PNIPAM	PMAA	Semi	Hydrogel	[44]
Acrylate/carbohydrate	PAA/Chitosan	PAA	Semi	Hydrogel	[45]
Urethane/styrene	PU/PS	PU	Semi	Thermoplastic	[46]
Sulfone/epoxy	PSf/DGEBA-DDS	PSf	Semi	Thermoplastic	[47]
Styrene/acrylate	PS/PMMA	PS	Semi	Gradient	[48]
Acrylate/vinyl alcohol	PVA/PAA	PAA	Full	Hydrogel	[49, 50]
Urethane/styrene	PU/PS	PU	Full	Latex	[51]
Acrylate/acrylate	PBA/PBMA	PBA	Full	-	[52]

Network I: the first network formed
FL = fullerenol; PAN = polyaniline; PMAA = poly(methacrylic acid); PAA = poly(acrylic acid); PS = polystyrene; PSf = polysulfone; DGEBA = diglycidyl ether of bisphenol A; DDS = diaminodiphenylsulfone; PMMA = poly(methyl methacrylate); PVA = poly(vinyl alcohol); PBA = poly(butylene adipate); PBMA = poly(butyl methacrylate)

8.4.3 Semi-IPN

Semi-interpenetrating networks (semi-IPN) consist of one or more networks and one or more linear or branched polymers characterised by the penetration on a molecular scale of at least one of the networks by at least some of the linear or branched macromolecules (**Figure 8.3**).

These polymer networks are distinguished from full interpenetrating polymer networks because the constituent linear or branched polymers may be separated from the constituent polymer networks without breaking chemical bonds, only by swelling in an appropriate

Figure 8.3 Schematic view of a semi-IPN system structure

solvent, so that they could be considered as intermediates between IPN and blends. Semi-IPN exhibit minor swelling capacity compared to the corresponding full-IPN.

Semi-IPN can be synthesised by both simultaneous and sequential methods. If a simultaneous method is used, they are also called pseudo-IPN. Because only one polymer is crosslinked, such networks are usually heterogeneous, weaker and less deformable, with separate phases and larger domain size compared to full-IPN. The preparation of semi-IPN can also be done through the super-critical carbon dioxide-assisted infusion strategy by the incorporation of crosslinking agents [53].

Selective crosslinking of poly(styrene-*co*-maleic anhydride) (SMA) with the formation of SMA/styrene-acrylonitrile copolymer (SAN) semi-IPN and induced phase separation was studied with NMR spin diffusion techniques [29]. An approximately linear dependence between the induced domain size (in the range 0-100 nm) and the amount of crosslinker, and an increase of the determined fraction of the SMA/SAN semi-IPN with the increasing of the amount of crosslinker were found. It was also established that the morphology upon crosslinking at elevated temperatures is determined by the crosslinking rate, the diffusion rate of SAN through the increasingly crosslinked and immobilised SMA, and by the entropic factors of SAN that can choose between different mesh sizes in the SMA matrix. So, if the crosslinking rate were extremely high, the homogeneous SAN/SMA system becomes fixed and, if the crosslinking rate was slow compared with the demixing process, a very coarse demixed system would result. Crosslinking processes can be efficiently controlled for the morphological evolution on a length scale of up to 100 nm. A semi-IPN of organic-inorganic hybrid type was made from linear poly(methyl methacrylate) (PMMA) and a silicone network formed by a self-condensation reaction among the polymethylphenylsiloxane (PMPS) pendent reactive methoxysilane groups [54]. DSC analysis shows one single, particularly broad T_g, shifted inward compared with the T_g of the pure components. The investigations of PMMA/PMPS microstructure via solid-state NMR show interfacial heterogeneity on the length scale of 21 nm.

The semi-IPN preparation could be an appropriate method to obtain various biodegradable materials by interpenetrating linear natural polymers (usually carbohydrates) into a synthetic polymer network. Such semi-IPN materials are those obtained by crosslinking the castor oil-based polyurethane (PU) prepolymers and net-poly(ethylene glycol) (Net-PEG) in the presence of linear nitrokonjac glucomannan (NKGM) and bacterial poly(hydroxybutyrate) (PHB), respectively, [55, 56]. The PU/NKGM IPN exhibits a single broad a-relaxation peak, one lower glass transition temperature, and slightly lower elongation at break, higher optical transmittance and much higher tensile strength than PU. Despite the good miscibility between the components, soft and hard segments are phase separated. The optimum results were obtained with relatively low molecular weight NKGM, which plays an important role in plasticising, cure accelerating and strengthening of the semi-IPN product. The PHB/net-PEG semi-IPN show improved strength and toughness, associated with a crystallinity reduced below 24.1% in comparison with 67.7% for a linear PEG macromer.

The shelf-life performances of conductive species can be improved by hindering their local displacement by networking. So, the incorporation of conductive species or additives in a chemically modified network by means of a semi-IPN material may eliminate most of the potential migration within the material's matrix. Such matrix stabilisation enhancement was shown by the polyaniline/fullerenol-polyurethane (PAn/Fl-PU) and polyvinyl alcohol-glutaraldehyde/polyaniline (PVA-GA/PAn) semi-IPN [43, 57]. The PVA-GA/PAn semi-IPN also has excellent optical properties. The morphology of the phase domains changes from rice-grain to spherical shape as the degree of crosslinking increases. The conditions of reaction used in synthesis of some semi-IPN are summarised in **Table 8.3**.

Table 8.3 Semi-IPN preparation

Semi-IPN type	Network		Linear polymer	Reaction Conditions	Refs.
	Monomer (polymer)	Crosslinker			
SMA/SAN	SMA	MDA	SAN	Acetonic solutions/dried 48 h at 50 °C and reduced pressure	[29]
PMMA/PMPS	PMPS	-	PMMA	THF solutions/dried 24 h at 40-60 °C under vacuum/self-condensation crosslinking of casting films 72 h at 120 °C	[54]
PU/NKGM	PU/BD	DETA	NKGM	THF solutions/cast and cured 1.5 h at 55 °C	[55]
PHB/net-PEG	PEG macromer	UV	PHB	DCE solutions/dried 48 h at 25 °C under vacuum	[56]

SMA = *poly(styrene-co-maleic anhydride)*; SAN = *styrene-acrylonitrile*; MDA = *4,4'-methylenedianiline*; PMPS = *polymethylphenylsiloxane.*; NKGM = *nitrokonjac glucomannan*; BD = *1,4-butandiol*; DETA = *diethylenetriamine*; THF = *tetrahydrofuran*; PHB=*poly(hydroxybutyrate)*; Net-PEG = *net-poly(ethylene glycol)*; DCE = *1,2-dichloroethane*

8.4.4 Full-IPN

Full IPN represent the most complete case of an IPN according to its definition, in that all polymers are crosslinked in network form. The two networks are ideally juxtaposed, which generates a lot of entanglements and interaction between them [58] (**Figure 8.4**).

A higher degree of crosslinking restricts the mobility of polymer chains and suppresses the phase segregation and the size of the final morphology. Linear chains in polymer networks are diminished in size with increasing crosslink density until they collapse into aggregates, even at extremely low linear chain concentrations, affecting the homogeneity of polymer networks. In some cases a synergistic improvement of properties like toughness, elongation at break, tensile and impact strength, mechanical damping, etc., can be observed.

Figure 8.4 Schematic view of a full IPN system structure

When both polymers form structurally regulated IPN with the same composition over all, the full-IPN so formed is named homogenous IPN (homo-IPN). This could happen when it is very difficult to modify the position of one polymer chain as a consequence of a high crosslink density. So the second polymer chains will grow and interpenetrate the existing network, causing a forced compatibilisation of both polymers to take place [59]. On contrary, if the polymers are immiscible and the crosslink density of the first or more rapid formed polymer network is small, the growing chains of the other will push apart the already existing chains and a phase-separated IPN will be obtained.

A good opportunity for studying the intimate factors that affects the behaviour of full-IPN materials is done by the synthesis of different architectures based on the crosslinking of two polymers belonging to the same class of macromolecules. Duenas and co-workers [52] did work with PBA/poly butyl methacrylate (PBMA) IPN made by a sequential method, using dielectric and dynamo-mechanical techniques for the determination of miscibility and molecular mobility.

Surface segregation of components in semi- and full-IPN based on crosslinked PU, polyethyl acrylate (PEA), poly(butyl methacrylate) (PBMA) and poly(vinyl acetate) (PVAc) have been studied at the interfaces with solids of high and low free surface energy as function of the kinetic conditions of reaction and method of curing [37].

Sometimes, the distinction between semi- and full-IPN can only be made by the ratio of mixed polymers. Mao and co-workers [60] related such behaviour for calcium chloride crosslinked high acyl gellan/low acyl gellan gels IPN. Each acyl gellan polymer type that is in a higher amount, i.e., in a ratio of 75/25, forms a continuous gel matrix, so that the other one is dispersed discretely. However, at a high acyl gellan/low acyl gellan ratio of 50/50, both polymers form separate, continuous networks, resulting in full-IPN structures.

8.4.5 Main IPN Subtypes

The general classification of IPN materials as a function of the synthesis procedure (simultaneous or sequential) and number of networks formed (semi-IPN or full-IPN) give useful information about these systems, but are not too precise with relation to their specific properties.

For this reason, a further separation of IPN by the polymerisation method, composition and type of constitutive polymer species, gives a better distinction between IPN.

Such classification is shown in **Table 8.4** using examples of binary IPN systems:

Table 8.4 Subtypes of IPN		
Classification Criteria	**Example**	**IPN class**
1. Polymer type(s)	Thermoplastic	Thermoplastic
	Thermoplastic (both)	Elastomeric
	Hydrophilic	Hydrogel
	Grafted polymer(s)	Grafted
	Organic - inorganic	Hybrid
	Conductive	Conductive
2. Composition	Presence of fillers	Filled
	Uniform	Homogeneous
	Non-uniform, regular	Gradient
	Non-uniform, irregular	Heterogeneous
3. Polymerisation method	Emulsion polymerisation	Latex

Some of the most representative 'subtypes' of IPN materials will be discussed in the following sections.

8.4.5.1 Thermoplastic IPN

When networks are formed by physical crosslinks rather than chemical crosslinks between constituent polymers, an IPN is obtained, which flow at elevated temperature in a similar way to thermoplastic elastomers, while behaving like conventional thermoset IPN at their application temperature [7]. Usually, at least one component is a block copolymer and the other one a semi-crystalline or glassy polymer. Typical physical crosslinks include glassy blocks, ionic groups or crystalline segments, thermoplastic IPN being hybrids between IPN and polymer blends. Depending on the continuity and proportion of phases, this kind of IPN can exhibit a wide range of properties, from reinforced rubber to high impact plastics.

The elastomeric component could serve as network I in a sequential IPN synthesis, or in prepolymer form as one component of simultaneous IPN. Although all variety of elastomers could be used, (i.e., polypropylene, polyethylene [61-63], polyamide, polyimide [64], polytetrafluorethylene, polyvinylchloride (PVC), polysulfone [47, 65] etc.), the most common thermoplastic IPN are those based on polyurethane elastomers and rigid plastics, usually either polyacrylates [34, 35, 66-70], polyester [8, 71, 72], or polystyrene [46] (**Table 8.5, Table 8.6**).

Table 8.5 Thermoplastic IPN materials				
IPN components	**Example**	**Type**	**Method**	**Refs.**
Urethane/acrylate	PU/PMMA	Full	Simultaneous	[34]
Urethane/styrene	PU/PS	Full	Simultaneous	[34]
Urethane/unsaturated polyester	PU/UPE	Full	Simultaneous	[71]
Urethane/carbonate	PU/PADC	Full	Simultaneous	[75]
Sulfone/epoxy	PSf/DGEBA-DDS	Semi	Sequential	[47]
Ethylene/acrylate	PE/PDDMA-*co*-EMA	Semi	Sequential	[62]
Sulfone/epoxy	PESf/DGEBA-DADPM	Semi	Sequential	[65]
Ethylene/acrylate-styrene	PE/PBMA-*co*-PS	Semi	Sequential	[76]

UPE = unsaturated polyester; PADC = polyallyl diglycol carbonate;
PESf = polyethersulfone; PE = polyethylene; DADPM = diaminodiphenyl methane;
PDDMA-co-EMA = polyethylene(dodecyl methacrylate-co-ethyl methacrylate);
PBMA-co-PS = poly(butyl methacrylate)-co-polystyrene

Table 8.6 Some examples of full simultaneous IPN thermoplastics. PU-based systems							
Sim-IPN type	Network I			Network II			Refs.
	Monomer	Crosslinker	Initiator/ catalyst	Monomer	Crosslinker	Initiator/ catalyst	
PU/PMMA	PPG/ HMDI	TMP	DBTDL	MMA	TEGDMA	LPO	[34]
PU/PS	PPG/ HMDI	TMP	DBTDL	Styrene	DVB	LPO	[34]
PU/PADC	PPG/ HMDI	-	DBTDL	ADC	-	BP	[75]

HMDI = dicyclohexylmethane-4,4′-diisocyanate; TMP = trimethylolpropane; DBTDL = dibutyl tin dilaurate; TEGDMA = teraelthylene glycol dimethacrylate; LPO = lauroil peroxide; DVB=divinyl benzene; ADC = allyl diglycol carbonate; BP = benzyl peroxide

Generally, thermoplastic-IPN showed synergistic improvement of properties such as toughness, tensile and impact strength, elongation at break and mechanical damping. If the elastomer phase is both continuous and dominant these materials belong to the group of reinforced rubbers and if not, rubber reinforced high impact plastics are obtained.

When both IPN components have thermoplastic behaviour, the resulting thermoplastic material is usually called elastomeric IPN. An example is 2-hydroxyethyl methacrylate-terminated polyurethanes/polyetherurethanes (HPU/PEU) IPN [73]. Enhanced mechanical strength was also reported for semi-IPN composed from elastomers only, (i.e., thermoplastic PVC/urethane) [28, 74].

8.4.5.2 Grafted IPN

Grafted IPN are usually made by simultaneous synthesis and exhibit complex structure and morphologies; therefore it is difficult to control the length and the efficiency of the grafts. Degree of grafting and microstructure of the resulting graft IPN are affected by a series of interconnected factors like reactivity and amount of grafting agent and host polymer, network formation behaviour, grafting temperature, time and reaction rates.

In most cases, grafting reactions occur to a low extent during IPN synthesis as a secondary process. Grafting may be neglected only if the concentration of crosslinks significantly exceeds the concentration of graft sites and if the morphology and physical properties are unaffected.

Induced grafts may act as compatibilisers between phases. When the grafting reaction between the phases is promoted, phase separation is much weaker, and entirely suppressed in some cases. Thus, slight grafting could improve interfacial bonding and interpenetration between networks and thereby their physical and mechanical properties.

Sung [77, 78] and Han [79-81] have synthesised a series of graft-IPN with epoxy resins and polyurethane derivatives. These phase separated systems have better compatibility, toughness, high impact strength and tensile strength than the original polymers.

Comparative examples of grafted IPN belonging to a single class of epoxy/urethane systems are shown in **Table 8.7**.

Polyether and polyester type PU graft agents were used for synthesis of diglycidyl ether of bisphenol A/urethane modified bismaleimide (DGEBA/UBMI) [81], see **Table 8.7**. IPN systems with polyether type PU graft agent presents a heterogeneous phase with UBMI particles dispersed in the epoxy matrix, whereas those with polyester type PU graft agent may be homogeneous. Diglycidyl ethers of bisphenol A/polydimethylsiloxane/

Table 8.7 Example of grafting: grafted epoxy/urethane IPN				
Grafted IPN type	Grafting system	Crosslinker	Method	Refs.
DGEBA/PDMS/PPG	TDI/DGEBA/PPG	MDA	sequential	[78]
DGEBA/UBMI	PUp/DGEBA/PBA PUp/DGEBA/PPG	TDMP	simultaneous	[81]
DGEBA/UAR	UAR/DGEBA	DETA	simultaneous	[82, 83]

PDMS = polydimethylsiloxane; TDI = 2,4-toluene diisocyanate; MDA = 4,4´-methylene dianiline; UBMI = urethane modified bismaleimide; PUp = polyurethane prepolymer (MDI/PPG or PBA/HEMA); TDMP = 2,4,6-tri(dimethyl aminomethyl)phenol; UAR = urethane acrylate resin (TDI/PPO/HEMA/MMA); HEMA = hydroxyethyl methacrylate; MMA = methyl methacrylate; DETA = diethylene triamine

polypropylene glycol (DGEBA/PDMS/PPG) grafted IPN have broad and high T_g regions, so could be applied as effective absorbing materials. The degree of compatibility is strongly affected by the polysiloxane/polyurethane ratio. Grafted chain length could act as a regulator of phase separation and mechanical properties of graft IPN. Longer graft chains could serve as a plasticiser that increase the flexibility of the polymer network but also diminish the crosslinking density. The interpenetration between phases for epoxy/urethane acrylate resin (UAR) IPN was appreciably improved due to the excellent miscibility between the poly(oxypropylene) (PPO) grafts and PPO segments existing in the graft epoxy and the UAR network, respectively.

Using grafting techniques, luminophoric groups can be covalently linked to one IPN component, allowing dynamic and static photophysical and spectroscopic measurements of the degrees of domain interpenetration and permeability [76].

8.4.5.3 Hydrogel Type IPN

When one of the IPN components is a hydrophilic polymer the resulting system may form a hydrogel when soaked in water. The most suitable systems for obtaining latex and hydrogel IPN with advantageous properties was until now those containing at least one acrylate-type component (**Table 8.8**).

An IPN hydrogel sensitive to both pH and temperature, poly(methacrylic acid)/poly(N-isopropylacrylamide), (PMAA/PNIPAM), synthesised by a sequential method, exhibit

Table 8.8 Examples of hydrogel formation for the case of full-IPN systems							
Sim-IPN type	Network I			Network II			Refs.
	Monomer	Crosslinker	Initiator/ catalyst	Monomer	Crosslinker	Initiator/- catalyst	
PNIPAM/ Alginate	PNIPAM	BACA	APS/ TEMDA	Alginate	CaCl$_2$	-	[40]
PEGM/ Chitosan	PEGM	DMPAP	UV	Chitosan	GA	-	[41]
Seq-IPN type	Network I			Network II			Refs.
	Monomer	Crosslinker	Initiator/ catalyst	Monomer	Crosslinker	Initiator/- catalyst	
PMAA/ PNIPAM	MAA	TEGDMA- /UV	DMPAP/- UV	NIPAAm	TEGDMA/- UV	DMPAP/- UV	[44]
PVA/PAA	AA	MBAAM/- UV	DMPAP/- UV	PVA	Freeze- thawing	Freeze- thawing	[50]
PAAm-*co*- AA/PVA	AM/AA	MBAAm	-	PVA	GA	-	[91]
PNIPAM = N-isopropylacrylamide; BACA = N, N´-bis(acryloyl)cystamine; APS = ammonium persulfate; TEMDA = N,N,N´,N´ tetraethylmethylenediamine; DMPAP = 2,2-dimethoxy-2-phenylacetophenone; GA = glutaraldehyde; TEGDMA = tetraethylene glycol dimethyl acrylate; AM = acrylamide; AA = acrylic acid; MBAAm = N,N-methylene bis(acrylamide); MAA = methacrylic acid							

swelling transition at 31-32 °C, corresponding to the lower critical solution temperature of the PNIPAM network, and at pH~5.5, a similar value with the pK$_a$ of PMAA, so, in this case, the responses of each network are relatively independent from each other [44]. Permeation analysis of membranes constructed from PMAA/PNIPAM hydrogels with model drugs revealed significant size exclusion behaviour, strongly affected by pH and temperature conditions. For permeation studies on hydrogel IPN membranes the following equations were used:

$$-\frac{2A}{V} P \cdot t = \ln\left(1 - \frac{2C_t}{C_0}\right) \tag{8.16}$$

$$D_m = \frac{P \cdot l}{K_d} \tag{8.17}$$

$$K_d = \frac{C_m}{C_s} \tag{8.18}$$

where: P = membrane permeability coefficient;

A = effective aria of permeation;

V = volume of each half cell;

C_t = solute concentration in the receptor cell at time t;

C_0 = initial solute concentration of the donor cell;

D_m = diffusion coefficient of the membrane;

l = membrane thickness in the swollen state at constant pH and temperature;

K_d = solute partition coefficient;

C_m, C_s = concentrations of the solute in the membrane and in the surrounding solution at equilibrium.

Furthermore, it was shown that the semi-IPN hydrogels have improved mechanical properties compared to the homopolymer hydrogels, but without affecting the gel collapsing temperature [84]. Muniz and Geuskens [38, 39] have synthesised PAAm/PNIPAM semi-IPN hydrogels with good qualitative mechanical properties, even in the swollen state, and only low shrinking behaviour at temperatures above the lower critical solution temperature (32 °C) of the linear PNIPAM. This low shrinking behaviour is determined by the physical restrictions of the PAAm network, which prevent the PNIPAM chains from collapsing. The equilibrium swelling ratios markedly decrease by increasing the amount of the less hydrophilic PNIPAM chains (by 18% for 5 wt% PNIPAM in the PAAm gel), as result of higher polymer volume fraction. This could be due to the entanglements of the PNIPAM chains that may act inside the semi-IPN hydrogels.

The elastic modulus, E, and the apparent crosslinking density, v_e, for IPN hydrogels could be determined with the equations:

$$\sigma = \frac{f}{S_0} = E\left(\lambda - \lambda^{-2}\right) \tag{8.19}$$

$$\sigma = RT\left(\phi_{p,0}/\phi_p\right)^{2/3} \cdot \phi_p \cdot v_e \cdot \left(\lambda - \lambda^{-2}\right) \tag{8.20}$$

where: σ = applied stress (Pa/m²);

f = measured force;

S_0 = cross section of the undeformed swollen sample;

λ = relative deformation of the sample;

$\Phi_{p,0}$, Φ_p = polymer volume fractions of the gel in the relaxed state and in the swollen state, respectively.

The apparent crosslinking density and elastic modulus of PAAm/PNIPAM semi-IPN hydrogels for 25-40 °C domain were found to be greater than those of the crosslinked PAAm hydrogel network. These results were attributed to the presence of PNIPAM chains that increase the polymer volume fraction of swollen gel at temperatures below 32 °C and collapse around the PAAm networks above 32 °C, leading to much denser, rigid, reinforced gels.

The IPN hydrogels of PNIPAM with hydrophobic polymers can be used for surface modification of rigid hydrophobic polymers through coating [85]. Temperature/pH sensitive comb-type graft and core-shell type IPN hydrogels composed from PNIPAM and calcium-alginate were also prepared [40, 86].

Lee and Kim [49, 50, 87-89] synthesised electrical responsive PVA/PAA IPN hydrogels that exhibit different swelling patterns and permeabilities to solutes as function of pH, temperature, ionic strength of the external solution and ionic groups content.

The release of drugs incorporated into these hydrogels was strongly influenced by the degree of swelling and show pulsatile patterns as a response to pH and temperature. When a PVA/PAA IPN is swelled in a sodium chloride electrolyte solution and placed between a pair of electrodes, it exhibits bending behaviour upon applied electric field. The electric responses depend on applied voltage, charge density of ionic groups within the IPN, ionic strength of the electrolyte solution and greatly affect the drug release behaviour. The amount of loaded drug, increases with both IPN and drug ionisable groups.

To study the release of drugs as function of the electrokinetic processes, Lee and Kim [49, 50, 87-89] have determined the equilibrium conditions for PVA/PAA hydrogels through the osmotic pressure π, which is the sum of the osmotic pressure due to the rubber elasticity, π_1, the solubility of the solvent in the polymer chain, π_2, and the difference in ionic concentration between the gel and surrounding medium, π_3:

$$\pi = \frac{\left[\ln(1-v)+v+\chi v^2\right]RT}{V_1} + \frac{\left(v^{1/3}-\dfrac{v}{2}\right)RTv_e}{V_0} + \left(\sum C_i - \sum C_j\right)RT \qquad (8.21)$$

where: v = volume fraction of the polymer network;

χ = solubility parameter;

V_1 = molar volume of gel;

V_0 = volume of polymer network under dry conditions;

v_e = number of chains;

C_i, C_j = ionic concentrations.

At equilibrium, the osmotic pressure of the gel is equal to that of the surrounding aqueous solution:

$$\pi = \pi_0 = \pi_1 + \pi_2 + \pi_3$$

If the hydrogel components have different hydrophilic characteristics, preferential absorbance of water by the hydrophilic component could be used in IPN microstructure determination. Gallego Ferrer and co-workers [90] have used thermally stimulated depolarisation currents (TSDC) for studying a series of seq-IPN formed by a hydrophobic PEA network and a hydrophilic poly(hydroxyethyl acrylate) (PHEA) network, concluding that the PHEA component in the IPN is plasticised by water in essentially the same way as is pure PHEA. Therefore, the microstructure of the PEA/PHEA IPN will consist of a heterogeneous system of finely dispersed hydrophobic and hydrophilic domains, in which the water molecules randomly mixed with polymer chains reside in the latter.

The swelling behaviour of the hydrogel IPN could be influenced by the IPN composition, the degree of crosslinking and the swelling medium characteristics. For poly(acrylamide-co-acrylic acid)/poly(vinyl alcohol) hydrogel IPN, incorporation of acrylic acid raises the swelling ratio several times while crosslinking of PVA suppresses it, therefore they could be used as modulating factors. Furthermore, the IPN being ionic, swelling is affected by the pH and ionic strength of the medium [91].

Various chitosan-based hydrogel type IPN were also synthesised [41, 45, 92]. These IPN generally exhibit good mechanical properties and an equilibrium water content increasing with the degree of crosslinking of the networks from under 60% to over 95%. For a PEG/chitosan hydrogel full-IPN two T_g values were observed, with a value between those of the individual components, indicating the presence of phase separation. The tensile strength and elongation at break in the swollen state was in the range 0.06-0.18 MPa and 18%-48%, respectively.

8.4.5.4 Hybrid IPN

Hybrid organic-inorganic IPN have been formulated in an effort to improve material properties. However, this kind of IPN may have non-effective interfaces between phases, so it's necessary to use polymers that can form a variety of intermolecular bonds between phases, and also to use low molecular weight loadings of the inorganic phase [2].

Silica and silicon compounds are the main inorganic components used for making hybrid IPN. Silicon-containing organic/inorganic architectures manifest the desirable properties of both an organic polymer (light-weight, flexible, mouldable) and an inorganic network (high thermal stability and strength), but also offer the promise of producing low-cost high performance materials.

Hybrid semi-IPN that involve organic polymers covalently linked to a three-dimensional silica network, at low SiO_2 content, with improved mechanical properties, (i.e., storage modulus, impact and tensile strength, elongation at break), such as poly(2-hydroxyethyl acrylate), PHEA/SiO_2, poly(methyl acrylate-*co*-acrylic acid) (PMA-*co*-AA)/SiO_2, polyamide 6 (PA 6)/SiO_2, have been reported [93-95].

IPN hybrids of poly(*N,N'*-dimethylacrylamide) gel and polystyrene gel with silica gel were prepared by polymerisation of *N,N'*-dimethylacrylamide and *N,N'*-methylenebisacrylamide, respectively, by the polymerisation of styrene monomer in the presence of divinylbenzene in methanolic solutions of tetramethoxysilane (TMOS) [96].

The loss of organic elements could be prevented in both cases by the incorporation of crosslinking points in the organic network. The resulting homogeneous glassy materials present high surface areas, large pore volumes and a sharp distribution of pore size below 2 nm, indicating a molecular-level integration of the organic gel and silica gel in the IPN polymer hybrids.

The degree of swelling PNIPAM/silica IPN decreased continuously with increasing temperature. This behaviour was related to an endothermic peak of DSC analysis of the swollen hybrid, which corresponded to the dissociation of the hydrophobic interaction of PNIPA chain [97].

The organic–inorganic co-continuous hybrid-IPN composed of an epoxide–amine network and silica was prepared by both simultaneous and sequential procedures [98]. Reinforcing rubbery crosslinked epoxide with silica-siloxane structures formed *in situ* from tetraethoxysilane through sol-gel processes is reflected by the increase in modulus (by two orders of magnitude) for low silica content (< 10 vol%). Reinforcement efficiency depends on the reaction conditions, especially on the features of interphase formation. Acid catalysis of the sol-gel process may promote grafting between epoxide and silica phases, leading to a more uniform and finer structure with smaller silica domains.

The most homogeneous hybrid morphology with the smallest silica domains of size 10-20 nm appears in the sequential IPN, where the development of the silica structures is restricted by the rigid reaction medium of the preformed epoxide network.

Reversible, transparent and homogenous hybrid IPN of organic gel and silica gel were synthesised by Imai and co-workers [99, 100] from TMOS and poly(2-methyl-2-oxazoline) having coumarine or thymine moieties as reversibly photocrosslinkable side groups. Furthermore, when maleimide and furan groups were introduced in the side chain of poly(2-methyl-2-oxazoline), the resulting IPN was proved to be thermally reversible [101].

IPN structures can evolve through chelate complexation of linear polymer chains. Such materials were obtained by the treatment of alginate (Alg)/PVA films cast from mixed aqueous polymer solutions with calcium tetraborate solutions [102]. In this case IPN formation takes place by simultaneous occurrence of a chelate complexation of alginates with Ca^{2+} cations and a borate ion-aided crosslinking between PVA chains [102].

Conductive hybrid polyaniline (PAN)/styrene-isoprene-styrene (SIS) and polypyrrole (Ppy)/styrene-isoprene-styrene IPN were prepared by sequential crosslinking reactions of tetraethyl orthosilicate (TEOS) with silicon and acrylic acid grafted functional SIS triblock copolymer, and with PAN and Ppy doped with dodecylbenzenesulfonic acid (DBSA), respectively [103, 104]. The resulting IPN have both good conductivity and superior thermal stability. The amounts of water and TEOS used have a strong influence on conductivity obtained.

Both condensation and hydrosilylation reactions have been successfully applied to the synthesis of silicon-based IPN consisting of a stable Si-O and/or a Si-C linkage, i.e., a ladder silsesquioxane oligomer and a polycarbosilane were used as starting materials for IPN formation through hydrosilylation polymerisation of bifunctional Si-H and Si-vinyl monomers [105].

8.4.5.5 Latex IPN

In latex IPN both networks are included in a single latex particle, usually by emulsion polymerisation. In the sequential method, if the monomers corresponding to the second polymer react near the surface of the first polymer, a latex IPN with shell/core morphology will be obtained. The core could be a hard polymer and the shell a soft polymer, or *vice versa*. If the second monomer is not compatible with the first polymer a compatibiliser has to be added. In some cases both polymer networks can be preformed and subsequently mixed and crosslinked. A more homogenous morphology can be obtained if the second monomer can diffuse fast in the first polymer latex, and by simultaneous network formation. However, the domain size and hence the properties of latex IPN, such those of PU ionomers with acrylate and styrene polymers, can be easily controlled by managing the molecular weight between crosslinks [51]. Some PS-based IPN systems are presented in **Table 8.9.**

Table 8.9 Example of latex IPN: PS-based latexes formation					
Latex IPN System	Emulsion Method	Ist network	Dispersed Phase	Continuous Phase	Refs.
PUI/PS	Seed	PUI	-	-	[51]
PCP/PS	Seed	PCP	-	-	[107]
PMMA/PS	Membrane	-	PS/PMMA/LOH in DCM	PVA/SLS aqueous	[108]
PUI = polyurethane ionomer (IPDI/PPG/HEMA); IPDI = isophorone diisocyanate; PCP = polychloroprene; LOH = lauryl alcohol; SLS = sodium lauryl sulfate; DCM = dichloromethane					

The IPN core/shell polymers were found to be very good dampers due to their higher miscibility. The highest level of damping may be achieved by inverting core/shell particles with dual-phase continuity compared to normal core/shell particles, as was related for acrylate/SAN latex IPN with cores from PBA or poly(ethylhexyl methacrylate) (PEHMA)-based copolymers and glassy SAN shells [106]. The better miscibility of PBA-based copolymers was associated with their higher grafting efficiency.

8.4.5.6 Filled IPN

Introduction of inert fillers into the reaction system at the stage of crosslinked polymer formation essentially affects the reaction kinetics, inhibits the process of microphase separation, and modifies the viscoelastic properties of the system.

In order to improve seq-IPN material properties, the preformed crosslinked network could be mixed with fillers before swelling in the second monomer solution. Polyaniline incorporation (5%-20% per weight) as almost uniform dispersion in the sequential PU/PMA IPN matrix improves tensile strength and electrical properties, polyaniline acting as both reinforcing and conducting filler [109].

The tensile strength, flexural strength, hardness, tensile modulus, dynamic storage modulus and flexural modulus of PU/PVE filled IPN increase with kaolin content, reaching a maximum for 20%-25% filler [110]. An increase of the filler content lowered the impact strength and shifted the T_g to a higher temperature.

Based on studies on PU/PEA sim-IPN, Sergeeva and co-workers [20] concluded that introducing fillers into the IPN at their formation stage influences both the chemical crosslinking and the microphase separation. Thus, fillers will change the chemical reaction rates, the structure of networks being formed and the microphase separation rate, giving non-equilibrium structures with various segregation degrees by adsorption on one or several components. If there is an affinity between the filler and the IPN components, more thermodynamically stable could be realised [19].

In filled semi-IPN, such as those obtained through simultaneous formation of crosslinked PU and linear PBMA networks, diffusion may limit the initiation and chain termination reactions because of the local highly viscous medium generated by the filler [25]. Comparative examples of filled IPN, summarised for clarity only for the PU-based systems, are given in **Table 8.10**.

Table 8.10 Filled IPN. Comparative examples for the case of PU systems				
Filled IPN	**IPN type**	**IPN procedure**	**Fillers**	**Refs.**
PU/PEA	Full	Simultaneous	Aerosil-300, Al_2O_3 and PEA particles 3%, Carbon fibres 2%-10%	[19, 20]
PU/PBMA	Semi	Simultaneous	Talc 20%-40%, TEGMA 10%-40%	[25]
PU/PMA	Semi	Sequential	PAn 5%-20%	[109]
PU/PVE	Full	Simultaneous	Kaolin 20%-25%	[110]
PEA = polyethyl acrylate; PMA = polymethacrylate; PVE = polyvinyl ester				

8.5 Applications

IPN have improved properties compared to their constituent polymers. The wide range of these kinds of materials, the possibility of controlled synthesis and design, the long time resistance and stability, make them the most suitable solutions for various applications.

Classical ways for IPN use include sheet moulding (SMC) and reinforced injection moulding (RIM) compounds, and coatings and adhesives, damping, sound insulation and acoustic absorbing materials, ion exchange resins, permselective/permeable membranes, dental fillings, toughening of rubbery and plastic materials, and as impact modifiers for thermoset materials.

However, the emerging need for performance materials in all technological fields push the polymers and their derived materials far away from their traditional uses. In this way, IPN architectures could be found in applications like nonlinear optical materials, electro-conductive and medical devices, 'intelligent' polymers with improved biocompatibility for biomedical and pharmaceutical applications like molecular separation, enzyme activity controlling systems, controlled drug delivery, tissue culture substrates. Representative examples are given in **Table 8.11**.

Major advances were made in recent years in applying IPN type materials to biological applications, opening a wide and promising way for the future.

PAA-*co*-PEG/polyamide (PA) IPN thin films covalently grafted to metal oxides, silicon, quartz or glass prevents non-specific protein adsorption and specifically promotes cell attachment [117]. Furthermore modifications of these IPN by grafting peptides that mimic the cell binding, facilitate osteoblast adhesion, proliferation and formation of a mineralised tissue. Peptide-IPN could be also used in biomimetic surface engineering as a coating to improve the initial wound healing and stability of metallic and polymeric implants [118].

IPN materials having both high elasticity and high tensile strength, especially those containing PU or polysiloxane, are widely used in the manufacture of medical devices such as catheters, tubing, films, balloons and the like.

The most wanted IPN for biological applications are currently the hydrogel type semi-IPN, because of their specific properties, i.e., changing behaviours as response to variations in pH and temperature or to electric stimuli, swelling and shrinking capabilities, controlled permeability and better mechanical properties than each individual component hydrogel. Major applications are in the field of size exclusion and biomimetic membranes, encapsulation of bioactive species like drugs and enzymes, biosensors and controlled drug delivery.

PMAA/PNIPAM semi-IPN hydrogels with the highest permeability at the physiological status (37 °C, pH 7.4), useful as bioactive membranes and drug delivery systems, have been reported [44]. The electro sensitive PVA/PAA IPN hydrogels have great potential for uses in controlled drug release and could be a step forward in the challenge for obtaining synthetic skeletal muscles [50]. Other IPN hydrogels like chitosan/PAA could be also effective as wound dressing materials [45].

Thermoplastic IPN composed from castor oil-based polyurethane and methyl acrylate (MA) were coated onto the surface of regenerated cellulose film to obtain biodegradable water-resistant films with properties dependent on the MA content [113].

Semi-IPN with varying proportions of poly(ethylene oxide)dimethacrylate and poly(ethylene oxide) and primary, bovine articular chondrocytes were successfully implanted in mice for *in vivo* cartilage regeneration [119].

Table 8.11 Examples of IPN applications				
IPN Type	**IPN Category**	**Improved & Advantageous Properties**	**Application field**	**Refs**
PU/PA 6	Latex	- Damping and mechanical properties - High specific surface area	Damping materials	[67]
Epoxy/UAR	Graft	- Impact strength - Tensile strength	- Injection-moulding - Toughened epoxy network	[82, 83]
Alg/PNIPAM	Hydrogel	- pH/temperature sensitive	- Stimuli-responsive drug delivery systems - Biomimetic actuators	[86]
PAAm-*co*-AA/PVA	Hydrogel	- Swelling and mechanical properties	Super absorbent materials	[91]
PAn/Si-*g*-SIS; PPy/Si-*g*-SIS	Hybrid	- Thermal stability	Conductive films	[103, 104]
SAN/PBA	Latex, grafted	- Damping and mechanical properties - High specific surface area	Damping materials	[106]
PU/PMMA	Filled	- Tensile strength and conductivity due to the PAn filler	Conductive materials	[109]
Epoxy/PMMA	Full	- Homogeneous-phase morphology - High stability in the dipole orientation - Long-term stability of second harmonic coefficients	Nonlinear optic materials	[111]
PVA/PAA	Hydrogel	- Negatively charged with electrically modulated behaviour	Drug delivery systems	[112]

Table 8.11 Continued				
IPN Type	IPN Category	Improved & Advantageous Properties	Application field	Refs
PU/PMMA	Thermoplastic	- Tensile strength - Water resistance - Water vapour permeability - Light transmittance	Coatings onto RC - biodegradable water-resistant films	[113]
PU/VER	Thermoplastic	- Impact strength - Tensile strength	Injection-moulding	[114]
PEO-PPO-PEO /PCL	Hydrogel	- Temperature sensitive	Drug delivery systems	[115]
Chitosan/ glycine	Hydrogel	- pH sensitive swelling	Drug delivery systems	[116]
PA 6 = polyamide 6; Alg = alginate; PPy = polypyrrole; Si-g-SIS = silicon-grafted styrene-isoprene-styrene triblock copolymer; RC = regenerated cellulose; PEO = poly(ethylene oxide); PCL = poly(ε-caprolactone)				

An epoxy/PMMA homogeneous IPN with potential use as a nonlinear optical material (NLO), with high temporal stability at 25-100 °C) was reported [111].

Surprising, only a few IPN commercial products are explicitly named as IPN, therefore one can find true IPN-type materials under the guise of alloys, composites and even blends. One important consequence of this fact is a greater difficulty to try to make a commercial analysis of the market. From this point of view, IPN remain one of the few major groups of chemical products with significant discrepancies between industrial, commercial and scientific nomenclatures.

However, the large number of patents focused on them, shows the great global impact of IPN products associated with their broad, high technology applications. Some representative examples of patents in the field of IPN materials are reviewed in the **Tables 8.12 – 8.14.**

Table 8.12 Full IPN applications			
System	Applicants	Uses/Performance	Refs
ER/TAC or ER/TAIC	Akzo Nobel NV	- Electronics industry; - High glass transition and optimum peel strength.	[120, 121]
Blend/A or Blend/MA	Dentsply Research and Development	- Superior chemical and physical characteristics; - Applications as construction media; - Production of prosthetic dental appliances.	[122]
AM/AA	Research Development Corporation of Japan	- Novel IPN making possible to deliver drug with change of temperature.	[123]
Siloxane/PhP	Ameron International Corporation	- Improved properties of impact resistance, tensile strength, flexural modulus, and density.	[124]
Multifunctional molecules/ Crosslinking agents	Edwards Lifesciences Corporation	- Suited for providing medical devices with anti-thrombogenic coatings; - Entrapping of biologically active compounds.	[125]
UR/styrene derivative	Korea Institute Science Technology	Applications such as: - Artificial organs; - Surface finishing materials, which are in direct contact with blood.	[126]
(PDMS-b-PC)/PM-MA	Razor Associates, Inc.	- A method for producing a thermoformed gas permeable ocular lens, comprising thermoforming into the lens a thermoformable gas. - Permeable ocular lens composition.	[127]
Silane/ER or FR/BR or NBR or isoprene rubber	Korea Chemical Company, Ltd.	- Agents for encapsulating semiconductors.	[128]

Table 8.12 Continued			
System	Applicants	Uses/Performance	Refs
Acrylate/Urethane networks	Minnesota Mining and Manufacturing Company	- Protective and decorative film-based coatings for surfaces exposed to adverse environments, including outdoor weather, solvents, dirt, grease, marine environments.	[129]
PU/PVC	WR Grace & Company	- Superior properties as sealant, especially for automotive parts.	[130]
PI/PU	Schneider USA, Inc.	- Use in the manufacture of medical devices, such as catheters and catheter balloons.	[131]
A/PU lattices	Hi Tex, Inc.	The ability to transmit water vapours while being virtually totally water repellent.	[132]
Urethane/Epoxy/Silicone	Essex Specialty Production	Metal reinforcing patches, directly adherent to an oily metal surface such as an oily steel surface.	[133, 134]
Host polymer/Guest polymer	UOP, Inc.	- Water-insoluble proton-conducting polymers which may be formed into membranes; - Used in gas separation.	[135]
Soluble polymer/crosslinked polymer/ polymerisable compound	Dainippon Ink & Chemical, Inc; Kawamura Institute of Chemistry Research	A coating film having; - Sufficient hydrophilicity; - Causing no deterioration in adhesion and film strength even in wet conditions.	[136]
AP/ER	Akzo NV	The manufacture of new materials especially suited to be used in the electronics industry.	[137, 138]
TPE/NR and/or NBR	Universitat Chemnitz Technishe	Moulded products for agricultural or constructional applications subject to high mechanical and thermal load.	[139]
PU/VER	US Navy	Improved acoustic damping materials.	[140]

	Table 8.12 Continued		
System	Applicants	Uses/Performance	Refs
PU/WBP	Sherwin Williams Company	Superior film properties such as: - Improved MEK resistance; - Film hardness; - Water and alkali resistance; - Flexibility.	[141]
Fluorocarbon/SiE	Eastman Kodak Company	Releasing agent donor member for a toner fixing station in an electrophotographic printing apparatus.	[142]
PMMA/PU	Atochem	High impact strength cast sheet materials.	[143]
Silicone/Epoxy	State of Israel, Ministry of Defense Arnaments Development Authority	Improving optical fibre windings such as: - Stability and payout properties; - Improved packages.	[144]
PU(A)/PU(B)	H.B. Fuller Licensing & Financing, Inc	Uses in adhesives, coatings, binders, primers and sizers.	[145]
PDMS-b-PC/PMMA	Razor Associates, Inc	- An improved gas permeable contact lens composition; - Method of forming the contact lenses.	[146]
PPhZ/PS; PPhZ/PAN; PPhZPMMA; PPhZ/PAA; PPhZ/PDMS	The Penn State Research Foundation	Polyphosphazenes as components of interpenetrating polymer networks.	[147]
Polyurea/ Polyacrylic network	Johnson & Johnson Vision Products, Inc.	Contact or intraocular lenses having clarity, dimensional stability, oxygen permeability, wetability, and durability.	[148, 149]
First/second polymer/nonlinear optical component	University of Massachusetts Lowell	Forming an interpenetrating polymer network, which exhibits nonlinear optical properties.	[150]

Table 8.12 Continued			
System	Applicants	Uses/Performance	Refs
ER/CER	Minnesota Mining and Manufacturing Company	The IPN provide high temperature stable vibration damping materials, adhesives, binders for abrasives, and protective coatings.	[151]

ER = epoxy resin; TAC = triallyl cyanurate; TAIC = triallyl isocyanurate; A = acrylate; MA = methacrylate; AA = crylic acid; PhP = phenolic polymer; UR = urethane resin; [PDMS-b-PC]/PMMA = poly(dimethylsiloxane)-b-poly(carbonate)/poly(methyl methacrylate); FR = fenoxy resin; BR = butadiene rubber; NBR = acrylonitrile butadiene rubber; PI = polyisoprene; AP = allyl polymers; PVC = polyvinylchloride; TPE = thermoplastic elastomer; NR = natural rubber; WBP = waterborne polymer; SiE = silicone elastomer; PPhZ = polyphosphazenes; PAN = polyacrylonitrile; CER = cyanate ester resin

Table 8.13 Semi-IPN applications			
System	Applicants	Uses/performance	Refs
PGDO/ICS/CRA	Nisshin Spinning Corporation	High ionic conductivity; Excellent shape retention.	[152]
PTFE/SE	Tetratec Corporation	Improved physical properties as compared to extruded fibrillated polytetrafluoroethylene dispersion resin alone.	[153]
EGC-PhZ-MF/PLA-*co*-GA	The Penn State Research Foundation	Biomedical applications, including controlled drug delivery and tissue regeneration, and environmental applications.	[154]
PPF/VP	Cambridge Scientific, Inc.	Surgical plates and bone cements.	[155]
Flexible polymer/Fluorescent dye/Polymer phase	3M Innovative Properties Company	Useful in: - Fluorescent traffic signs - Safety devices - Pavement marking tape or paint.	[156]
PTFE/PDMS	Bio Medical Sciences, Inc.	- Suitable for the treatment of dermatological scars, such as those associated with traumatic or surgical injuries of the skin; - Improved physical integrity, durability, and elastic behaviour.	[157]
FP/ER	Minnesota Mining & Manufacturing	Useful as protective coatings, adhesives including adhesive tapes and in multilayer assemblies.	[158]
Thermosetting/ Thermoplastic polyimide	US Army	Improved processability, damage tolerance and mechanical performance.	[159]
Nylon 12/Modified PDMS	B.C. Arkles and R.A. Smith (US)	Improved chemical resistance, increased compressive or tensile strength and excellent temperature resistance and electrical properties; - Applications such as papermaking felts and wire and cable insulation.	[160]

		Table 8.13 Continued		
System	Applicants	Uses/performance	Refs
Polycyanurate/TPP/ polyaramide	Allied Corporation	Improved circuit board.	[161]
PI/PU	Schneider USA, Inc.	Use in the manufacture of medical devices.	[162]
PEOD/PEO; PEOD/SAD	- Massachusetts Institute of Technology; - University Technology Corporation; - The General Hospital Hospital Corporation	- A method for forming a tissue equivalent in a patient; - Compositions and photocrosslinkable polymeric hydrogels in medical treatments, especially joint resurfacing and plastic surgery and delivery of drugs.	[163]
Silicone sheet/ BMEP/ AEMAHCl/HMPP	Pharmacia & Upjohn AB	The process can be used to increase the ability of a surface modification agent, such as heparin, to adhere to the surface of the polymer substrate.	[164]
Poly (PCPP:SA)	Massachusetts Institute of Technology	Compositions useful as bone cements, dental materials, fillers, tissue implants, and as bone grafts, pins, screws, plates.	[165]
Thermoplastic/ Thermosetting resins	Minnesota Mining and Manufacturing Company	New semi-IPN prepared by polymerisation of a thermosetting resin in the presence of a fully pre-polymerised thermoplastic polymer.	[166]

PGDO = polyglycidol; CRA =crosslinking agents; ICS = ion-conductive salt; PTFE = polytetrafluoroethylene; EGC-PhZ-MF = ethyl glycinato-substituted polyphosphazene with p-methylphenoxy as co-substituent; PLA-co-GA = poly(lactide-co-glycolide); PPF = polypropylene fumarate; VP = vinyl pyrrolidone; FP = fluoropolymer; TPP = thermoplastic polymer; PEOD = methacrylated mixed anhydride of succinic acid and PEO and dimethacrylate; SAD = succinic dimethacrylate; BMEP = bis(2-methacryloxyethyl) phosphate; AEMAC = 2-aminoethyl methacrylate hydrochloride; HMPP = 2-hydroxy-2-methyl-1-phenylproponone; SA = sebacic acid; PCPP = (p-carboxy phenoxy) propane

Table 8.14 Applications of other IPN				
Type	System	Applicants	Uses/performance	Refs
Hetero-IPN	PEO or PDMAA-*co*-S/S or AMA or PTMEG	University of Utah	Used in the controlled release of drugs.	[167]
Hydrophilic semi-IPN		Dainippon Ink & Chemical, Inc.; Kawamura Institute of Chemical Research	A coating film having sufficient hydrophilicity and film strength even in wet conditions.	[168]
Hydrogel semi-IPN	Chitosan/ PEO	Northeastern University	Used for site-specific drug delivery in the gastro-intestinal tract.	[169]
Hydrogel IPN	AMA/ VM/ CRA/PVP or PEOz	Sunsoft Corporation	A material having: - Low degree of surface friction and dehydration rate; - High degree of biodeposit resistance. - Used for the contact lens and other medical devices.	[170]
Hybrid-IPN	PDMS/ PTFE	Biomed Sciences, Inc.	Non-adherent covering for fragile and sensitive wounds on polyurethane foam support.	[171]
Hybrid-IPN	PU/ACP	Reichhold, Inc.	Air curable waterborne urethane-acrylic hybrid polymers suitable for coatings.	[172]
Latex-IPN	CD/HVM/ MVM/ PVCM	Dow Chemical Company	Stable to heat and light.	[173]
Semi-2-IPN	PISO2/ ATPISO2	NASA	Improved strength, adhesion, and processability.	[174]
Silicone IPN		LNP Corporation	Reduced shrinkage and warpage and more isotropic mould shrinkage than conventional fibre reinforced thermoplastics.	[175]

	Table 8.14 Continued			
Type	System	Applicants	Uses/performance	Refs
Sequential IPN	PDMS-PM-AA	-	Relates to a method of making IPN that aids in the control of the IPN morphology.	[176]
Ionic semi-IPN or ionic IPN	Copoly (TMAEAC /allylamine); copoly(TM-AEAC/ N-decylallyl amine)	GelTex Pharmaceuticals, Inc.	- Removing bile salts from a patient. - Effective for an IPN comprising a cationic polymer.	[177]
Semi- and Full IPN	Highly crosslinked polyimides/ bis-maleimides	The USA as represented by the Administrator of the NASA	A process for controlling the degree of phase separation and improving toughness, micro-cracking resistance and thermal-mechanical performance of high temperature IPN or semi-IPN.	[178]
Silicone-thermo-plastic semi-IPN	Hybrid silicone/ Unsaturated silicone	Huls America, Inc.	A composition and a method for improving the impact resistance of thermoplastics.	[179]
Gel-IPN	Polyacrylic acid/Polyac-rylamide	Massachusetts Institute of Technology	The phase-transition gels and methods of forming phase-transition gels, which undergo a significantly large volume, change at a desired phase-transition condition in response to a stimulus.	[180]
Simul-taneous semi-IPN	Thermid RTM series/ NR-150B2	The USA as represented by the Administrator of the NASA	Useful as moulding compounds, adhesives, and polymer matrix composites for the electronics and aerospace industries.	[181]

PDMAA-co-S = poly(N,N´ dimethyl acrylamide-co-styrene); S = styrene; AMA = alkyl methacrylate; PTMEG = polytetramethylene ether glycol; VM = vinyl monomer: PEOz = poly-2-ethyl-2-oxazoline; ACP = acrylic copolymer: CD = conjugated diene; HVM = hydrophobic vinyl monomer; PVCM = polyvinyl crosslinking monomer; PISO2 = polyimidesulfone; ATPISO2 = acetylene-terminated polyimidesulfone; TMAEAC = 2-trimethylammonioethylacrylate chloride; Thermid RTM Series = acetylene-endcapped polyimides for resin transfer moulding; NR-150B2 = thermoplastic polyimide

8.6 Conclusions

Interpenetrated polymer networks are advanced materials with complex, rigid architecture, resulting from the irreversible interlocking of the growing chains of each individual polymer network, without any chemical bonding between them.

Reciprocally permanent mutual entanglements prevent total phase separation and stabilise the morphology holding together these networks. So, IPN control the phase separation in a multi-component polymer mixture by the formation of individual polymer networks, allowing multi-phase materials to be made, which have useful mechanical, permeation, optical, and biological properties. However, the suppression of phase-separation is not absolutely complete in practice.

This particular use of networks in IPN synthesis allows the making of materials with bicontinuous structure by forced compatibilisation of incompatible polymers and from polymers insoluble in solvents, giving a clear distinction between IPN and blends. Obviously, IPN materials are also different from alloys and composites that involve only mixing of at least partially miscible polymers by physical means.

The IPN morphology and properties depend and could be modulated by a large group of factors, such as: method of synthesis, polymer type, amount and ratio, kinetics of polymerisation and network formation, crosslinker, catalyst and initiator type and amount, additives, cure and post-cure conditions, compatibility of the monomers and polymer systems used, crosslinking density in each of polymer networks and crosslinking degree reached in IPN and inter-network grafting extent.

IPN analysis can be done with various techniques, such as: DSC, DMA, DMTA, M-TDSC, FT-IR, NMR, DPC, SEM, TEM, SANS, SAXS and AFM.

Two major methods are used for making IPN: simultaneous and sequential. If only one polymer is crosslinked the resulting material is a semi-IPN, while a full-IPN is made when all species are in network form.

Function of criteria like phase segregation extent, type of polymer network, number and types of crosslinks that occur, and the final product properties, both semi- and full-IPN major types could be further split into thermoplastic, hydrogel, latex, grafted, hybrid, filled, conductive, elastomeric, gradient and homo-IPN subtypes.

All polymers have advantageous and disadvantageous properties. For example, acrylics have good UV resistance and colour clarity, but they make poor bonds to wet surfaces, are brittle, flammable and shrink. Epoxy resins have good chemical resistance and good adhesion on both dry and wet surfaces, but are brittle and poor resistant against UV.

Polyurethanes are flexible and tough, but allow moisture penetration, can blister and bubble. Silicones have good flexibility and heat resistance, but they make poor bonds with surfaces and other polymers.

By making IPN systems from such starting polymers, one has the opportunity of obtaining materials with a range of properties that overcome the disabilities of individual polymer components and perhaps generating a synergistic effect on one or more of the properties. Along with the improvements of some components properties, minimising of unwanted ones and synergistic findings, the IPN-type structure itself gives to this kind of material the gift of long life, stability and resistance to physical and chemical agents.

References

1. J.R. Millar, *Journal of Chemical Society*, 1960, **263**, 1311.

2. L.H. Sperling, *Interpenetrating Polymer Networks and Related Materials*, Plenum Press, New York, NY, USA, 1981.

3. *IPNs Around the World, Science and Engineering*, Eds., S.C. Kim and L.H. Sperling, Wiley, New York, NY, USA, 1997.

4. *Advances in Interpenetrating Polymer Networks*, 2nd Edition, Eds., D. Klempner and K.C. Frisch, Technomic, Lancaster, PA, USA, 1994.

5. *Interpenetrating Polymer Networks*, Eds., D. Klempner, L.H. Sperling and L.A. Utracki, Advances in Chemistry Series No.239, ACS, Washington, DC, USA, 1994.

6. Y.S. Lipatov and L.M. Sergeeva, *Interpenetrating Polymer Networks*, Naukova Dumka, Kiev, Ukraine, 1979.

7. R. Bischoff and S.E. Cray, *Progress in Polymer Science*, 1999, **24**, 185.

8. X. Ramis, A. Cadenato, J.M. Morancho and J.M. Salla, *Polymer*, 2001, **42**, 9469.

9. S.C. Kim, D. Klempner, K.C. Frisch and H.L. Frisch, *Macromolecules*, 1977, **10**, 1187.

10. D.J. Hourston and F.U. Schäfer, *Polymer*, 1996, **37**, 3521.

11. L.E. Nielsen and R.F. Landel, *Mechanical Properties of Polymers and Composites*, 2nd Edition, Marcel Dekker, Inc, New York, NY, USA, 1994.

12. D.J. Hourston and F.U. Schäfer, *Journal of Applied Polymer Science*, 1996, **62**, 2025.

13. R. Hu, V.L. Dimonie, M.S. El-Aasser, R.A. Pearson, A. Hiltner, S.G. Mylonakis and L.H. Sperling, *Journal of Polymer Science, Part B, Polymer Physics*, 1997, **35**, 1501.

14. Y.C. Chern, S.M. Tseng and K.H. Hsieh, *Journal of Applied Polymer Science*, 1999, **74**, 335.

15. M.J. Brekner, H.A. Schneider and H.J. Cantow, *Polymer*, 1988, **29**, 2085.

16. B. Budiansky, *Journal of Mechanical Physics Solids*, 1965, **13**, 223.

17. I. Soos, Y. Nagase, W.I. Lenggoro and H. Kodaka, *Journal of Polymer Engineering*, 1996, **16**, 73.

18. C. Clark-Monks and B. Ellis, *Journal of Polymer Science, Polymer Physics*, 1973, **11**, 2089.

19. Y.S. Lipatov, L.V. Karabanova and I.M. Sergeeva, *Polymer International* 1994, **34**, 7.

20. L.M. Sergeeva, S.I. Skiba and L.V. Karabanova, *Polymer International*, 1996, **39**, 317.

21. P. Hedvig, *Dielectric Spectroscopy of Polymers*, Adam Hilger, Bristol, UK, 1977.

22. A. Kanapitsas, P. Pissis, L. Karabanova, L. Sergeeva and L. Apekis, *Polymer Gels & Networks*, 1998, **6**, 83.

23. J.H. Ambrus, C.T. Moynihan and P.B. Macedo, *Journal of Physical Chemistry*, 1992, **176**, 3287.

24. S. Havriliak and S.J. Negami, *Journal of Polymer Science, Polymer Symposium*, 1966, **14**, 89.

25. Y.S. Lipatov, T.T. Alekseeva, V.F. Rosovitsky and N.V. Babkina, *Polymer International*, 1995, **37**, 97.

26. P.J. Flory, *Principles of Polymer Chemistry*, Cornell University Press, Ithaca, NY, USA, 1953.

27. H.L. Frisch, K.C. Frisch and D. Klempner, *Polymer Engineering and Science*, 1974, **14**, 648.

28. C.U. Pittman Jr., X. Xu, L. Wang and H. Toghiani, *Polymer*, 2000, **41**, 5405.

29. F.M. Mulder, W. Heinen, M. van Duin, J. Lugtenburg and H.J.M. de Groot, *Macromolecules*, 2000, **33**, 5544.

30. M. Schulz and B. Paul, *Physical Review B*, 1998, 58, 17.

31. P. Zhou, Q. Xu and H.L. Frisch, *Macromolecules*, 1994, **27**, 938.

32. M. Junker, I. Alig, H.L. Frisch, G. Fleischer and M. Schulz, *Macromolecules*, 1997, 30, 2085.

33. Y.C. Chou and L.J. Lee, *Polymer Engineering and Science*, 1994, **34**, 1239.

34. V. Mishra, F.E. Du Prez, E. Gosen, E.J. Goethals and L.H. Sperling, *Journal of Applied Polymer Science*, 1995, **58**, 331.

35. L.H. Sperling and V. Mishra, *Macromolecular Symposia*, 1997, **118**, 363.

36. D. Jia, L. Chen, B. Wu and M. Wang in *Advances in Interpenetrating Polymer Networks*, 1st Edition, Eds., D. Klempner and K. C. Frisch, Technomic, Lancaster, PA, USA, 1989.

37. Y.S. Lipatov and G.M. Semenovich, *Polymer*, 1999, **40**, 6485.

38. E.C. Muniz and G. Geuskens, *Journal of Membrane Science*, 2000, **172**, 287.

39. E.C. Muniz and G. Geuskens, *Macromolecules*, 2001, **34**, 4480.

40. T.G. Park and H.K. Choi, *Macromolecular Rapid Communication*, 1998, **19**, 167.

41. S.J. Lee, S.S. Kim and Y.M. Lee, *Carbohydrate Polymers*, 2000, **41**, 197.

42. L.A. Utracki in *Interpenetrating Polymer Networks*, Eds., D. Klempner, L.H. Sperling and L.A. Utracki, Advances in Chemistry Series, No.239, ACS, Washington, DC, USA, 1994.

43. L.Y. Chiang, L.Y. Wang, C.S. Kuo, J.G. Lin and C.Y. Huang, *Synthetic Metals*, 1997, **84**, 721.

44. J. Zhang and N.A. Peppas, *Macromolecules*, 2000, **33**, 102.

45. J.W. Lee, S.Y. Kim, S.S. Kim, Y.M. Lee, K.H. Lee and S.J. Kim, *Journal of Applied Polymer Science*, 1999, **73**, 113.

46. Siddaramaiah, P. Mallu and A. Varadarajulu, *Polymer Degradation and Stability*, 1999, **63**, 305.

47. H.S. Min and S.C. Kim, *Polymer Bulletin*, 1999, **42**, 221.

48. G. Akovali, *Journal of Applied Polymer Science*, 1999, 73, 1721.

49. Y.M. Lee, S.H. Kim and C.S. Cho, *Journal of Applied Polymer Science*, 1996, **62**, 301.

50. S.Y. Kim, H.S. Shin, Y.M. Lee and C.N. Jeong, *Journal of Applied Polymer Science*, 1999, **73**, 1675.

51. B.K. Kim, *Macromolecular Symposia*, 1997, **118**, 195.

52. J.M.M. Duenas, D.T. Escuriola, G. Gallego Ferrer, M.M. Pradas, J.L.G. Ribelles, P. Pissis and A. Kyritsis, *Macromolecules*, 2001, **34**, 5525.

53. P. Rajagopalan and T.J. McCarthy, *Macromolecules*, 1998, **31**, 4791.

54. S.S. Hou and P.L. Kuo, *Polymer*, 2001, **42**, 9505.

55. S. Gao and L. Zhang, *Macromolecules*, 2001, **34**, 2202.

56. J. Hao and X. Deng, *Polymer*, 2001, **42**, 4091.

57. W.M. de Azevedo, J.M. de Souza and J.V. de Melo, *Synthetic Metals*, 1999, **100**, 241.

58. L.H. Sperling in *Polymeric Materials Encyclopedia*, Volume 5, Ed., J.C. Salamone, CRC Press, Boca Raton, FL, USA, 1996.

59. D.G. Fradkin, J.N. Foster, L.H. Sperling and D.A. Thomas, *Polymer Engineering and Science*, 1996, **26**, 730.

60. R. Mao, J. Tang and B.G. Swanson, *Carbohydrate Polymers*, 2000, **41**, 331.

61. M. Song, D.J. Hourston, M. Reading, H.M. Pollock and A. Hammiche, *Journal of Thermal Analysis and Calorimetry*, 1999, **56**, 991.

62. J. Hu, U. Schulze and J. Pionteck, *Polymer*, 1999, **40**, 5279.

63. R. Greco, M. Iavarone, A. Fiedlerova and E. Borsig, *Journal of Macromolecular Science - Pure and Applied Chemistry*, 2000, **37**, 433.

64. K.O. Gaw and M. Kakimoto, *Progress in Polyimide Chemistry I, Advances in Polymer Science Series*, 1999, **140**, 107.

65. I. Alig, W. Jenninger, and J.E.K. Schawe, *Thermochimica Acta*, 1999, **330**, 167.

66. V. Mishra and L.H. Sperling, *Journal of Polymer Science*, 1996. **34**, 883.

67. X.Q. Yu, G. Gao, J.Y. Wang, F. Li and X.Y. Tang, *Polymer International*, 1999, **48**, 805.

68. Y. Rharbi, A. Yekta, M.A. Winnik, R.J. DeVoe and D. Barrera, *Macromolecules*, 1999, **32**, 3241.

69. S.H. Wang, S. Zawadzki, and L. Akcelrud, *Materials Research*, 2001, **4**, 27.

70. K. Prashantha, K.V.K. Pai, B.S. Sherigara and S. Prasannakumar, *Bulletin of Material Science*, 2001, **24**, 535.

71. L.H. Fan, C.P. Hu and S.K. Ying, *Polymer Engineering and Science*, 1997, **37**, 2.

72. G.Y. Wang, Y.L. Wang and C.P. Hu, *European Polymer Journal*, 2000, **36**, 735.

73. T.T. Hsieh, K.H. Hsieh, G.P. Simon and C. Tiu, *Polymer*, 1999, **40**, 3153.

74. S.N. Jaisankar, A. Anandprabu, Y. Lakshminarayana and G. Radhakrishnan, *Journal of Materials Science*, 2000, **35**, 1065.

75. S. Dadbin, R. P. Burford and R. P. Chapin, *Polymer Gels and Networks*, 1995, **3**, 179.

76. C. Kosa, M. Danko, A. Fiedlerova, P. Hrdlovic, E. Borsig and R.G. Weiss, *Macromolecules*, 2001, **34**, 2673.

77. P.H. Sung and W.G. Wu, *European Polymer Journal*, 1994, **30**, 905.

78. P.H. Sung and C.Y. Lin, *European Polymer Journal*, 1997, **33**, 231.

79. J.L. Han and Y.C. Chem, K,Y. Li and K.H. Hsieh, *Journal of Applied Polymer Science*, 1998, **70**, 529.

80. J.L. Han and K.Y. Li, *Polymer Journal*, 1999, **31**, 401.

81. K.H. Hsieh, J.L. Han, C.T. Yu and S.C. Fu, *Polymer*, 2001, **42**, 2491.

82. F.J. Hua and C.P. Hu, *European Polymer Journal*, 1999, **35**, 103.

83. F.J. Hua and C.P. Hu, *European Polymer Journal*, 2000, **36**, 27.

84. B.C. Shin, M.S. Jhon, H.B. Lee, and S.K. Yuk, *European Polymer Journal*, 1998, **34**, 1675.

85. A. Gutowska, *Macromolecular Symposia*, 1997, **116**, 545.

86. H.K. Ju, S.Y. Kim and Y.M. Lee, *Polymer*, 2001, **42**, 6851.

87. H.S. Shin, S.Y. Kim and Y.M. Lee, *Journal of Applied Polymer Science*, 1997, **65**, 685.

88. H.S. Shin, S.Y. Kim, Y.M. Lee, K.H. Lee, S.J. Kim and C. Rogers, *Journal of Applied Polymer Science*, 1998, **69**, 479.

89. S.Y. Kim and Y.M. Lee, *Journal of Applied Polymer Science*, 1999, **74**, 1752.

90. G. Gallego Ferrer, M.M. Pradas, J.L.G. Ribelles and P. Pissis, *Proceedings of the Workshop Dielectrica 98*, Monte de Caparica, Portugal, 1998.

91. D. Dhara, C.K. Nisha and P.R. Chatterji, *Journal of Macromolecular Science - Pure and Applied Chemistry*, 1999, **A36**, 197.

92. S.M. Cho, S.Y. Kim, Y.M. Lee, Y.K. Sung and C.S. Cho, *Journal of Applied Polymer Science*, 1999, **73**, 2151.

93. B.M. Novak and C. Davies, *Macromolecules*, 1991, **24**, 5481.

94. F. Yang, Y. Ou and Z. Yu, *Journal of Applied Polymer Science*, 1998, **69**, 355.

95. W. Yin, J. Li, J. Wu and T. Gu, *Journal of Applied Polymer Science*, 1997, **64**, 903.

96. R. Tamaki, T. Horiguchi and Y. Chujo, *Bulletin of the Chemical Society of Japan*, 1998, **71**, 2749.

97. Y. Imai, N. Yoshida, K. Naka and Y. Chujo, *Polymer Journal*, 1999, **31**, 258.

98. L. Matejka, K. Dusek, J. Plestil, J. Kriz and F. Lednicky, *Polymer*, 1999, **40**, 171.

99. Y. Imai, K. Naka and Y. Chujo, *Polymer Journal*, 1998, **30**, 990.

100. Y. Imai, T. Ogoshi, K. Naka and Y. Chujo, *Polymer Bulletin*, 2000, **45**, 9.

101. Y. Imai, H. Itoh, K. Naka and Y. Chujo, *Macromolecules*, 2000, **33**, 4343.

102. K. Miura, N. Kimura, H. Suzuki, Y. Miyashita and Y. Nishio, *Carbohydrate Polymers*, 1999, **39**, 139.

103. L.H. Gan, Y.Y. Gan and W.S. Yin, *Polymer International*, 1999, **48**, 1160.

104. L.H. Gan, Y.Y. Gan and W.S. Yin, *Polymer*, 1999, **40**, 4035.

105. M. Tsumura and T. Iwahara, *Journal of Applied Polymer Science*, 2000, **78**, 724.

106. M.S. El-Aasser, R. Hu, V.L. Dimonie and L.H. Sperling, *Colloids and Surfaces A: Physicochemical and Engineering Aspects*, 1999, **153**, 241.

107. C.D. Vo and R.P. Burford, *Journal of Applied Polymer Science*, 1999, **74**, 622.

108. G.H. Ma, M. Nagai and S. Omi, *Journal of Colloid and Interface Science*, 1999, **214**, 264.

109. T. Jeevananda, M. Begum and Siddaramaiah, *European Polymer Journal,* 2001, **37**, 1213.

110. C.H. Chen, W.J. Chen, M.H. Chen and Y.M. Li, *Journal of Applied Polymer Science*, 1999, **71**, 1977.

111. L.Z. Zhang, Z.G. Cai, Q.D. Ying and Z.X. Liang, *Chinese Journal of Polymer Science*, 1999, **17**, 435.

112. S.Y. Kim and Y.M. Lee, *Journal of Applied Polymer Science*, 1999, **74**, 1752.

113. J.H. Yu, Y.M. Du and F.Y. Cheng, *Journal of Macromolecular Science - Pure and Applied Chemistry*, 1999, **36**, 1259.

114. L.H. Fan, C.P. Hu and S.K. Ying, *Polymer Engineering and Science*, 1997, **37**, 338.

115. S.Y. Kim, J.C. Ha and Y.M. Lee, *Journal of Controlled Release*, 2000, **65**, 345.

116. K.C. Gupta and M.N.V.R. Kumar, *Journal of Applied Polymer Science*, 2000, **76**, 672.

117. J.P. Bearinger, D.G. Castner and K.E. Healy, *Journal of Biomaterials Science, Polymer Edition*, 1998, **9**, 629.

118. K.E. Healy, *Current Opinion in Solid State & Materials Science*, 1999, **4**, 381.

119. J. Elisseeff, K. Anseth, D. Sims, W. McIntosh, M. Randolph, M. Yaremchuk and R. Langer, *Plastic and Reconstructive Surgery*, 1999, **104**, 1014.

120. J.A.J. Schutyser and A.J.W. Buser, inventors; Akzo Nobel NV, assignee; World Patent 9,506,075, 1995.

121. J.A.J. Schutyser, A.J.W. Buser, and A. Steenbergen, inventors; Akzo Nobel NV, assignee; US Patent 5,728,468, 1998.

122. L. Tateosian and F.D. Roemer, inventors; Dentsply Research & Devevelopment, assignee; US Patent 4,698,373, 1987.

123. M. Okano, N. Ogata, K. Sanai, A. Maruyama, H. Katono and Y. Sakurai, inventors; Research Development Corporation of Japan, assignee; Japanese Patent 3,079,608, 1991.

124. Kane J.F. and N.R. Mowrer, inventors; Ameron International Corporation, assignee; US Patent 5,736,619, 1998.

125. S-D. Tong, L-C. Hsu and C.B. Hu, inventors; Edwards Lifesciences Corporation, assignee; World Patent 0,108,718, 2001.

126. S-C. Kim, H-W. Roh, M-J. Song and Y-C. Shin, inventors; Korea Institute Science Technology, assignee; US Patent 6,156,344, 2000.

127. D.R. Ingenito, H.F. Rugge, D.S. Soane and W.L. Sturm, inventors; Rasor Associates, Inc, assignee; US Patent 5,986,001 1999.

128. C-T. Lee, M-Y. Lee and C-J. Ryu, inventors; Korea Chemical Company Ltd, assignee; Korean Patent 9,104,647, 1991.

129. D.A. Barrera, inventor; Minnesota Mining and Manufacturing Company, assignee; US Patent 5,965,256, 1999.

130. N.E. Blank, R.C. Hartwig and C. Vu, inventors; W.R. Grace and Company, assignee; US Patent 5,091,455, 1992.

131. L.H. Sperling, C.J. Murphy and V. Mishra, inventors; Schneider USA, Inc., assignee; US Patent 5,783,633, 1998.

132. K.C. Frisch, P. Geng and H.X. Xiao, inventors; Hi Tex Inc, assignee; US Patent 5,747,392, 1998.

133. S.D. Rizk, N.B. Shah and J.W. Powers, inventors; Essex Specialty Production, assignee; US Patent 4,842,938, 1989.

134. S.D. Rizk and N.B. Shah, inventors; Essex Specialty Production, assignee; US Patent 4,766,183, 1988.

135. J.J. Zupancic, R.J. Swedo and S. Petty-Weeks, inventors; UOP, Inc., assignee; US Patent 4,708,981, 1987.

136. T. Anazawa and T. Takada, inventors; Dainippon Ink & Chemical, Inc. and Kawamura Institute of Chemistry Research, assignees; Japanese Patent 11,209,648, 1999.

137. J.A.J. Schutyser, H.J. Slots, A.J.W. Buser and P.H. Zuuring, inventors; Akzo NV, assignee; European Patent 0,417,837, 1991.

138. J.A.J. Schutyser, H.J. Slots, A.J.W. Buser and P.H. Zuuring, inventors; Akzo NV, assignee; European Patent 0,413,386, 1991.

139. J. Jentzsch, H. Michael, E. Herscher and D. Kaune, inventors; Universitat Chemnitz Technishe, assignee; German Patent 4,102,237, 1992.

140. U.A. Sorathia, W.L. Yeager and T.L. Dapp, inventors; US Navy, assignee; US Patent 5,237,018, 1993.

141. R.R. Tomko, inventor; Sherwin Williams Company, assignee; US Patent 6,166,127, 2000.

142. S.V. Davis, J. E. Mathers, B.H. Mills, B. Tan and J-H. Chen, inventors; Eastman Kodak Company, assignee; US Patent 6,067,438, 2000.

143. M. Avenel, inventor; Atochem, assignee; US Patent 5,539,053, 1996.

144. H. Birnholz and A. Buchman, inventors; State of Israel, Ministry of Defense Arnaments Development Authority, assignee; US Patent 6,103,375, 2000.

145. Y. Duan, Y. Wei, Y. Zhu, S.E. Stammler, B.L. Marty, G. Haider, R.R. Davies and M. Maksymkiw, inventors; H.B. Fuller Licensing & Financing, Inc., assignee; US Patent 6,017,998, 2000.

146. D.R. Ingenito, H.F. Rugge, D.S. Soane and W.L. Sturm, inventors; Rasor Associates, Inc., assignee; US Patent 5,789,483, 1998.

147. H.R. Allcock, K. Visscher and Y.B. Kim, inventors; The Penn State Research Foundation, assignee; US Patent 5,747,60, 1998.

148. G.A. Hill, K.C. Frisch, V. Sendijarevic and S-W. Wong, inventors; Johnson & Johnson Vision Products, Inc., assignee; US Patent 5,674,942, 1997.

149. G.A. Hill, K.C. Frisch, V. Sendijarevic and S-W. Wong, inventors; Johnson & Johnson Vision Products, Inc., assignee; US Patent 5,656,210, 1997.

150. S.K.Tripathy, R-J. Jeng, J. Kumar, S. Marturunkakul and J.I. Chen, inventors; University of Massachusetts Lowell, assignee; US Patent 5,532,320, 1996.

151. I. Gorodisher and M.C. Palazzotto, inventors; Minnesota Mining and Manufacturing Company, assignee; US Patent 5,494,981, 1996.

152. T. Sato, inventor; Nisshin Spinning, assignee; European Patent 1,057,869, 2000.

153. J.A. Dillon and M.E. Dillon, inventors; Tetratec Corporation, assignee; US Patent 4,945,125, 1990.

154. H. Allcock, S. Ibim, A. Ambrosio, M. Kwon and C. Laurencin, inventors; Penn State Research Foundation assignee; US Patent 6,077,916, 2000.

155. Y-Y. Hsu, D.L. Wise, J.D. Gresser and D.J. Trantolo, inventors; Cambridge Scientific, Inc., assignee; US Patent 6,071,982, 2000.

156. R.E. Harelstad, L.A. Pavelka, D.A., Barrera and W.D Joseph, inventors; 3M Innovative Properties Company, assignee; US Patent 6,001,936, 1999.

157. M.E. Dillon, inventor; Bio Medical Sciences, Inc., assignee; US Patent 5,980,923, 1999.

158. M.A. Perez, W.D. Coggio, M.C. Palazzotto and D.S. Parker, inventors; Minnesota Mining & Manufacturing, assignee; World Patent 9,808,906, 1998.

159. R.H. Pater, inventor; US Army assignee; US Patent 5,492,979, 1996.

160. B.C. Arkles and R.A. Smith, inventors; no assignee; US Patent 4,970,263, 1990.

161. E.S. Hsiue, D. Ziatyk, G.R Stone and B.T. Debona, inventors; Allied Corporation, assignee; US Patent 4,623,577, 1986.

162. L.H. Sperling, C.J. Murphy and V. Mishra, inventors; Schneider USA, Inc., assignee; US Patent 5,786,426, 1998.

163. R.S. Langer, J.H. Elisseeff, K. Anseth and D. Sims, inventors; Massachusetts Institute of Technology, University Technology Corporation and The General Hospital Hospital Corporation, assignees; US Patent 6,224,893, 2001.

164. Y. Wang, R. van Boxtel and S.Q. Zhou, inventors; Pharmacia & Upjohn AB, assignee; US Patent 6,011,082, 2000.

165. V.R. Shastri, R.S. Langer and P.J. Tarcha, inventors; Massachusetts Institute of Technology, assignee; US Patent 5,837,752, 1998.

166. M.A. Perez, D.A. Ylitalo, T.M. Clausen, R J. DeVoe, K.E. Kinzer and M.D. Swan, inventors; Minnesota Mining & Manufacturing Company, assignee; US Patent 5,709,948, 1998.

167. Y.H. Bae, T. Okano and S.W. Kim, inventors; University of Utah, assignee; US Patent 4,931,287, 1990.

168. N. Shirai, T. Takada and T. Anazawa, inventors; Dainippon Ink & Chemical, Inc. and Kawamura Institute of Chemical Research, assignees; Japanese Patent 11,293,147, 1999.

169. M.M. Amiji, inventor; Northeastern University, assignee; US Patent 5,904,927, 1999.

170. T. Nguyen, H. Hu, E. Rossberg, H. Tran and C.R. Briggs, inventors; Sunsoft Corporation, assignee; World Patent 0,002,937, 2000.

171. M.E. Dillon, inventor; Biomed Sciences, Inc., assignee; World Patent 0,149,228, 2001.

172. G. Petschke and S. Yang, inventors; Reichhold, Inc., assignee; US Patent 6,239,209, 2001.

173. No inventor; The Dow Chemical Company, assignee; Spanish Patent 8,701,208, 1987.

174. A.O. Egli and T.L. St Clair, inventors; NASA, assignee; US Patent 4,695,610, 1987.

175. J.M. Crosby and M.K. Hutchins, inventors; LNP Corporation, assignee; US Patent 4,695,602, 1987.

176. J. Turner and Y.L. Cheng, inventors; US Patent 6,331,578, 2001.

177. W.H. Mandeville, III, S.R. Holmes-Farley, T.X. Neenan and G.M. Whitesides, inventors; GelTex Pharmaceuticals, Inc. assignee; US Patent 5,925,379, 1999.

178. R.H. Pater and M.G. Hansen, inventors; The USA as represented by the Administrator of the NASA, assignee; US Patent 5,648,432, 1997.

179. M. Zolotnitsky, inventor; Hüls America Inc., assignee; US Patent 5,648,426, 1997.

180. T. Tanaka, F. Ilmain, E. Kokufuta and M. Annaka, inventors; Massachusetts Institute of Technology, assignee; US Patent 5,580,929, 1996.

181. R.H. Pater, inventor; The USA as represented by the Administrator of the NASA, assignee; US Patent 5,492,979, 1996.